BIOLOGY | CELL BIOLOGY AND GENETICS

THIRTEENTH EDITION

CECIE STARR   RALPH TAGGART   CHRISTINE EVERS   LISA STARR

BROOKS/COLE
CENGAGE Learning

Australia • Brazil • Japan • Korea • Mexico • Singapore • Spain •   United Kingdom • United States

W9-ANM-333

BROOKS/COLE
CENGAGE Learning·

Cell Biology and Genetics
Biology: The Unity and Diversity of Life,
Thirteenth Edition
**Cecie Starr, Ralph Taggart, Christine Evers,
Lisa Starr**

Senior Acquisitions Editor, Life Sceinces:
   Peggy Williams

Publisher: Yolanda Cossio

Assistant Editor: Shannon Holt

Editorial Assistant: Sean Cronin

Media Editor: Lauren Oliveira

Marketing Manager: Tom Ziolkowski

Marketing Coordinator: Jing Hu

Marketing Communications Manager: Linda Yip

Content Project Manager: Hal Humphrey

Design Director: Rob Hugel

Art Director: John Walker

Print Buyer: Karen Hunt

Rights Acquisitions Specialist: Dean Dauphinais

Production Service: Grace Davidson &
   Associates

Text Designer: John Walker

Photo Researcher: Myrna Engler Photo
   Research Inc.

Text Researcher: Pablo D'Stair

Copy Editor: Anita Wagner

Illustrators: Gary Head, ScEYEnce Studios,
   Lisa Starr

Cover Image: The diversity of body forms
   among flower mantids, wolves, and sea
   anemones conceals an underlying unity. All
   are animals, and thus belong to the same
   branch of life that you do.

   Top: Bob Jensen Photography; middle, Jeff
   Vanuga/Corbis; bottom: John Easley.

Compositor: Lachina Publishing Services

For product information and technology assistance, contact us at
Cengage Learning Customer & Sales Support, 1-800-354-9706.

For permission to use material from this text or product,
submit all requests online at **www.cengage.com/permissions**.
Further permissions questions can be emailed to
permissionrequest@cengage.com.

Library of Congress Control Number: 2011938380

ISBN-13: 978-1-111-57985-2
ISBN-10: 1-111-57985-7

Brooks/Cole
20 Davis Drive
Belmont, CA 94002
USA

Cengage Learning is a leading provider of customized learning solutions with
office locations around the globe, including Singapore, the United Kingdom,
Australia, Mexico, Brazil, and Japan. Locate your local office at:
www.cengage.com/global.

Cengage Learning products are represented in Canada by Nelson Education, Ltd.

To learn more about Brooks/Cole visit **www.cengage.com/brookscole**.

Purchase any of our products at your local college store or at our preferred
online store **www.CengageBrain.com**.

Printed in Canada
1 2 3 4 5 6 7 15 14 13 12 11

CONTENTS IN BRIEF

DETAILED CONTENTS

INTRODUCTION

UNIT I   PRINCIPLES OF CELLULAR LIFE

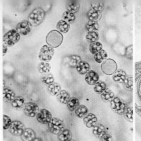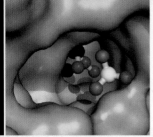

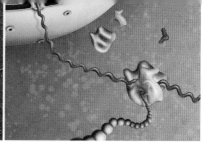

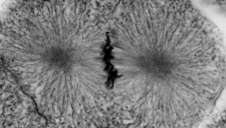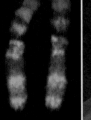

Preface

This edition of *Biology: The Unity and Diversity of Life* includes a wealth of new information about recent discoveries in biology (details can be found in the *Power Bibliography*, which lists journal articles and other references we used in the revision process). Descriptions of current research, along with photos and videos of scientists who carry it out, underscore the concept that science is an ongoing endeavor carried out by a diverse community of people. We discuss not only what was discovered, but also how the discoveries were made, how our understanding has changed over time, and what remains undiscovered.

At the same time, we provide a thorough, accessible introduction to well-established concepts and principles that underpin modern biology. We examine every topic from an evolutionary perspective and emphasize the connections between all forms of life. Throughout the book, we revised text and diagrams to make difficult concepts easier for students to grasp.

We also continue to focus on real world applications of the topics we discuss. This edition provides expanded coverage of human health and diseases. It also covers in more detail the many ways in which human activities can alter the environment and threaten both human health and the biodiversity of our planet.

CHANGES TO THIS EDITION

Learning Road Map   Each chapter opens with a *Learning Road Map* that places the chapter in a larger context. It notes which previously discussed material the chapter draws upon, highlights key concepts discussed within the chapter, and looks ahead to later sections that refer back to the chapter's concepts. For added impact, each *Learning Road Map* overlays an intriguing full-page opening photo. The photo reappears within the chapter, together with a caption that explains its content.

Opening Essay/Essay Revisited   We now devote the first section of each chapter to an essay about an environmental or health application of that chapter's theme. The final section revisits this topic as a reminder of the chapter's relevance to the real world.

Section-Based Links to Earlier Concepts   *Links to Earlier Concepts* appear at the start of most sections. These links emphasize connections between biological topics and facilitate review of material that has not been fully mastered. For example, the heading of the section about membrane potential refers back to earlier coverage of potential energy and of transport proteins.

Section-Based Glossaries   In addition to a full glossary of terms at the end of the book, each section now has a *Section-Based Glossary* that includes all boldface terms for that section. A student can simply glance at this glossary for a quick check of a term's meaning.

Step-by-Step Graphics Integrated With Text   We have long explained biological processes using figures accompanied by step-by-step captions. In this edition, we've also integrated stepwise descriptions of many graphics into the text, using colored, numbered balls to make the connections.

Visual Chapter Summary   To further enhance visual learning, we've added graphics to each section of the chapter summary. As before, each chapter summary includes all key terms from that chapter in boldface.

Chapter-Specific Changes   Every chapter was extensively revised for clarity; this edition has more than 200 new photos and over 250 new or updated figures. A section-by-section guide to content and figures is available upon request, but we summarize the highlights here.

- *Chapter 1, Invitation to Biology* Revised and expanded coverage of critical thinking, scientific process, and philosophy of science. Material on classification systems moved to this chapter.
- *Chapter 2, Life's Chemical Basis* New opening essay discusses mercury toxicity and prevalence; revisions emphasize electron behavior in atoms as it relates to ions and bonding.
- *Chapter 3, Molecules of Life* Importance of protein structure now exemplified with prions.
- *Chapter 4, Cell Structure and Function* Expanded discussion and new photos of archeans.
- *Chapter 5, Ground Rules of Metabolism* Chapter now includes material on cell membrane function; new section on cofactors, including ATP.
- *Chapter 6, Where It Starts—Photosynthesis* Biofuels essay revised to include food supply debate; revisited section on climate change and CO_2 now includes atmospheric change after photosynthesis evolution.
- *Chapter 7, How Cells Release Chemical Energy* Introductory section now discusses the relationship between the evolution of oxygenic photosynthesis and aerobic respiration. Third stage art revised to connect with summary illustration.
- *Chapter 8, DNA Structure and Function* This chapter was moved forward in book to allow a more complete explanation of genes and inheritance; now includes material on chromosome structure and mutations. New opener essay about cloned 9/11 rescue dog Trakr.
- *Chapter 9, From DNA to Protein* Introductory essay revised to include RIPs other than ricin; material on sickle cell anemia and other effects of beta globin mutations added.
- *Chapter 10, Gene Control* Theme of evolutionary connections strengthened throughout; male sex determination and riboswitches added to existing sections; new epigenetics section added.
- *Chapter 11, How Cells Reproduce* New fluorescence micrograph of HeLa cells shows how defects in cell cycle controls lead to cancer; section on telomeres added.
- *Chapter 12, Meiosis and Sexual Reproduction* Artwork updated; new fluorescence micrograph illustrates checkpoint proteins in meiosis.
- *Chapter 13, Observing Patterns in Inherited Traits* Opening essay is now on inheritance of CF mutation. Environmental effects section rewritten to include epigenetics and pelage cycle.

- *Chapter 14, Chromosomes and Human Inheritance* Opening essay on human skin color moved to this chapter. Chapter reorganized to strengthen introduction to human genetics; genetic screening section updated; many photos replaced with newer versions.
- *Chapter 15, Studying and Manipulating Genomes* New opening section on personal genetic testing; updated genomics section includes expanded material on DNA profiling. New brainbow photo.
- *Chapter 16, Evidence of Evolution* New graphics include a page from Darwin's evolution journal and updated paleogeography illustrations.
- *Chapter 17, Processes of Evolution* New illustrated examples include allopatric speciation in snapping shrimp, sympatric speciation in Lake Victoria cichlids, ring species; coevolution of an ant and a butterfly.
- *Chapter 18, Organizing Information About Species* Expanded and updated phylogeny/cladistics section; *Hox* gene comparison material updated with new examples and photos.
- *Chapter 19, Life's Origin and Early Evolution* Opening essay focuses on astrobiology. Added micrographs and discussion of bacteria with internal membranes. Improved figure illustrates step on the road to life.
- *Chapter 20, Viruses, Bacteria, and Archaea* Opening essay discusses SIV and evolution of HIV. More detailed depiction of HIV replication. Coverage of prions moved to Chapter 3. New figure shows viral recombination.
- *Chapter 21, Protists—The Simplest Eukaryotes* New opening essay about malaria. New step-by-step figure illustrating *Plasmodium* life-cycle. Clearer, simpler illustration of *Chlamydamonas* life cycle. Choanoflagellates now introduced here.
- *Chapter 22, The Land Plants* Essay about Nobel Prize winner W. Mathai's work. Life cycle figures revised.
- *Chapter 23, Fungi* New opening essay about the threat wheat stem rust poses to food supplies. Life cycle figures revised for clarity. Additional photos of fungal diversity.
- *Chapter 24, Animal Evolution—Invertebrates* Improved or new graphics of animal evolutionary tree, types of body cavities, crab life cycle, grasshopper anatomy.
- *Chapter 25, Animal Evolution—Vertebrates* New evolutionary tree diagram, graphic of tunicate. Discussion of fish evolution revised. New graphic illustrates major mammal orders. Material on human evolution moved to separate chapter.
- *Chapter 26, Human Evolution* Opening essay describes recent evidence that Neanderthals mated with *Homo sapiens*. The chapter covers primate groups and the latest discoveries regarding human ancestry.
- The chapter *Plants and Animals—Common Challenges* has been deleted and material previously covered therein is now integrated into other chapters.
- *Chapter 27, Plant Tissues* New introductory essay on carbon sequestration by plants; reorganized tissue section now includes thumbnail photos; many revised illustrations and new micrographs.
- *Chapter 28, Plant Nutrition and Transport* Plant nutrient table revised to include functions; soil water uptake artwork revised; new artwork and micrographs of xylem tubes; expanded discussion of stomata function.
- *Chapter 29, Life Cycles of Flowering Plants* Material on early plant development, senescence, and dormancy moved to this chapter and expanded. Many new photos.
- *Chapter 30, Communication Strategies in Plants* Extensively revised and expanded to reflect paradigm shifts driven by recent research. New introductory essay on health benefits of cocoa; new section introducing plant hormones added; material on major hormones expanded and reorganized by section; new section on plant stress responses added.
- *Chapter 31, Animal Tissues and Organ Systems* New graphics depicting levels of organization, endocrine versus exocrine glands. Added discussion of carcinomas, vitiligo, loss of body hair in human evolution.
- *Chapter 32, Neural Control* Added latest research on effects of Ecstasy use. Additional info about the role of neuroglia in disease. Split-brain research deleted.
- *Chapter 33, Sensory Perception* Updated discussion of olfaction with new figure. Increased coverage of diversity of visual systems. More about effects of noise pollution.
- *Chapter 34, Endocrine Control* New graphic of human endocrine system. Added section about thymus. Added information about pthalates as endocrine disruptors.
- *Chapter 35, Structural Support and Movement* New essay about effects of myostatin. Improved figures depicting muscle contraction.
- *Chapter 36, Circulation* Updated art throughout. Added illustration of a stent.
- *Chapter 37, Immunity* Innate responses material expanded to two sections that include updated illustrations and photos; also added neutrophil nets. Expanded sections on immune dysfunction, vaccines.
- *Chapter 38, Respiration* New essay about carbon monoxide poisoning. New graphic of CO_2 transport.
- *Chapter 39, Digestion and Nutrition* New essay about the importance of gut bacteria. Discussion of stomach and small intestine revised to improve flow.
- *Chapter 40, Maintaining the Internal Environment* New figures for urine formation and renin–angiotensin–aldosterone system. New coverage of hibernation.
- *Chapter 41, Animal Reproductive Systems* New graphics depicting sperm formation, female genitals. Section about FSH and twins deleted.
- *Chapter 42, Animal Development* Updated essay with disucssion of "octomom" and early opposition to IVF.
- *Chapter 43, Animal Behavior* Added discussions of kinesis and taxis, epigenetic effects, behavioral plasticity.
- *Chapter 44, Population Ecology* New essay about Canada goose population explosion. Human population material updated.
- *Chapter 45, Community Ecology* New studies of competition cited. Revised discussion of disturbance and exotic species.
- *Chapter 46, Ecosystems* New essay on phosphate pollution. Section on carbon cycle revised and updated.
- *Chapter 47, The Biosphere* Added information about latest deep ocean life. New biome photos.
- *Chapter 48, Human Impacts on the Biosphere* Biological magnification, ozone depletion and pollution now covered in this section

STUDENT AND INSTRUCTOR RESOURCES

Test Bank   Nearly 4,000 test items, ranked according to difficulty and consisting of multiple-choice (organized by section heading), selecting the exception, matching, labeling, and short answer exercises. Includes selected images from the text. Also included in Microsoft® Word format on the PowerLecture DVD.

ExamView®   Create, deliver, and customize tests (both print and online) in minutes with this easy-to-use assessment and tutorial system. Each chapter's end-of-chapter material is also included.

Instructor's Resource Manual   Includes chapter outlines, objectives, key terms, lecture outlines, suggestions for presenting the material, classroom and lab enrichment ideas, discussion topics, paper topics, possible answers to critical thinking questions, answers to data analysis activities, and more. Also included in Microsoft® Word format on the PowerLecture DVD.

Student Interactive Workbook   Labeling exercises, self-quizzes, review questions, and critical thinking exercises help students with retention and better test results.

PowerLecture   This convenient tool makes it easy for you to create customized lectures. Each chapter includes the following features, all organized by chapter: lecture slides, all chapter art and photos, bonus photos, animations, videos, Instructor's Manual, Test Bank, Examview testing software, and JoinIn polling and quizzing slides. This single disc places all the media resources at your fingertips.

The Brooks/Cole Biology Video Library 2010 featuring BBC Motion Gallery   Looking for an engaging way to launch your lectures? The Brooks/Cole series features short high-interest segments: Pesticides: Will More Restrictions Help or Hinder?; A Reduction in Biodiversity; Are Biofuels as Green as They Claim?; Bone Marrow as a New Source for the Creation of Sperm; Repairing Damaged Hearts with Patients' Own Stem Cells; Genetically Modified Virus Used to Fight Cancer; Seed Banks Helping to Save Our Fragile Ecosystem; The Vanishing Honeybee's Impact on Our Food Supply.

Webtutors for WebCT and BlackBoard   Jump-start your course with customizable, rich, text-specific content. Whether you want to Web-enable your class or put an entire course online, WebTutor delivers. WebTutor offers a wide array of resources including media assets, quizzing, web links, exercises, flashcards, and more. Visit webtutor.cengage.com to learn more. New to this edition are pop-up tutors, which help explain key topics with short video explanations.

Biology CourseMate   Cengage Learning's Biology CourseMate brings course concepts to life with interactive learning, study, and exam preparation tools that support the printed textbook, or the included ebook. With Course-Mate, professors can use the included Engagement Tracker to assess student preparation and engagement. Use the tracking tools to see progress for the class as a whole or for individual students.

To get access, visit Cengage Brain.com

http://www.cengage.com/qr/9

Premium eBook   This complete online version of the text is integrated with multimedia resources and special study features, providing the motivation that so many students need to study and the interactivity they need to learn. New to this edition are pop-up tutors, which help explain key topics with short video explanations.

ACKNOWLEDGMENTS

Writing, revising, and illustrating a biology textbook is a major undertaking for two full-time authors, but our efforts constitute only a small part of what is required to produce and sell this one. We are truly fortunate to be part of a huge team of very talented people who are as committed as we are to creating and disseminating an exceptional science education product.

Biology is not dogma; paradigm shifts are a common outcome of the fantastic amount of research in the field. Thus, ideas about what material should be taught and how best to present that material to students changes even from one year to the next. It is only with the ongoing input of our many academic reviewers and advisors (see *opposite page*) that we can continue to tailor this book to the needs of instructors and students while integrating new information and models. We continue to learn from and be inspired by these dedicated educators. A special thank-you goes to Michael Plotkin for his thoughtful, thorough input on current paradigms in evolutionary biology.

On the production side of our team, the indispensable Grace Davidson orchestrated the flow of countless files, photos, and illustrations while managing schedules, budgets, and whatever else happened to be on fire at the time. Grace, thank you for your continued patience and dedication. Photoresearch for this book presents a special challenge because we often request images from unusual or difficult-to-reach sources. Paul Forkner, thank you for your persistence and determination to track down many of the exceptional photos that make this book unique. Copyeditor Anita Wagner and proofreader Diane Miller, your valuable suggestions kept our text clear and concise.

At Cengage Learning, the brilliant John Walker created this book's highly appealing new design. Yolanda Cossio, thank you for continuing to support us and for encouraging our efforts to innovate and improve. Peggy Williams, we are as always grateful for your enthusiastic, thoughtful guidance, and for your many travels (and travails) on behalf of our books.

Thanks to Hal Humphrey our Cengage Production Manager, Tom Ziolkowski our Marketing Manager, Lauren Oliveira who creates our exciting technology package, Shannon Holt who managed all the print supplements for this edition, and Editorial Assistant Sean Cronin.

LISA STARR AND CHRISTINE EVERS
August 20011

Marc C. Albrecht
University of Nebraska at Kearney

Ellen Baker
Santa Monica College

Sarah Follis Barlow
Middle Tennessee State University

Tesfaye Belay
Bluefield State College

Michael C. Bell
Richland College

Lois Brewer Borek
Georgia State University

Robert S. Boyd
Auburn University

Uriel Angel Buitrago-Suarez
Harper College

Matthew Rex Burnham
Jones County Junior College

P.V. Cherian
Saginaw Valley State University

Larissa Crawford Clark
*Arkansas State University –
Newport*

Warren Coffeen
Linn Benton

Luigia Collo
Universita' Degli Studi Di Brescia

David T. Corey
Midlands Technical College

David F. Cox
Lincoln Land Community College

Kathryn Stephenson Craven
*Armstrong Atlantic State
University*

Jessica Crowe
Valdosta State University

Juville Dario-Becker
Central Virginia Community College

Sondra Dubowsky
Allen County Community College

Peter Ekechukwu
Horry-Georgetown Technical College

Daniel J. Fairbanks
Brigham Young University

Lewis A. Foster
Martin Methodist College

Mitchell A. Freymiller
University of Wisconsin - Eau Claire

Raul Galvan
South Texas College

Nabarun Ghosh
West Texas A&M University

Julian Granirer
URS Corporation

Stephanie G. Harvey
Georgia Southwestern State University

James A. Hewlett
Finger lakes community College

James Holden
Tidewater Community College - Portsmouth

Helen James
Smithsonian Institution

David Leonard
*Hawaii Department of Land and Natural
Resources*

Steve Mackie
Pima West Campus

Cindy Malone
California State University - Northridge

Kathleen A. Marrs
*Indiana University - Purdue University
Indianapolis*

Emilio Merlo-Pich
GlaxoSmithKline

Carlos Morales
Ocean County College

Charlotte Otts
New Mexico State University

Michael Plotkin
Mt. San Jacinto College

Michael D. Quillen
Maysville Community and Technical College

Wenda Ribeiro
Thomas Nelson Community College

Margaret G. Richey
Centre College

Jennifer Curran Roberts
Lewis University

Frank A. Romano, III
Jacksonville State University

Cameron Russell
*Tidewater Community College -
Portsmouth*

Al Schlundt
Faulkner University

Robin V. Searles-Adenegan
Morgan State University

Bruce Shmaefsky
Kingwood College

Bruce Stallsmith
University of Alabama - Huntsville

Linda Smith Staton
*Pollissippi State Technical
Community College*

Peter Svensson
West Valley College

Lisa Weasel
Portland State University

Diana C. Wheat
Linn-Benton Community College

Claudia M. Williams
Campbell University

Martin Zahn
Thomas Nelson Community College

LEARNING ROADMAP

Where you have been Whether or not you have studied biology, you already have an intuitive understanding of life on Earth because you are part of it. Every one of your experiences with the natural world—from the warmth of the sun on your skin to the love of your pet—contributes to that understanding.

Where you are now

The Science of Nature
We can understand life by studying it at many levels, starting with its component atoms and molecules and extending to interactions of organisms with their environment.

Life's Unity
All living things must have ongoing inputs of energy and raw materials; all sense and respond to change; and all have DNA that guides their development and functioning.

Life's Diversity
Observable characteristics vary tremendously among organisms. Various classification systems help us keep track of the differences.

How We Study Life
Carefully designed experiments help researchers unravel cause-and-effect relationships in complex natural systems.

What Science Is (and Is Not)
Science addresses only testable ideas about observable events and processes. It does not address the untestable, including beliefs and opinions.

Where you are going This book parallels nature's levels of organization, from atoms to the biosphere. Learning about the structure and function of atoms and molecules will prime you to understand how living cells work. Learning about processes that keep a single cell alive can help you understand how multicelled organisms survive. Knowing what it takes for organisms to survive can help you see why and how they interact with one another and their environment.

1.1 The Secret Life of Earth

In this era of detailed satellite imagery and cell phone global positioning systems, could there possibly be any places left on Earth that humans have not yet explored? Actually, there are plenty of them. As recently as 2005, for example, helicopters dropped a team of scientists into the middle of a vast and otherwise inaccessible cloud forest covering the top of New Guinea's Foja Mountains. Within a few minutes, the explorers realized that their landing site, a dripping, moss-covered swamp, had been untouched by humans. Team member Bruce Beehler remarked, "Everywhere we looked, we saw amazing things we had never seen before. I was shouting. This trip was a once-in-a-lifetime series of shouting experiences."

How did the explorers know they had landed in uncharted territory? For one thing, the forest was filled with plants and animals previously unknown even to native peoples that have long inhabited other parts of the region. During the next month, the team members discovered many new species, including a rhododendron plant with flowers the size of a plate and a frog the size of a pea. They also came across hundreds of species that are on the brink of extinction in other parts of the world, and some that supposedly had been extinct for decades. The animals had never learned to be afraid of humans, so they could easily be approached. A few were discovered as they casually wandered through campsites (**Figure 1.1**).

New species are discovered all the time, often in places much more mundane than Indonesian cloud forests. How do we know what species a particular organism belongs to? What is a species, anyway, and why should discovering a new one matter to anyone other than a scientist? You will find the answers to such questions in this book. They are part of the scientific study of life, **biology**, which is one of many ways we humans try to make sense of the world around us.

Trying to understand the immense scope of life on Earth gives us some perspective on where we fit into it. For example, hundreds of new species are discovered every year, but about 20 species become extinct every minute in rain forests alone—and those are only the ones we know about. The current rate of extinctions is about 1,000 times faster than normal, and human activities are responsible for the acceleration. At this rate, we will never know about most of the species that are alive on Earth today. Does that matter? Biologists think so. Whether or not we are aware of it, humans are intimately connected with the world around us. Our activities are profoundly changing the entire fabric of life on Earth. These changes are, in turn, affecting us in ways we are only beginning to understand.

Ironically, the more we learn about the natural world, the more we realize we have yet to learn. But don't take our word for it. Find out what biologists know, and what they do not, and you will have a solid foundation upon which to base your own opinions about how humans fit into this world. By reading this book, you are choosing to learn about the human connection—your connection—with all life on Earth.

biology The scientific study of life.

Figure 1.1 Explorers found hundreds of rare species and dozens of new ones during recent survey expeditions to the Foja Mountain cloud forest (*top*).

Bottom, Paul Oliver discovered this tree frog (*Litoria*) perched on a sack of rice during a particularly rainy campsite lunch. The explorers dubbed the new species "Pinocchio frog" after the Disney character because the male frog's long nose inflates and points upward during times of excitement.

1.2 Life Is More Than the Sum of Its Parts

- Biologists study life by thinking about it at different levels of organization.
- The quality of life emerges at the level of the cell.

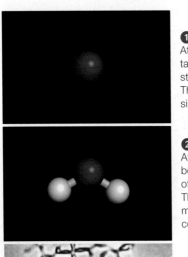

1 atom
Atoms are fundamental units of all substances, living or not. This is a model of a single oxygen atom.

2 molecule
Atoms joined in chemical bonds. This is a model of a water molecule. The molecules of life are much larger and more complex than water.

3 cell
The cell is the smallest unit of life. Some, like this plant cell, live and reproduce as part of a multicelled organism; others do so on their own.

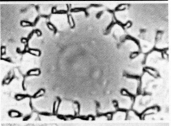

4 tissue
Organized array of cells and substances that interact in a collective task. This is epidermal tissue on the surface of a flower petal.

5 organ
Structural unit of interacting tissues. Flowers are the reproductive organs of many plants.

6 organ system
A set of interacting organs. The shoot system of this poppy plant includes its above-ground parts: leaves, flowers, and stems.

Figure 1.2 Animated Levels of life's organization.

Biologists study all aspects of life, past and present. What, exactly, is the property we call "life"? We may never actually come up with a good definition, because living things are too diverse, and they consist of the same basic components as nonliving things. When we try to define life, we end up only identifying properties that differentiate living from nonliving things.

Complex properties, including life, often emerge from the interactions of much simpler parts. For an example, take a look at these drawings:

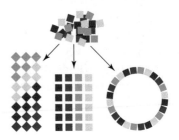

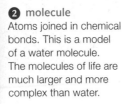

The property of "roundness" emerges when the parts are organized one way, but not other ways. Characteristics of a system that do not appear in any of the system's components are called **emergent properties**. The idea that structures with emergent properties can be assembled from the same basic building blocks is a recurring theme in our world, and also in biology.

Through the work of biologists, we are beginning to understand a pattern in life's organization. That organization occurs in successive levels, with new emergent properties appearing at each level (**Figure 1.2**).

Life's organization starts with interactions among atoms. **Atoms** are fundamental units of matter, building blocks of all substances **1**. Atoms join as **molecules** **2**. There are no atoms unique to living things, but there are unique molecules. In today's world, only living things make the "molecules of life," which are lipids, proteins, DNA, RNA, and complex carbohydrates. The emergent property of "life" appears at the next level, when many molecules of life become organized as a cell **3**. A **cell** is the smallest unit of life. Cells survive and replicate themselves using energy, raw materials, and information in their DNA. Some cells live and reproduce independently. Others do so as part of a multicelled organism. An **organism** is an individual that consists of one or more cells. A poppy plant is an example of a multicelled organism **7**.

In most multicelled organisms, cells make up tissues **4**. The cells of a **tissue** are typically specialized, and they are organized in a particular pattern. The arrangement allows the cells to collectively perform a special function such as protection from injury (dermal tissue), movement (muscle tissue), and so on. An **organ** is an organized array of tissues that collectively carry out a particular task or set of tasks **5**. For example, a flower is an organ of reproduction in plants; a heart, an organ that pumps blood in animals. An **organ system** is a set of organs and tissues that interact to keep the individual's body working properly **6**. Examples of organ systems include the aboveground parts of a plant (the shoot system), and the heart and blood vessels of an animal (the circulatory system).

A **population** is a group of individuals of the same type, or species, living in a given area **8**. An example would be all of the California poppies in California's Antelope Valley Poppy Reserve. At the next level, a **community** consists of all populations of all species in a given area. The Antelope Valley Reserve community includes California poppies and all other plants, animals, microorganisms, and so on **9**. Communities may be large or small, depending on the area defined.

The next level of organization is the **ecosystem**, or a community interacting with its physical and chemical environment **10**. The most inclusive level, the **biosphere**, encompasses all regions of Earth's crust, waters, and atmosphere in which organisms live **11**.

atom Fundamental building block of all matter.
biosphere All regions of Earth where organisms live.
cell Smallest unit of life.
community All populations of all species in a given area.
ecosystem A community interacting with its environment.
emergent property A characteristic of a system that does not appear in any of the system's component parts.
molecule An association of two or more atoms.
organ In multicelled organisms, a grouping of tissues engaged in a collective task.
organ system In multicelled organisms, set of organs engaged in a collective task that keeps the body functioning properly.
organism Individual that consists of one or more cells.
population Group of interbreeding individuals of the same species that live in a given area.
tissue In multicelled organisms, specialized cells organized in a pattern that allows them to perform a collective function.

Take-Home Message

How do living things differ from nonliving things?

» All things, living or not, consist of the same building blocks: atoms. Atoms join as molecules.

» The unique properties of life emerge as certain kinds of molecules become organized into cells.

» Higher levels of life's organization include multicelled organisms, populations, communities, ecosystems, and the biosphere.

» Emergent properties occur at each successive level of life's organization.

7 multicelled organism Individual that consists of one or more cells. Cells of this California poppy plant are part of its two organ systems: aboveground shoots and belowground roots.

8 population Group of single-celled or multicelled individuals of a species in a given area. This population of California poppy plants is in California's Antelope Valley Poppy Reserve.

9 community All populations of all species in a specified area. These plants are part of the Antelope Valley Poppy Reserve community.

10 ecosystem A community interacting with its physical environment through the transfer of energy and materials. Sunlight and water sustain the natural community in the Antelope Valley.

11 biosphere The sum of all ecosystems: every region of Earth's waters, crust, and atmosphere in which organisms live. The biosphere is a finite system, so no ecosystem in it can be truly isolated from any other.

1.3 How Living Things Are Alike

- Continual inputs of energy and the cycling of materials maintain life's complex organization.
- Organisms sense and respond to change.
- All organisms use information in the DNA they inherited from their parent or parents to develop and function.

A Producers harvest energy from the environment. Some of that energy flows from producers to consumers.

sunlight energy

Producers
plants and other
self-feeding organisms

B Nutrients that become incorporated into the cells of producers and consumers are eventually released by decomposition. Some cycle back to producers.

Consumers
animals, most fungi,
many protists, bacteria

C All of the energy that enters the world of life eventually flows out of it, mainly as heat released back to the environment.

Figure 1.3 Animated The one-way flow of energy and cycling of materials in the world of life. The photo (*top*) shows a producer acquiring energy and nutrients from the environment, and consumers acquiring energy and nutrients by eating the producer.

Even though we cannot define "life," we can intuitively understand what it means because all living things share a set of key features. All require ongoing inputs of energy and raw materials; all sense and respond to change; and all have DNA that guides their functioning.

Organisms Require Energy and Nutrients

Not all living things eat, but all require energy and nutrients on an ongoing basis. Both are essential to maintain life's organization and functioning. **Energy** is the capacity to do work. A **nutrient** is a substance that an organism needs for growth and survival but cannot make for itself.

Organisms spend a lot of time acquiring energy and nutrients. However, what type of energy and nutrients are acquired varies considerably depending on the type of organism. The differences allow us to classify all living things into two categories: producers and consumers. **Producers** make their own food using energy and simple raw materials they get directly from their environment. Plants are producers that use the energy of sunlight to make sugars from water and carbon dioxide (a gas in air), a process called **photosynthesis**. By contrast, **consumers** cannot make their own food. They get energy and nutrients by feeding on other organisms. Animals are consumers. So are decomposers, which feed on the wastes or remains of other organisms. The leftovers of consumers' meals end up in the environment, where they serve as nutrients for producers. Said another way, nutrients cycle between producers and consumers.

Energy, however, is not cycled. It flows through the world of life in one direction: from the environment, through organisms, and back to the environment. The flow of energy maintains the organization of each individual, and it is also the basis of how organisms interact with one another and their environment. It is a one-way flow because with each transfer, some energy escapes as heat. Cells do not use heat to do work. Thus, all of the energy that enters the world of life eventually leaves it, permanently (**Figure 1.3**).

Organisms Sense and Respond to Change

An organism cannot survive for very long in a changing environment unless it adapts to the changes. Thus, every living thing has the ability to sense and respond to conditions both inside and outside of itself (**Figure 1.4**). For example, after you eat, the sugars from your meal enter your bloodstream. The added sugars set in

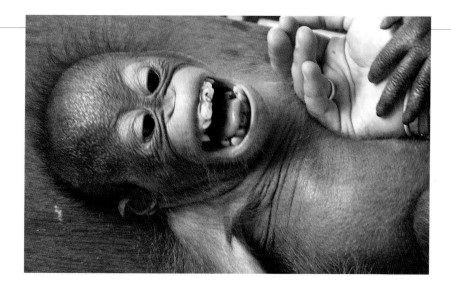

Figure 1.4 Organisms sense and respond to stimulation. This baby orangutan is laughing in response to being tickled. Apes and humans make different sounds when being tickled, but the airflow patterns are so similar that we can say apes really do laugh.

motion a series of events that causes cells throughout the body to take up sugar faster, so the sugar level in your blood quickly falls. This response keeps your blood sugar level within a certain range, which in turn helps keep your cells alive and your body functioning.

The fluid in your blood is part of your body's internal environment, which consists of all body fluids outside of cells. Unless that internal environment is kept within certain ranges of composition, temperature, and other conditions, your body cells will die. By sensing and adjusting to change, you and all other organisms keep conditions in the internal environment within a range that favors cell survival. **Homeostasis** is the name for this process, and it is one of the defining features of life.

Organisms Use DNA

With little variation, the same types of molecules perform the same basic functions in every organism. For example, information encoded in an organism's **DNA** (deoxyribonucleic acid) guides the ongoing metabolic activities that sustain the individual through its lifetime. Such activities include **development**: the process by which the first cell of a new individual becomes a multicelled adult; **growth**: increases in cell number, size, and volume; and **reproduction**: processes by which individuals produce offspring.

Individuals of every natural population are alike in certain aspects of their body form and behavior, an outcome of shared information encoded in DNA. Orangutans look like orangutans and not like caterpillars because they inherited orangutan DNA, which differs from caterpillar DNA in the information it carries. **Inheritance** refers to the transmission of DNA to offspring. All organisms receive their DNA from one or more parents.

DNA is the basis of similarities in form and function among organisms. However, the details of DNA molecules differ, and herein lies the source of life's diversity. Small variations in the details of DNA's structure give rise to differences among individuals, and also among types of organisms. As you will see in later chapters, these differences are the raw material of evolutionary processes.

consumer Organism that gets energy and nutrients by feeding on tissues, wastes, or remains of other organisms.
development Multistep process by which the first cell of a new individual becomes a multicelled adult.
DNA Deoxyribonucleic acid; carries hereditary information that guides development and functioning.
energy The capacity to do work.
growth In multicelled species, an increase in the number, size, and volume of cells.
homeostasis Set of processes by which an organism keeps its internal conditions within tolerable ranges.
inheritance Transmission of DNA to offspring.
nutrient Substance that an organism needs for growth and survival, but cannot make for itself.
photosynthesis Process by which producers use light energy to make sugars from carbon dioxide and water.
producer Organism that makes its own food using energy and simple raw materials from the environment.
reproduction Processes by which individuals produce offspring.

Take-Home Message

How are all living things alike?

» A one-way flow of energy and a cycling of nutrients sustain life's organization.

» Organisms sense and respond to conditions inside and outside themselves. They make adjustments that keep conditions in their internal environment within a range that favors cell survival, a process called homeostasis.

» Organisms develop and function based on information encoded in their DNA, which they inherit from their parents. DNA is the basis of similarities and differences in form and function.

1.4 How Living Things Differ

■ There is great variation in the details of appearance and other observable characteristics of living things.

Living things differ tremendously in their observable characteristics. Various classification schemes help us organize what we understand about the scope of this variation, which we call Earth's **biodiversity**.

For example, organisms can be classified into broad groups depending on whether they have a **nucleus**, which is a sac with two membranes that encloses and protects a cell's DNA. **Bacteria** (singular, bacterium) and **archaea** (singular, archaeon) are two types of organisms whose DNA is not contained within a nucleus. All bacteria and archaea are single-celled, which means each organism consists of one cell (Figure 1.5A,B). As a group, they are also the most diverse organisms: Different kinds are producers or consumers in nearly all regions of the biosphere. Some inhabit such extreme environments as frozen desert rocks, boiling sulfurous lakes, and nuclear reactor waste. The first cells on Earth may have faced similarly hostile challenges to survival.

Traditionally, organisms without a nucleus have been called prokaryotes, but this designation is an informal one. Despite their similar appearance, bacteria and archaea are less related to one another than we had once thought. Archaea are actually more closely related to **eukaryotes**, organisms whose DNA is contained within a nucleus. Some eukaryotes live as individual cells; others are multicelled (Figure 1.5C). Eukaryotic cells are typically larger and more complex than bacteria or archaea.

Structurally, **protists** are the simplest eukaryotes. As a group they vary a great deal, from single-celled consumers to giant, multicelled producers.

Most **fungi** (singular, fungus), such as the types that form mushrooms, are multicelled eukaryotes. Many are decomposers. All are consumers that secrete substances that break down food outside of the body. Their cells then absorb the released nutrients.

animal Multicelled consumer that develops through a series of stages and moves about during part or all of its life cycle.
archaeon Member of a group of single-celled organisms that lack a nucleus but are more closely related to eukaryotes than to bacteria.
bacterium Member of the most diverse and well-known group of single-celled organisms that lack a nucleus.
biodiversity Scope of variation among living organisms.
eukaryote Organism whose cells characteristically have a nucleus.
fungus Single-celled or multicelled eukaryotic consumer that digests material outside its body, then absorbs released nutrients.
nucleus Double-membraned sac that encloses a cell's DNA.
plant A multicelled, typically photosynthetic producer.
protist Member of a diverse group of simple eukaryotes.

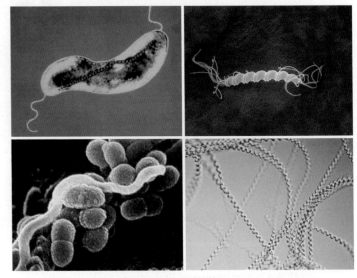

A Bacteria are the most numerous organisms on Earth. *Clockwise from upper left*, a bacterium with a row of iron crystals that acts like a tiny compass; a common resident of cat and dog stomachs; spiral cyanobacteria; types found in dental plaque.

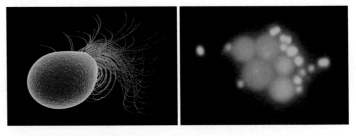

B Archaea resemble bacteria, but are more closely related to eukaryotes. *Left*, an archaeon from volcanic ocean sediments. *Right*, two types of archaea from a seafloor hydrothermal vent.

Figure 1.5 Animated Representatives of life's diversity.

Plants are multicelled eukaryotes that live on land or in freshwater environments. Nearly all are photosynthetic producers. Besides feeding themselves, plants and other photosynthesizers also serve as food for most of the other organisms in the biosphere.

Animals are multicelled consumers that ingest tissues or juices of other organisms. Herbivores graze, carnivores eat meat, scavengers eat remains of other organisms, parasites withdraw nutrients from the tissues of a host, and so on. Animals develop through a series of stages that lead to the adult form. All kinds actively move about during at least part of their lives.

Take-Home Message

How do organisms differ from one another?

» Organisms differ in their details; they show tremendous variation in observable characteristics, or traits.

» We divide Earth's biodiversity into broad groups based on traits such as having a nucleus or being multicellular.

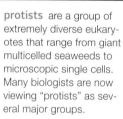

protists are a group of extremely diverse eukaryotes that range from giant multicelled seaweeds to microscopic single cells. Many biologists are now viewing "protists" as several major groups.

plants are multicelled eukaryotes, most of which are photosynthetic. Nearly all have roots, stems, and leaves. Plants are the primary producers in land ecosystems.

fungi are eukaryotes. Most are multicelled. Different kinds are parasites, pathogens, or decomposers. Without decomposers such as fungi, communities would be buried in their own wastes.

animals are multicelled eukaryotes that ingest tissues or juices of other organisms. All actively move about during at least part of their life.

C **Eukaryotes** are single-celled or multicelled organisms whose DNA is contained within a nucleus.

1.5 Organizing Information About Species

- Each type of organism, or species, is given a unique name.
- We define and group species based on shared traits.

Each time we discover a new **species**, or unique kind of organism, we name it. **Taxonomy**, a system of naming and classifying species, began thousands of years ago. However, naming species in a consistent way did not become a priority until the eighteenth century. At the time, European explorers and naturalists who were just discovering the scope of life's diversity started having more and more trouble communicating with one another because species often had multiple names.

For example, the dog rose (a plant native to Europe, Africa, and Asia) was alternately known as briar rose, witch's briar, herb patience, sweet briar, wild briar, dog briar, dog berry, briar hip, eglantine gall, hep tree, hip fruit, hip rose, hip tree, hop fruit, and hogseed—and those are only the English names! Species often had multiple scientific names too, in Latin that was descriptive but often cumbersome. The scientific name of the dog rose was *Rosa sylvestris inodora seu canina* (odorless woodland dog rose), and also *Rosa sylvestris alba cum rubore, folio glabro* (pinkish white woodland rose with smooth leaves).

An eighteenth-century naturalist, Carolus Linnaeus, standardized a two-part naming system that we still use today. By the Linnaean system, every species is given a unique two-part scientific name. The first part is the name of the **genus** (plural, genera), a group of species that share a unique set of features. The second part is the **specific epithet**. Together, the genus name plus the specific epithet designate one species. Thus, the dog rose now has one official name, *Rosa canina*, that is recognized worldwide.

Genus and species names are always italicized. For example, *Panthera* is a genus of big cats. Lions belong to the species *Panthera leo*. Tigers belong to a different species in the same genus (*Panthera tigris*), and so do leopards (*P. pardus*). Note how the genus name may be abbreviated after it has been spelled out once.

Today we rank species into ever more inclusive categories. Each rank, or **taxon** (plural, taxa), is a group of organisms that share a unique set of features. The categories above species—genus, family, order, class, phylum, kingdom, and domain—are the higher taxa (**Figure 1.6**). Each higher taxon consists of a group of the next lower taxon. Using this system, we can sort all life into a few categories (**Figure 1.7**).

A Rose by Any Other Name . . .

The individuals of a species share a unique set of observable characteristics, or **traits**. For example, giraffes normally have very long necks, brown spots

domain	Eukarya	Eukarya	Eukarya	Eukarya	Eukarya
kingdom	Plantae	Plantae	Plantae	Plantae	Plantae
phylum	Magnoliophyta	Magnoliophyta	Magnoliophyta	Magnoliophyta	Magnoliophyta
class	Magnoliopsida	Magnoliopsida	Magnoliopsida	Magnoliopsida	Magnoliopsida
order	Apiales	Rosales	Rosales	Rosales	Rosales
family	Apiaceae	Cannabaceae	Rosaceae	Rosaceae	Rosaceae
genus	*Daucus*	*Cannabis*	*Malus*	*Rosa*	*Rosa*
species	*carota*	*sativa*	*domestica*	*acicularis*	*canina*
common name	wild carrot	marijuana	apple	prickly rose	dog rose

Figure 1.6 Taxonomic classification of five species that are related at different levels. Each species has been assigned to ever more inclusive groups, or taxa: in this case, from genus to domain.
Figure It Out: Which of the plants shown here are in the same order?

Answer: Marijuana, apple, prickly rose, and dog rose are all in the order Rosales.

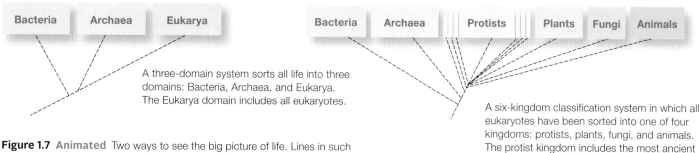

A three-domain system sorts all life into three domains: Bacteria, Archaea, and Eukarya. The Eukarya domain includes all eukaryotes.

Figure 1.7 **Animated** Two ways to see the big picture of life. Lines in such diagrams indicate evolutionary connections. Compare **Figure 1.6**.

A six-kingdom classification system in which all eukaryotes have been sorted into one of four kingdoms: protists, plants, fungi, and animals. The protist kingdom includes the most ancient multicelled and all single-celled eukaryotes.

on white coats, and so on. These are morphological traits (*morpho*– means form). Individuals of a species also share physiological traits, and they respond the same way to certain stimuli, as when hungry giraffes feed on tree leaves. These are behavioral traits.

A species is assigned to higher taxa based on some subset of heritable traits it shares with other species. That assignment may change as we discover more about the species and the traits involved. For example, Linnaeus grouped plants by the number and arrangement of reproductive parts, a scheme that resulted in odd pairings such as castor-oil plants with pine trees. Having more information today, we place these plants in separate phyla.

Traits vary a bit within a species, such as eye color does in people. However, there are often tremendous differences between species. Think of petunias and whales, beetles and emus, and so on. Such species look very different, so it is easy to tell them apart. Species that share a more recent ancestor may be much more difficult to distinguish (**Figure 1.8**).

How do we know whether similar-looking organisms belong to different species? The short answer is that we rely on whatever information we have. For example, early naturalists studied anatomy and distribution—essentially the only methods available at the time. Thus, species were named and classified according to what they looked like and where they lived. Today's biologists have at their disposal an array of techniques much more sophisticated than those of the eighteenth century. They are able to study and compare traits that the early naturalists did not even know about, including biochemical ones such as DNA sequences. As you will see in Section 18.4, such comparisons are deepening our understanding of the shared evolutionary history among species.

Evolutionary biologist Ernst Mayr defined a species as one or more groups of individuals that potentially can interbreed, produce fertile offspring, and do not interbreed with other groups. This "biological species concept" is useful in many cases, but it is not universally applicable. For example, we may never know whether separate populations could interbreed even if they did get together. As another example, populations often continue to interbreed even as they diverge, so the exact moment at which two populations become two species is often impossible to pinpoint. We return to speciation and how it occurs in Chapter 17, but for now it is important to remember that a "species" is a convenient but artificial construct of the human mind.

Figure 1.8 Four butterflies, two species: Which are which?

The *top* row shows two forms of the species *Heliconius melpomene*; the *bottom* row, two forms of *Heliconius erato*.

H. melpomene and *H. erato* never cross-breed. Their alternate but similar patterns of coloration evolved as a shared warning signal to predatory birds that these butterflies taste terrible.

genus A group of species that share a unique set of traits; also the first part of a species name.
species Unique type of organism.
specific epithet Second part of a species name.
taxon A group of organisms that share a unique set of features.
taxonomy The science of naming and classifying species.
trait An observable characteristic of an organism or species.

Take-Home Message

How do we keep track of all the species we know about?

» Each species has a unique, two-part scientific name.

» Classification systems group species on the basis of shared, inherited traits.

1.6 The Science of Nature

■ Judging the quality of information before accepting it is called critical thinking.

■ Scientists practice critical thinking by testing predictions about how the natural world works.

Thinking About Thinking

Most of us assume that we do our own thinking, but do we, really? You might be surprised to find out how often we let others think for us. For instance, a school's job, which is to impart as much information as possible to students, meshes perfectly with a student's job, which is to acquire as much knowledge as possible. In this rapid-fire exchange of information, it is sometimes easy to forget about the quality of what is being exchanged. Anytime you accept information without questioning it, you let someone else think for you.

Critical thinking is the deliberate process of judging the quality of information before accepting it. "Critical" comes from the Greek *kriticos* (discerning judgment). When you use critical thinking, you move beyond the content of new information to consider supporting evidence, bias, and alternative interpretations. How does the busy student manage this? Critical thinking does not necessarily require extra time, just a bit of extra awareness. There are many ways to do it. For example, you might ask yourself some of the following questions while you are learning something new:

> What message am I being asked to accept?
> Is the message based on facts or opinion?
> Is there a different way to interpret the facts?
> What biases might the presenter have?
> How do my own biases affect what I'm learning?

A Sequencing the human genome.

Figure 1.9 Examples of research in the field of biology.

Such questions are a way of being conscious about learning. They can help you decide whether to allow new information to guide your beliefs and actions.

How Science Works

Critical thinking is a big part of **science**, the systematic study of the observable world and how it works (**Figure 1.9**). A scientific line of inquiry usually begins with curiosity about something observable, such as, say, a decrease in the number of birds in a particular area. Typically, a scientist will read about what others have discovered before making a **hypothesis**, a testable explanation for a natural phenomenon. An example of a hypothesis would be, "The number of birds is decreasing because the number of cats is increasing." Making a hypothesis this way is an example of **inductive reasoning**, which means arriving at a conclusion based on one's observations. Inductive reasoning is the way we come up with new ideas about groups of objects or events.

A **prediction**, or statement of some condition that should exist if the hypothesis is correct, comes next. Making predictions is called the if–then process, in which the "if" part is the hypothesis, and the "then" part is the prediction. Using a hypothesis to make a prediction is a form of **deductive reasoning**, or logical process of using a general premise to draw a conclusion about a specific case.

Next, a scientist will devise ways to test a prediction. Tests may be performed on a **model**, or analogous system, if working with an object or event directly is not possible. For example, animal diseases are often used as models of similar human diseases. Careful

Table 1.1 The Scientific Method
1. Observe some aspect of nature.
2. Think of an explanation for your observation (in other words, form a hypothesis).
3. Test the hypothesis.
a. Make a prediction based on the hypothesis.
b. Test the prediction using experiments or surveys.
c. Analyze the results of the tests (data).
4. Decide whether the results of the tests support your hypothesis or not (form a conclusion).
5. Report your results to the scientific community.

B Looking for fungi in atmospheric dust.

C Improving the efficiency of biofuel production from agricultural wastes.

D Studying the ecological benefits of weedy buffer zones on farms.

observations are one way to test predictions that flow from a hypothesis. So are **experiments**: tests designed to support or falsify a prediction. A typical experiment explores a cause-and-effect relationship.

Researchers investigate causal relationships by changing or observing **variables**, which are characteristics or events that can differ among individuals or over time. An **independent variable** is defined or controlled by the person doing the experiment. A **dependent variable** is an observed result that is supposed to be influenced by the independent variable. For example, an independent variable in an investigation of hangover preventions may be the administration of artichoke extract before alcohol consumption. The dependent variable in this experiment would be the severity of the forthcoming hangover.

Biological systems are complex, with many interacting variables. It can be difficult to study one variable separately from the rest. Thus, biology researchers often test two groups of individuals simultaneously. An **experimental group** is a set of individuals that have a certain characteristic or receive a certain treatment. This group is tested side by side with a **control group**, which is identical to the experimental group except for one independent variable: the characteristic or the treatment being tested. Any differences in experimental results between the two groups should be an effect of changing the variable.

Test results—**data**—that are consistent with the prediction are evidence in support of the hypothesis. Data inconsistent with the prediction are evidence that the hypothesis is flawed and should be revised.

A necessary part of science is reporting one's results and conclusions in a standard way, such as in a peer-reviewed journal article. The communication gives other scientists an opportunity to evaluate the information for themselves, both by checking the conclusions drawn and by repeating the experiments.

Forming a hypothesis based on observation, and then systematically testing and evaluating the hypothesis are collectively called the **scientific method** (Table 1.1).

control group In an experiment, a group of individuals who are not exposed to the independent variable being tested.
critical thinking Judging information before accepting it.
data Experimental results.
deductive reasoning Using a general idea to make a conclusion about a specific case.
dependent variable In an experiment, a variable that is presumably affected by the independent variable being tested.
experiment A test designed to support or falsify a prediction.
experimental group In an experiment, a group of individuals who are exposed to an independent variable.
hypothesis Testable explanation of a natural phenomenon.
independent variable Variable that is controlled by an experimenter in order to explore its relationship to a dependent variable.
inductive reasoning Drawing a conclusion based on observation.
model Analogous system used for testing hypotheses.
prediction Statement, based on a hypothesis, about a condition that should exist if the hypothesis is correct.
science Systematic study of the observable world.
scientific method Making, testing, and evaluating hypotheses.
variable In an experiment, a characteristic or event that differs among individuals or over time.

Take-Home Message

What is science?

» The scientific method consists of making, testing, and evaluating hypotheses. It is a way of critical thinking, or systematically judging the quality of information before allowing it to guide one's beliefs and actions.

» Experiments measure how changing an independent variable affects a dependent variable.

1.7 Examples of Experiments in Biology

■ Researchers unravel cause-and-effect relationships in complex natural processes by changing one variable at a time.

There are many different ways to do research, particularly in biology. Some biologists do surveys; they observe without making hypotheses. Some make hypotheses and leave the experimentation to others. However, despite a broad range of subject matter, scientific experiments are typically designed in a consistent way. Experimenters try to change one independent variable at a time, and see what happens to a dependent variable.

To give you a sense of how biology experiments work, we summarize two published studies here.

Potato Chips and Stomachaches

In 1996 the U.S. Food and Drug Administration (the FDA) approved Olestra®, a fat replacement manufactured from sugar and vegetable oil, as a food additive. Potato chips were the first Olestra-containing food product on the market in the United States. Controversy about the food additive soon raged.

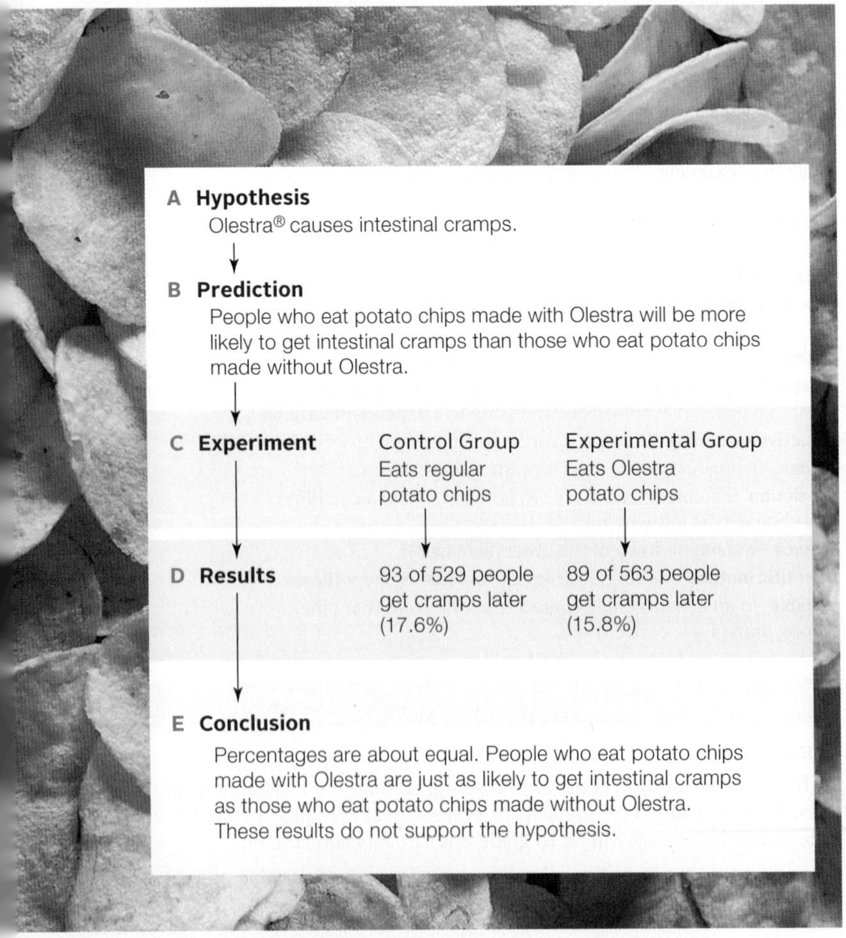

Many people complained of intestinal problems after eating the chips, and thought that the Olestra was at fault. Two years later, researchers at the Johns Hopkins University School of Medicine designed an experiment to test whether Olestra causes cramps.

The researchers predicted that if Olestra indeed causes cramps, then people who eat Olestra will be more likely to get cramps than people who do not. To test the prediction, they used a Chicago theater as a "laboratory." They asked 1,100 people between the ages of thirteen and thirty-eight to watch a movie and eat their fill of potato chips. Each person got an unmarked bag that contained 13 ounces of chips.

In this experiment, the individuals who got Olestra-containing potato chips constituted the experimental group, and individuals who got regular chips were the control group. The independent variable was the presence or absence of Olestra in the chips.

A few days after the experiment was finished, the researchers contacted all of the people and collected any reports of post-movie gastrointestinal problems. Of 563 people making up the experimental group, 89 (15.8 percent) complained about cramps. However, so did 93 of the 529 people (17.6 percent) making up the control group—who had eaten the regular chips.

People were about as likely to get cramps whether or not they ate chips made with Olestra. These results did not support the prediction, so the researchers concluded that eating Olestra does not cause cramps (**Figure 1.10**).

Butterflies and Birds

The peacock butterfly is a winged insect named for the large, colorful spots on its wings. In 2005, researchers reported the results of experiments investigating whether certain behaviors help peacock butterflies defend themselves against insect-eating birds. The researchers began with two observations. First, when a peacock butterfly rests, it folds its wings, so only the dark underside shows (**Figure 1.11A**). Second, when a butterfly sees a predator approaching, it repeatedly flicks its wings open and closed, while also moving the hindwings in a way that produces a hissing sound and a series of clicks.

A Hypothesis
Olestra® causes intestinal cramps.

B Prediction
People who eat potato chips made with Olestra will be more likely to get intestinal cramps than those who eat potato chips made without Olestra.

C Experiment	Control Group Eats regular potato chips	Experimental Group Eats Olestra potato chips
D Results	93 of 529 people get cramps later (17.6%)	89 of 563 people get cramps later (15.8%)

E Conclusion
Percentages are about equal. People who eat potato chips made with Olestra are just as likely to get intestinal cramps as those who eat potato chips made without Olestra. These results do not support the hypothesis.

Figure 1.10 The steps in a scientific experiment to determine if Olestra causes cramps. A report of this study was published in the *Journal of the American Medical Association* in January 1998. **Figure It Out:** What was the dependent variable in this experiment?

Answer: Whether or not a person got cramps

Figure 1.11 Peacock butterfly defenses against predatory birds. (**A**) With wings folded, a resting peacock butterfly looks a bit like a dead leaf. (**B**) When a bird approaches, the butterfly repeatedly flicks its wings open and closed, a behavior that exposes brilliant spots and produces hissing and clicking sounds.

Researchers tested whether the butterfly's behavior deters blue tits (**C**). They painted over the spots of some butterflies, cut the sound-making part of the wings on other butterflies, and did both to a third group; then the biologists exposed each butterfly to a hungry bird.

The results, listed in **Table 1.2**, support the hypotheses that peacock butterfly spots and sounds can deter predatory birds.

Figure It Out: What percentage of butterflies with no spots and no sound survived the test?   Answer: 20 percent

Wing Spots	Wing Sound	Total Number of Butterflies	Number Eaten	Number Survived
Spots	Sound	9	0	9 (100%)
No spots	Sound	10	5	5 (50%)
Spots	No sound	8	0	8 (100%)
No spots	No sound	10	8	2 (20%)

Table 1.2 Results of Peacock Butterfly Experiment*

* *Proceedings of the Royal Society of London, Series B* (2005) 272: 1203–1207.

The researchers were curious about why the peacock butterfly flicks its wings. After they reviewed earlier studies, they came up with two hypotheses that might explain the wing-flicking behavior:

1. Although wing-flicking probably attracts predatory birds, it also exposes brilliant spots that resemble owl eyes (**Figure 1.11B**). Anything that looks like owl eyes is known to startle small, butterfly-eating birds, so exposing the wing spots might scare off predators.
2. The hissing and clicking sounds produced when the peacock butterfly moves its hindwings may be an additional defense that deters predatory birds.

The researchers used their hypotheses to make the following predictions:

1. If peacock butterflies startle predatory birds by exposing their brilliant wing spots, then individuals with wing spots will be less likely to get eaten by predatory birds than those without wing spots.
2. If peacock butterfly sounds deter predatory birds, then sound-producing individuals will be less likely to get eaten by predatory birds than silent individuals.

The next step was the experiment. The researchers used a marker to paint the wing spots of some but-

terflies black, and scissors to cut off the sound-making part of the hindwings of others. A third group had their wing spots painted and their hindwings cut. The researchers then put each butterfly into a large cage with a hungry blue tit (**Figure 1.11C**) and watched the pair for thirty minutes.

Table 1.2 lists the results of the experiment. All of the butterflies with unmodified wing spots survived, regardless of whether they made sounds. By contrast, only half of the butterflies that had spots painted out but could make sounds survived. Most of the silenced butterflies with painted-out spots were eaten quickly. The test results confirmed both predictions, so they support the hypotheses. Predatory birds are indeed deterred by peacock butterfly sounds, and even more so by wing spots.

Take-Home Message

Why do biologists perform experiments?

» Natural processes are often very complex and influenced by many interacting variables.

» Experiments help researchers unravel causes of complex natural processes by focusing on the effects of changing a single variable.

1.8 Analyzing Experimental Results

■ Biology researchers often experiment on subsets of a group. Results from such an experiment may differ from results of the same experiment performed on the whole group.

■ Science is, ideally, a self-correcting process because scientists check one another's work.

Sampling Error

Researchers can rarely observe all individuals of a group. For example, the explorers you read about in Section 1.1 did not—and could not—survey every uninhabited part of the Foja Mountains. The cloud forest itself cloaks more than 2 million acres, so surveying all of it would take unrealistic amounts of time and effort. Besides, tromping about even in a small area can damage delicate forest ecosystems.

Given such limitations, researchers often look at subsets of an area, a population, or an event. They test or survey the subset, then use the results to make generalizations. However, generalizing from a subset is risky because a subset may not be representative of the whole.

For example, the golden-mantled tree kangaroo was first discovered in 1993 on a single forested mountaintop in New Guinea. For more than a decade, the species was never seen outside of that habitat, which is getting smaller every year because of human activities. Thus, the golden-mantled tree kangaroo was considered to be one of the most endangered animals on the planet. Then, in 2005, the New Guinea explorers discovered that this kangaroo species is fairly common in the Foja Mountain cloud forest (**Figure 1.12**). As a result,

Figure 1.12 Biologist Kris Helgen holds a golden-mantled tree kangaroo he found during the 2005 expedition to the Foja Mountains cloud forest in New Guinea.

This kangaroo species is extremely rare in other regions of the world, so it was thought to be critically endangered prior to the 2005 survey expedition.

biologists now believe its future is secure, at least for the moment.

Making generalizations from testing or surveying a subset is risky because of sampling error. **Sampling error** is a difference between results obtained from testing a subset of a group, and results from testing the whole group. Sampling error may be unavoidable, as illustrated by the example of the golden-mantled tree kangaroo. However, knowing how it can occur helps researchers design their experiments to minimize it. For example, sampling error can be a substantial problem with a small subset (**Figure 1.13**), so experimenters try to start with a relatively large sample, and they typically repeat their experiments.

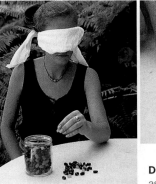

A Natalie, blindfolded, randomly plucks a jelly bean from a jar. The jar contains 120 green and 280 black jelly beans, so 30 percent of the jelly beans in the jar are green, and 70 percent are black.

B The jar is hidden from Natalie's view before she removes her blindfold. She sees one green jelly bean in her hand and assumes that the jar must hold only green jelly beans.

C Still blindfolded, Natalie randomly picks out 50 jelly beans from the jar. She ends up picking out 10 green and 40 black ones.

D The larger sample leads Natalie to assume that one-fifth of the jar's jelly beans are green (20 percent) and four-fifths are black (80 percent). This sample more closely approximates the jar's actual green-to-black ratio of 30 percent to 70 percent. The more times Natalie repeats the sampling, the greater the chance she has of guessing the actual ratio.

Figure 1.13 How sample size affects sampling error.

To understand why such practices reduce the risk of sampling error, think about what happens each time you flip a coin. There are two possible outcomes: The coin lands heads up, or it lands tails up. Thus, the chance that the coin will land heads up is one in two (1/2), which is a proportion of 50 percent. However, when you flip a coin repeatedly, it often lands heads up, or tails up, several times in a row. With just 3 flips, the proportion of times that heads actually land up may not even be close to 50 percent. With 1,000 flips, the proportion of times that the coin lands heads up is likely to be close to 50 percent.

In cases like flipping a coin, it is possible to calculate **probability**: the measure, expressed as a percentage, of the chance that a particular outcome will occur. That chance depends on the total number of possible outcomes. For instance, imagine that 10 million people enter a random drawing to win a car. There are 10 million possible outcomes, so each person has the same probability of winning the item: 1 in 10 million, or (an extremely improbable) 0.00001 percent.

Analysis of experimental data often includes calculations of probability. If a result is very unlikely to have occurred by chance alone, it is said to be **statistically significant**. In this context, the word "significant" does not refer to the result's importance. It means that a complicated statistical analysis shows the result has a very low probability (typically, less than a 5 percent chance) of being skewed by sampling error. In science, every result—even a statistically significant one—has a probability of being incorrect.

Variation in a set of data is often shown as error bars on a graph (**Figure 1.14**). Depending on the graph, error bars may indicate variation around an average for one sample set, or the difference between two sample sets.

Bias in Interpreting Results

Experimenting with a single variable apart from all others is not often possible, particularly when studying humans. For example, remember that the people who participated in the Olestra experiment were chosen randomly, which means the study was not controlled for gender, age, weight, medications taken, and so on. Such variables may well have influenced the results.

probability The chance that a particular outcome of an event will occur; depends on the total number of outcomes possible.
sampling error Difference between results derived from testing an entire group of events or individuals, and results derived from testing a subset of the group.
statistically significant Refers to a result that is statistically unlikely to have occurred by chance.

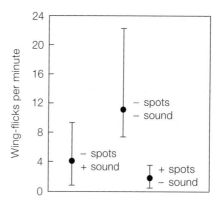

Figure 1.14 Example of error bars in a graph. This particular graph was adapted from the peacock butterfly research described in Section 1.7.

The researchers recorded the number of times each butterfly flicked its wings in response to an attack by a bird. The dots represent average frequency of wing flicking for each sample set of butterflies. The error bars that extend above and below the dots indicate the range of values—the sampling error.

Figure It Out: What was the fastest rate at which a butterfly with no spots or sound flicked its wings?

Answer: 22 times per minute

Humans are by nature subjective, and scientists are no exception. Experimenters risk interpreting their results in terms of what they want to find out. That is why they often design experiments to yield quantitative results, which are counts or some other data that can be measured or gathered objectively. Such results minimize the potential for bias, and also give other scientists an opportunity to repeat the experiments and check the conclusions drawn from them.

This last point gets us back to the role of critical thinking in science. Scientists expect one another to recognize and put aside bias in order to test their hypotheses in ways that may prove them wrong. If a scientist does not, then others will, because exposing errors is just as useful as applauding insights. The scientific community consists of critically thinking people trying to poke holes in one another's ideas. Their collective efforts make science a self-correcting endeavor.

Take-Home Message

How do scientists avoid potential pitfalls of sampling error and bias when doing research?

» Researchers minimize sampling error by using large sample sizes and by repeating their experiments.

» A statistical analysis can show the probability that a result has occurred by chance alone.

» Science is a self-correcting process because it is carried out by an aggregate community of people systematically checking one another's ideas.

1.9 The Nature of Science

- Scientific theories are our best descriptions of reality.
- Science helps us to be objective about our observations, in part because it is limited to the observable.

Suppose a hypothesis stands even after years of tests. It is consistent with all data ever gathered, and it has helped us make successful predictions about other phenomena. When a hypothesis meets these criteria, it is considered to be a **scientific theory** (Table 1.3).

To give an example, all observations to date have been consistent with the hypothesis that matter consists of atoms. Scientists no longer spend time testing this hypothesis for the compelling reason that, since we started looking 200 years ago, no one has discovered matter that doesn't consist of atoms. Thus, scientists use the hypothesis, now called atomic theory, to make other hypotheses about matter and the way it behaves.

Scientific theories are our best descriptions of reality. However, they can never be proven absolutely, because to do so would necessitate testing under every possible circumstance. For example, in order to prove atomic theory, the atomic composition of all matter in the universe would have to be checked—an impossible task even if someone wanted to try.

Like all hypotheses, a scientific theory can be disproven by a single observation or result that is inconsistent with it. For example, if someone discovers a form of matter that does not consist of atoms, atomic theory would have to be revised. The potentially falsifiable nature of scientific theories means that science has a built-in system of checks and balances. A theory is revised until no one can prove it to be incorrect. For example, the theory of evolution, which states that

change occurs in a line of descent over time, still holds after a century of observations and testing. As with all other scientific theories, no one can be absolutely sure that it will hold under all possible conditions, but it has a very high probability of not being wrong. Few other theories have withstood as much scrutiny.

You may hear people apply the word "theory" to a speculative idea, as in the phrase "It's just a theory." This everyday usage of the word differs from the way it is used in science. Speculation is an opinion, belief, or personal conviction that is not necessarily supported by evidence. A scientific theory is different. By definition, it is supported by a large body of evidence, and it is consistent with all known facts.

A scientific theory also differs from a **law of nature**, which describes a phenomenon that has been observed to occur in every circumstance without fail, but for which we do not have a complete scientific explanation. The laws of thermodynamics, which describe energy, are examples. We know how energy behaves, but not exactly why it behaves the way it does.

The Limits of Science

Science helps us be objective about our observations in part because of its limitations. For example, science does not address many questions, such as "Why do I exist?" Answers to such questions can only come from within as an integration of the personal experiences and mental connections that shape our consciousness. This is not to say subjective answers have no value, because no human society can function for long unless its individuals share standards for making judgments,

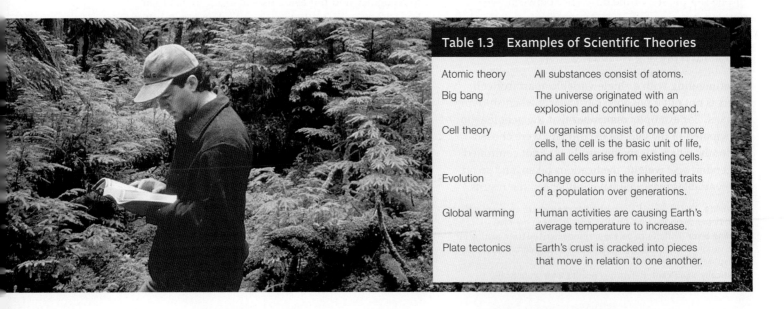

Table 1.3 Examples of Scientific Theories

Atomic theory	All substances consist of atoms.
Big bang	The universe originated with an explosion and continues to expand.
Cell theory	All organisms consist of one or more cells, the cell is the basic unit of life, and all cells arise from existing cells.
Evolution	Change occurs in the inherited traits of a population over generations.
Global warming	Human activities are causing Earth's average temperature to increase.
Plate tectonics	Earth's crust is cracked into pieces that move in relation to one another.

The Secret Life of Earth (revisited)

Of an estimated 100 billion species that have ever lived, at least 100 million are still with us. That number is only an estimate because we are still discovering them. For example, a mouse-sized opossum and a cat-sized rat turned up on a return trip to the Foja Mountains. Other recent surveys have revealed a new species of leopard in Borneo; lemurs and sucker-footed bats in Madagascar; birds in the Philippines; a jackal in Egypt; monkeys in Tanzania, Brazil, and India;

Figure 1.15 The discoverer of a new species typically has the honor of naming it. Dr. Jason Bond holds a new species of spider he discovered in California in 2008. Bond named the spider *Aptostichus stephencolberti*, after TV personality Stephen Colbert.

insects and spiders in California; a giant crayfish in Tennessee; a rat-eating plant in the Philippines; carnivorous sponges near Antarctica; whales, sharks, giant jellylike animals, fishes, and other aquatic wildlife; and scores of plants and single-celled organisms. Most were discovered by biologists simply trying to find out what lives where.

Biologists discover thousands of new species per year (**Figure 1.15**). Each is a reminder that we do not yet know all of the organisms living on our own planet. We don't even know how many to look for. The vast information about the 1.8 million species we do know about changes so quickly that collating it has been impossible—until recently. A web site titled the Encyclopedia of Life is intended to be an online reference source and database of species information maintained by collaborative effort. See its progress at www.eol.org.

How would you vote? Substantial populations of some species currently listed as endangered may exist in unexplored areas. Should we wait to protect endangered species until all of Earth has been surveyed?

---

even if they are subjective. Moral, aesthetic, and philosophical standards vary from one society to the next, but all help people decide what is important and good. All give meaning to our lives.

Neither does science address the supernatural, or anything that is "beyond nature." Science neither assumes nor denies that supernatural phenomena occur, but scientists often cause controversy when they discover a natural explanation for something that was thought to have none. Such controversy often arises when a society's moral standards are interwoven with its understanding of nature.

For example, Nicolaus Copernicus concluded in 1540 that Earth orbits the sun. Today that idea is generally accepted, but during Copernicus's time the prevailing belief system had Earth as the immovable center of the universe. In 1610, astronomer Galileo Galilei published evidence for the Copernican model of the solar system, an act that resulted in his imprisonment. He was publicly forced to recant his work, spent the rest of his life under house arrest, and was never allowed to publish again.

law of nature Generalization that describes a consistent natural phenomenon for which there is incomplete scientific explanation.
scientific theory Hypothesis that has not been disproven after many years of rigorous testing.

As Galileo's story illustrates, exploring a traditional view of the natural world from a scientific perspective is often misinterpreted as a violation of morality. As a group, scientists are no less moral than anyone else. However, they follow a particular set of rules that do not necessarily apply to others: Their work concerns only the natural world, and their ideas must be testable in ways others can repeat.

Science helps us communicate our experiences without bias. As such, it may be as close as we can get to a universal language. We are fairly sure, for example, that the laws of gravity apply everywhere in the universe. Intelligent beings on a distant planet would likely understand the concept of gravity. We might well use gravity or another scientific concept to communicate with them, or anyone, anywhere. The point of science, however, is not to communicate with aliens. It is to find common ground here on Earth.

Take-Home Message

Why does science work?

» Checks and balances inherent in the scientific process help researchers to be objective about their observations.

» Because a scientific theory is revised until no one can prove it wrong, it is our best way of describing reality.

Summary

Section 1.1 Biology is the systematic study of life. We have encountered only a fraction of the organisms that live on Earth, in part because we have explored only a fraction of its inhabited regions.

Section 1.2 Biologists think about life at different levels of organization. **Emergent properties** appear at successively higher levels. Life emerges at the cellular level. All matter consists of **atoms**, which combine as **molecules**. **Organisms** are individuals that consist of one or more **cells**. Cells of larger multicelled organisms are organized as **tissues**, **organs**, and **organ systems**. A **population** is a group of individuals of a species in a given area; a **community** is all populations of all species in a given area. An **ecosystem** is a community interacting with its environment. The **biosphere** includes all regions of Earth that hold life.

Section 1.3 Life has underlying unity in that all living things have similar characteristics: (1) All organisms require **energy** and **nutrients** to sustain themselves. **Producers** harvest energy from the environment to make their own food by processes such as **photosynthesis**; **consumers** eat other organisms, or their wastes and remains. (2) Organisms keep the conditions in their internal environment within ranges that their cells tolerate—a process called **homeostasis**. (3) **DNA** contains information that guides an organism's form and function, which include **development**, **growth**, and **reproduction**. The passage of DNA from parents to offspring is called **inheritance**.

Section 1.4 The many types of organisms that currently exist on Earth differ greatly in details of body form and function. **Biodiversity** is the sum of differences among living things. **Bacteria** and **archaea** are all single-celled, and their DNA is not contained within a **nucleus**. **Eukaryotes** (**protists**, **plants**, **fungi**, and **animals**) can be single-celled or multicelled. Their DNA is contained within a nucleus.

Section 1.5 Each type of organism has a two-part name. The first part is the **genus** name. When combined with the **specific epithet**, it designates a particular **species**. With **taxonomy**, species are ranked into ever more inclusive **taxa** on the basis of shared **traits**.

Section 1.6 Critical thinking, the self-directed act of judging the quality of information as one learns, is an important part of **science**. Generally, a researcher observes something in nature, uses **inductive reasoning** to form a **hypothesis** (testable explanation) for it, then uses **deductive reasoning** to make a **prediction** about what might occur if the hypothesis is correct. Predictions are tested with observations, **experiments**, or both. Experiments typically are performed on an **experimental group** as compared with a **control group**, and sometimes on **models**. Conclusions are drawn from experimental results, or **data**. A hypothesis that is not consistent with data is modified. Making, testing, and evaluating hypotheses are the **scientific method**.

Biological systems are usually influenced by many interacting **variables**. Experiments test how an **independent variable** influences a **dependent variable**.

Section 1.7 Scientific approaches differ, but experiments are typically designed in a consistent way. A researcher changes an independent variable, then observes the effects of the change on a dependent variable. This practice allows the researcher to study a cause-and-effect relationship in a complex natural system.

Section 1.8 A small sample size increases the potential for **sampling error** in experimental results. In such cases, a subset may be tested that is not representative of the whole. Researchers design experiments carefully to minimize sampling error and bias, and they use **probability** rules to check the **statistical significance** of their results. Scientists check and test one another's work, so science is ideally a self-correcting process.

Section 1.9 Science helps us be objective about our observations because it is concerned only with testable ideas about observable aspects of nature. Opinion and belief have value in human culture, but they are not addressed by science. A **scientific theory** is a long-standing hypothesis that is useful for making predictions about other phenomena. It is our best way of describing reality. A **law of nature** describes something that occurs without fail, but our scientific explanation of why it occurs is incomplete.

Self-Quiz   *Answers in Appendix III*

1. _____ are fundamental building blocks of all matter.
   a. Atoms   c. Cells
   b. Molecules   d. Organisms

2. The smallest unit of life is the _____ .
   a. atom   c. cell
   b. molecule   d. organism

3. All organisms require _____ and _____ to sustain themselves.
   a. DNA; energy   c. nutrients; energy
   b. cells; raw materials d. DNA; cells

4. _____ is a process that maintains conditions in the internal environment within ranges that cells can tolerate.

5. DNA _____ .
   a. guides development   c. is transmitted from
      and functioning   parents to offspring
   b. is the basis of traits d. all of the above

Peacock Butterfly Predator Defenses

The photographs on the *right* represent the experimental and control groups used in the peacock butterfly experiment discussed in Section 1.7.

See if you can identify the experimental groups, and match them up with the relevant control group(s).

Hint: Identify which variable is being tested in each group (each variable has a control).

A Wing spots painted out

B Wing spots visible; wings silenced

C Wing spots painted out; wings silenced

D Wings painted but spots visible

E Wings cut but not silenced

F Wings painted, spots visible; wings cut, not silenced

---

6. A process by which an organism produces offspring is called _____ .

7. _____ is the transmission of DNA to offspring.

8. _____ move around for at least part of their life.
   a. Organisms   c. Consumers
   b. Animals   d. Cells

9. A butterfly is a(n) _____ (choose all that apply).
   a. organism   e. consumer
   b. domain   f. producer
   c. animal   g. hypothesis
   d. eukaryote   h. trait

10. A bacterium is _____ (choose all that apply).
    a. an organism   c. an animal
    b. single-celled d. a eukaryote

11. A long-standing hypothesis that is used to make predictions about other phenomena is called a _____ .

12. Science addresses only that which is _____ .
    a. alive   c. variable
    b. observable   d. indisputable

13. A control group is _____ .
    a. a set of individuals that have a certain characteristic or receive a certain treatment
    b. the standard against which an experimental group is compared
    c. the experiment that gives conclusive results

14. Fifteen randomly selected students are found to be taller than 6 feet. The researchers concluded that the average height of a student is greater than 6 feet. This is an example of _____ .
    a. experimental error   c. sampling bias
    b. sampling error   d. experimental bias

15. Match the terms with the most suitable description.
    ___ emergent property   a. statement of what a hypothesis leads you to expect
    ___ species   b. type of organism
    ___ scientific theory   c. occurs at a higher organizational level
    ___ hypothesis   d. time-tested hypothesis
    ___ prediction   e. testable explanation
    ___ probability   f. measure of chance

Critical Thinking

1. A person is declared to be dead upon the irreversible cessation of spontaneous body functions: brain activity, or blood circulation and respiration. However, only about 1% of a person's cells have to die in order for all of these things to happen. How can someone be dead when 99% of his or her cells are still alive?

2. Explain the difference between a one-celled organism and a single cell of a multicelled organism.

3. Why would you think twice about ordering from a cafe menu that lists the genus name but not the specific epithet of its offerings? *Hint:* Look up *Homarus americanus, Ursus americanus, Ceanothus americanus, Bufo americanus, Lepus americanus,* and *Nicrophorus americanus.*

4. Once there was a highly intelligent turkey that had nothing to do but reflect on the world's regularities. Morning always started out with the sky turning light, followed by the master's footsteps, which were always followed by the appearance of food. Other things varied, but food always followed footsteps. The sequence of events was so predictable that it eventually became the basis of the turkey's theory about the goodness of the world. One morning, after more than 100 confirmations of the goodness theory, the turkey listened for the master's footsteps, heard them, and had its head chopped off.

   Any scientific theory is modified or discarded upon discovery of contradictory evidence. The absence of absolute certainty has led some people to conclude that "facts are irrelevant—facts change." If that is so, should we stop doing scientific research? Why or why not?

5. In 2005, researcher Woo-suk Hwang reported that he had made immortal stem cells from human patients. His research was hailed as a breakthrough for people affected by degenerative diseases, because stem cells may be used to repair a person's own damaged tissues. Hwang published his results in a peer-reviewed journal. In 2006, the journal retracted his paper after other scientists discovered that Hwang's group had faked their data.

   Does the incident show that results of scientific studies cannot be trusted? Or does it confirm the usefulness of a scientific approach, because other scientists discovered and exposed the fraud?

2 Life's Chemical Basis

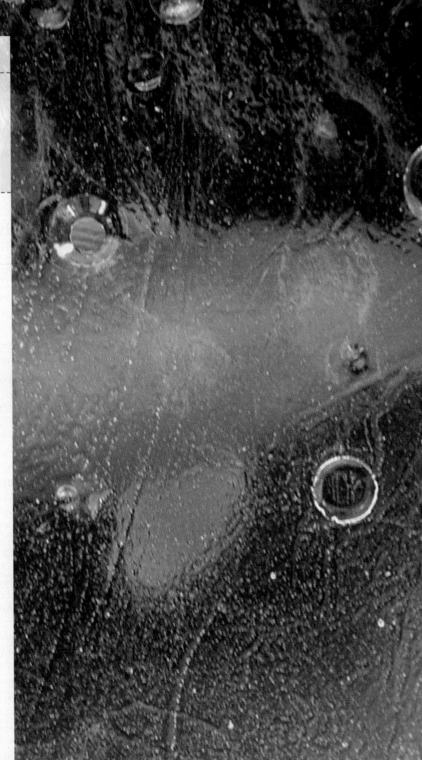

LEARNING ROADMAP

Where you have been In this chapter, you will explore the first level of life's organization—atoms—as you encounter the first example of how the same building blocks, arranged different ways, form different products (Section 1.2). You will also see one aspect of homeostasis, the process by which organisms keep themselves in a state that favors cell survival (Section 1.3).

Where you are now

Atoms and Elements
Atoms, the building blocks of all matter, differ in their numbers of protons, neutrons, and electrons. Atoms of an element have the same number of protons.

Why Electrons Matter
Whether and how an atom interacts with other atoms depends on the number of electrons it carries. An atom with an unequal number of electrons and protons is an ion.

Atoms Bond
Atoms of many elements interact by acquiring, sharing, and giving up electrons. Interacting atoms may form ionic, covalent, or hydrogen bonds.

Water
Hydrogen bonding among individual molecules gives water properties that make life possible: temperature stabilization, cohesion, and the ability to dissolve many other substances.

Hydrogen Power
Most of the chemistry of life occurs in a narrow range of pH, so most fluids inside organisms are buffered to stay within that range.

Where you are going Electrons will come up again as you learn how energy drives metabolism, especially in photosynthesis (Chapter 6) and respiration (Chapter 7). Hydrogen bonding is critical for the molecules of life (3.4, 3.6–3.8); the properties of water, for membranes (Section 4.2 and Chapter 5), plant nutrition and transport (28.4), and temperature regulation (40.9). You will also see how radioisotopes are used to date rocks and fossils (16.6), and the dangers of free radicals (8.6) and acid rain (48.5).

2.1 Mercury Rising

Actor Jeremy Piven, best known for his Emmy-winning role on the television series *Entourage*, began starring in a Broadway play in 2008. He quit suddenly after two shows, citing medical problems. Piven explained that he was suffering from mercury poisoning caused by eating too much sushi. The play's producers and his co-actors were skeptical. The playwright ridiculed Piven, saying he was leaving to pursue a career as a thermometer. But mercury poisoning is no laughing matter.

Mercury is a naturally occurring metal. Most of it is safely locked away in rocky minerals, but volcanic activity and other geologic processes release it into the atmosphere. So do human activities, especially burning coal (**Figure 2.1**). Once airborne, mercury can drift long distances before settling to Earth's surface. There, microbes combine it with carbon to form a substance called methylmercury.

Mercury does not cross skin or mucous membranes very well, but methylmercury does so easily. In water, it ends up in the tissues of aquatic organisms. All fish and shellfish contain methylmercury. Humans contain it too, mainly as a result of eating seafood.

Once mercury enters the body's internal environment, it damages the nervous system, brain, kidneys, and other organs. A dose as low as 3 micrograms per kilogram of body weight (about 200 micrograms for an average-sized adult) can cause tremors, itching or burning sensations, and loss of coordination. Exposure to larger amounts can result in thought and memory impairment, coma, and death. The developing brain is particularly sensitive to mercury because the metal interferes with nerve formation. Thus, mercury is acutely toxic in infants, and it causes long-term neurological effects in children. Mercury in a pregnant woman's blood passes to her unborn child, along with a legacy of permanent developmental problems.

The U.S. Food and Drug Administration requires that foods contain less than 1 part per million of mercury, and for the most part they do. However, it takes months or even years for mercury to be cleared from the body, so it can build up to high levels if even small amounts are ingested on a regular basis. That is why large predatory fish have a lot of mercury in their tissues. It is also why the U.S. Environmental Protection Agency recommends that adults ingest less than 0.1 microgram of mercury per kilogram of body weight per day. For an average-sized person, that limit works out to be about 7 micrograms per day, which is not a big amount if you eat seafood. A two-ounce piece of sushi tuna typically contains about 40 micrograms of mercury, and the occasional piece has many times that amount. It doesn't matter if the fish is raw, grilled, or canned, because mercury is unaffected by cooking. Eat a medium-sized tuna steak, and you could be getting more than 700 micrograms of mercury along with it.

With this chapter, we turn to the first of life's levels of organization: atoms. Interactions between atoms make the molecules that sustain life, and also some that destroy it.

Figure 2.1 Atmospheric fallout from coal-fired power plant emissions is now the biggest cause of mercury pollution. Mercury can accumulate to toxic levels in the tissues of tuna and other large predatory fish.

2.2 Start With Atoms

- The behavior of elements, which make up all living things, starts with the structure of individual atoms.
- The number of protons in the atomic nucleus defines the element, and the number of neutrons defines the isotope.
- Link to atoms 1.2

Figure 2.2 Atoms consist of electrons moving around a core, or nucleus, of protons and neutrons.

Models such as this diagram cannot show what atoms really look like. Electrons zoom around in fuzzy, three-dimensional spaces about 10,000 times bigger than the nucleus.

- ● proton
- ◐ neutron
- ○ electron

Life's unique characteristics start with the properties of different **atoms**, tiny particles that are building blocks of all substances. Even though atoms are about 20 million times smaller than a grain of sand, they consist of even smaller subatomic particles. Positively charged **protons** (p$^+$) and uncharged **neutrons** occur in an atom's core, or **nucleus**. Negatively charged **electrons** (e$^-$) move around the nucleus (**Figure 2.2**). **Charge** is an electrical property: opposite charges attract, and like charges repel.

Atoms differ in their number of subatomic particles. The number of protons in an atom's nucleus is called the **atomic number**, and it determines the type of atom, or element. **Elements** are pure substances, each consisting only of atoms that have the same number of protons in their nucleus. For example, the atomic number of carbon is 6, so all atoms with six protons in their nucleus are carbon atoms, no matter how many electrons or neutrons they have. A chunk of carbon consists only of carbon atoms, and all of those atoms have six protons.

The same elements that make up a living body also occur in nonliving things, but their proportions differ. For example, a human body contains a much larger proportion of carbon atoms than sand or seawater. Why? Unlike sand or seawater, a body consists of a very high proportion of the molecules of life, which in turn consist of a high proportion of carbon atoms.

Knowing the numbers of electrons, protons, and neutrons in atoms helps us predict how elements will behave. In 1869, chemist Dmitry Mendeleyev arranged the elements known at the time by their chemical properties (**Figure 2.3**). The arrangement, which he called the **periodic table**, turned out to be by atomic number, even though subatomic particles would not be discovered until the early 1900s. Gaps in the table allowed Mendeleyev to predict the existence of elements that had not yet been discovered.

In the periodic table, each element is represented by a symbol that is typically an abbreviation of the element's Latin or Greek name. For instance, Pb (lead) is short for *plumbum*; the word "plumbing" is related—ancient Romans made their water pipes with lead. Carbon's symbol, C, is from *carbo*, the Latin word for coal (which is mostly carbon).

Figure 2.3 Animated The periodic table and its creator, Dmitry Mendeleyev. Until he came up with the table, Mendeleyev was known mainly for his extravagant hair (he cut it only once per year).

Atomic numbers are shown above the element symbols. Some of the symbols are abbreviations for their Latin names. For instance, Pb (lead) is short for *plumbum*; the word "plumbing" is related—ancient Romans made their water pipes with lead. Appendix IV shows the table in more detail.

atom Particle that is a fundamental building block of all matter.
atomic number Number of protons in the atomic nucleus; determines the element.
charge Electrical property. Opposite charges attract, and like charges repel.
electron Negatively charged subatomic particle that occupies orbitals around an atomic nucleus.
element A pure substance that consists only of atoms with the same number of protons.
isotopes Forms of an element that differ in the number of neutrons their atoms carry.
mass number Total number of protons and neutrons in the nucleus of an element's atoms.
neutron Uncharged subatomic particle in the atomic nucleus.
nucleus Core of an atom; occupied by protons and neutrons.
periodic table Tabular arrangement of the elements by atomic number.
proton Positively charged subatomic particle that occurs in the nucleus of all atoms.
radioactive decay Process by which atoms of a radioisotope emit energy and/or subatomic particles when their nucleus spontaneously disintegrates.
radioisotope Isotope with an unstable nucleus.
tracer Substance with a detectable component, such as a molecule labeled with a radioisotope.

1 H																	2 He
3 Li	4 Be											5 B	6 C	7 N	8 O	9 F	10 Ne
11 Na	12 Mg											13 Al	14 Si	15 P	16 S	17 Cl	18 Ar
19 K	20 Ca	21 Sc	22 Ti	23 V	24 Cr	25 Mn	26 Fe	27 Co	28 Ni	29 Cu	30 Zn	31 Ga	32 Ge	33 As	34 Se	35 Br	36 Kr
37 Rb	38 Sr	39 Y	40 Zr	41 Nb	42 Mo	43 Tc	44 Ru	45 Rh	46 Pd	47 Ag	48 Cd	49 In	50 Sn	51 Sb	52 Te	53 I	54 Xe
55 Cs	56 Ba	71 Lu	72 Hf	73 Ta	74 W	75 Re	76 Os	77 Ir	78 Pt	79 Au	80 Hg	81 Tl	82 Pb	83 Bi	84 Po	85 At	86 Rn
87 Fr	88 Ra	103 Lr	104 Rf	105 Db	106 Sg	107 Bh	108 Hs	109 Mt	110 Ds	111 Rg	112 Uub	113 Uut	114 Uuq	115 Uup	116 Uuh		118 Uuo

57 La	58 Ce	59 Pr	60 Nd	61 Pm	62 Sm	63 Eu	64 Gd	65 Tb	66 Dy	67 Ho	68 Er	69 Tm	70 Yb
89 Ac	90 Th	91 Pa	92 U	93 Np	94 Pu	95 Am	96 Cm	97 Bk	98 Cf	99 Es	100 Fm	101 Md	102 No

Isotopes and Radioisotopes

Atoms of an element do not differ in the number of protons, but they can differ in the number of other subatomic particles. Those that differ in the number of neutrons are called **isotopes**. We define isotopes by their **mass number**, which is the total number of protons and neutrons in their nucleus. Mass number is written as a superscript to the left of an element's symbol. For example, the most common isotope of carbon has six protons and six neutrons, so it is designated ^{12}C, or carbon 12. The other naturally occurring isotopes of carbon are ^{13}C (six protons, seven neutrons), and ^{14}C (six protons, eight neutrons).

Carbon 14 is a **radioisotope**, or radioactive isotope. Atoms of a radioisotope have an unstable nucleus that breaks down spontaneously. As a nucleus breaks down, it emits radiation—subatomic particles, energy, or both—a process called **radioactive decay**. The atomic nucleus cannot be altered by ordinary means, so radioactive decay is unaffected by external factors such as temperature, pressure, or whether the atoms are part of molecules.

Each radioisotope decays at a predictable rate into predictable products. For example, when carbon 14 decays, one of its neutrons splits into a proton and an electron. The nucleus emits the electron as radiation. Thus, an atom with eight neutrons and six protons (^{14}C) becomes an atom with seven neutrons and seven protons, which is nitrogen (^{14}N):

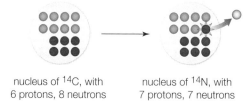

nucleus of ^{14}C, with
6 protons, 8 neutrons

nucleus of ^{14}N, with
7 protons, 7 neutrons

This process is so predictable that we can say with certainty that about half of the atoms in any sample of ^{14}C will be ^{14}N atoms after 5,730 years. Researchers use the predictability of radioactive decay when they estimate the age of a rock or fossil by measuring its isotope content (we return to this topic in Section 16.6).

All isotopes of an element generally have the same chemical properties regardless of the number of neutrons in their atoms. The consistent chemical behavior means that organisms use atoms of one isotope the same way that they use atoms of another. Thus, radioisotopes can be used to study biological processes. A **tracer** is any substance with a detectable component, such as a molecule in which an atom (such as ^{12}C) has been replaced with a radioisotope (such as ^{14}C).

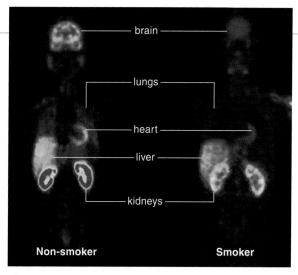

Figure 2.4 Animated PET scans. The result of a PET scan is a digital image of a process in the body's interior. These two PET scans show the activity of a molecule called MAO-B in the body of a non-smoker (*left*) and a smoker (*right*). The activity is color-coded from *red* (highest activity) to *purple* (lowest). Low MAO-B activity is associated with violence, impulsiveness, and other behavioral problems.

A radioactive tracer can be delivered into a biological system such as a cell, body, or ecosystem, and then followed as it moves through the system with instruments that detect radiation.

Radioactive tracers are widely used in research. A famous example is a series of experiments carried out by Melvin Calvin and Andrew Benson. These researchers synthesized carbon dioxide with ^{14}C, then let green algae take up the radioactive gas. Using instruments that detect electrons emitted by the radioactive decay of ^{14}C, they tracked carbon through steps by which the algae—and all plants—make sugars.

Radioisotopes have medical applications as well. For example, PET (short for positron-emission tomography) helps us "see" a functional process inside the body. By this procedure, a radioactive sugar or other tracer is injected into a patient, who is then moved into a scanner. Inside the patient's body, cells with differing rates of activity take up the tracer at different rates. The scanner detects radioactive decay wherever the tracer is, then translates that data into an image (**Figure 2.4**).

Take-Home Message

What are the basic building blocks of all matter?

» All matter consists of atoms, tiny particles that in turn consist of electrons moving around a nucleus of protons and neutrons.

» An element is a pure substance that consists only of atoms with the same number of protons. Isotopes are forms of an element that have different numbers of neutrons.

» Unstable nuclei of radioisotopes disintegrate spontaneously (decay) at a predictable rate to form predictable products.

2.3 Why Electrons Matter

- Whether an atom will interact with other atoms depends on how many electrons it has.
- Link to building blocks of life 1.2

Electrons are really, really small. How small are they? If they were as big as apples, you would be about 3.5 times taller than our solar system is wide. Simple physics explains the motion of, say, an apple falling from a tree, but electrons are so tiny that such everyday physics does not explain their behavior.

For example, electrons carry energy, but only in incremental amounts. An electron can gain energy only by absorbing the exact amount needed to boost it to the next energy level. Likewise, it can lose energy only by emitting the exact difference between two energy levels. This concept will be especially important to remember in later chapters, when you learn about how cells harvest and release energy.

A typical atom has approximately the same number of electrons as protons. The atoms of many elements have a lot of electrons zipping around the same nucleus. However, despite moving at nearly the speed of light (300,000 kilometers per second, or 670 million miles per hour), electrons that are part of an atom never collide. Why not? Electrons avoid one another because they occupy different orbitals, which are defined volumes of space around the nucleus.

To understand how orbitals work, imagine that an atom is a multilevel apartment building with a nucleus in the basement. Each "floor" of the building corresponds to a certain energy level, and each has a certain number of "rooms" (orbitals) available for rent.

Two electrons can occupy each room. Pairs of electrons populate rooms from the ground floor up; in other words, they fill orbitals from lower to higher energy levels. The farther an electron is from the nucleus in the basement, the greater its energy. An electron can move to a room on a higher floor if an energy input gives it a boost, but it immediately emits the extra energy and moves back down.

Shell models help us visualize how electrons populate atoms (**Figure 2.5**). In this model, nested "shells" correspond to successively higher energy levels. Thus, each shell includes all of the rooms on one floor of our atomic apartment building.

We draw a shell model of an atom by filling shells with electrons (represented as balls or dots), from the innermost shell out, until there are as many electrons as the atom has protons. There is only one room on the first floor, one orbital at the lowest energy level

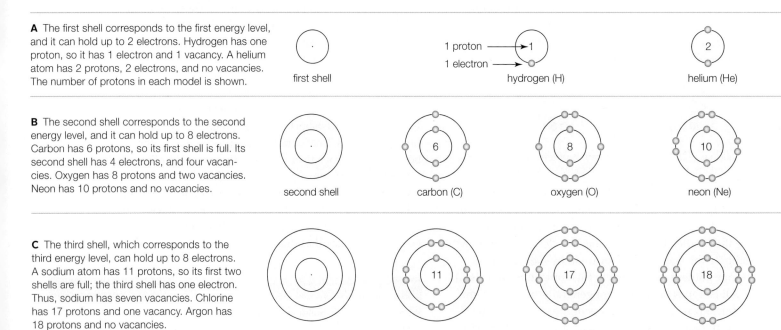

A The first shell corresponds to the first energy level, and it can hold up to 2 electrons. Hydrogen has one proton, so it has 1 electron and 1 vacancy. A helium atom has 2 protons, 2 electrons, and no vacancies. The number of protons in each model is shown.

first shell

1 proton
1 electron
hydrogen (H)

helium (He)

B The second shell corresponds to the second energy level, and it can hold up to 8 electrons. Carbon has 6 protons, so its first shell is full. Its second shell has 4 electrons, and four vacancies. Oxygen has 8 protons and two vacancies. Neon has 10 protons and no vacancies.

second shell

carbon (C)

oxygen (O)

neon (Ne)

C The third shell, which corresponds to the third energy level, can hold up to 8 electrons. A sodium atom has 11 protons, so its first two shells are full; the third shell has one electron. Thus, sodium has seven vacancies. Chlorine has 17 protons and one vacancy. Argon has 18 protons and no vacancies.

third shell

sodium (Na)

chlorine (Cl)

argon (Ar)

Figure 2.5 Animated Shell models. Each circle (shell) represents all orbitals at one energy level. A model is filled with electrons from the innermost shell out, until there are as many electrons as protons. Atoms with vacancies (room for additional electrons) in their outermost shell tend to get rid of them.

Figure It Out: Which of these models have unpaired electrons in their outer shell?

Answer: Hydrogen, carbon, oxygen, sodium, and chlorine

(Figure 2.5A). It fills up first. In hydrogen, the simplest atom, one electron occupies that room. Helium, with two protons, has two electrons that fill the room—and the first shell. In larger atoms, more electrons rent the second-floor rooms (Figure 2.5B). When the second floor fills, more electrons rent third-floor rooms (Figure 2.5C), and so on.

About Vacancies

When an atom's outermost shell is filled with electrons, we say that it has no vacancies. Any atom is in its most stable state when it has no vacancies. Helium, neon, and argon are examples of elements with no vacancies. Atoms of these elements are chemically stable, which means they have no tendency to interact with other atoms. Thus, these elements occur most frequently in nature as solitary atoms.

By contrast, when an atom's outermost shell has room for another electron, it has a vacancy. Atoms with vacancies tend to get rid of them by interacting with other atoms; in other words, they are chemically active. For example, the sodium atom (Na) depicted in Figure 2.5C has one electron in its outer (third) shell, which can hold eight. With seven vacancies, we can predict that this atom is chemically active.

In fact, this particular sodium atom is not just active, it is extremely so. Why? The shell model shows that a sodium atom has an unpaired electron, but electrons like to occupy orbitals in pairs. Thus, with some exceptions, solitary atoms that have unpaired electrons are not very stable. These atoms are called **free radicals**. Most free radicals have such a strong tendency to interact with other atoms that they exist only briefly in that state. As you will see in later chapters, such interactions make free radicals dangerous to life.

A sodium atom can easily lose its one unpaired electron. After that happens, the atom's second shell —which is full of electrons—is its outermost. Thus, no vacancies remain. This is a very stable configuration for sodium, so it is not surprising that the vast majority of sodium atoms on Earth have 11 protons and 10 electrons. Atoms like this, with an unequal number of protons and electrons, are called **ions**. Ions carry a net (or overall) charge. The negative charge of an elec-

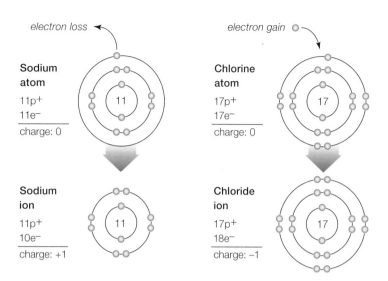

A A sodium atom (Na) becomes a positively charged sodium ion (Na⁺) when it loses the electron in its third shell. The atom's full second shell is now its outermost, so it has no vacancies.

B A chlorine atom (Cl) becomes a negatively charged chloride ion (Cl⁻) when it gains an electron and fills the vacancy in its third, outermost shell.

Figure 2.6 Animated Ion formation. **Figure It Out:** Does a chloride ion have an unpaired electron?

Answer: No

tron is the same magnitude as the positive charge of a proton, so the two charges cancel one another. Thus, an atom with the same number of electrons and protons carries no net charge. Sodium ions (Na⁺) offer an example of how atoms gain a positive charge by losing an electron (Figure 2.6A).

Other atoms gain a negative charge by accepting an electron. For example, an uncharged chlorine atom has 17 protons and 17 electrons. Its outer shell can hold eight electrons, but it has only seven. This atom has one vacancy and one unpaired electron, so we can predict—correctly—that it is chemically very active. A chlorine atom easily fills its third shell by accepting an electron. When that happens, the atom becomes a chloride ion (Cl⁻) with 17 protons, 18 electrons, and a net negative charge (Figure 2.6B).

free radical Atom with an unpaired electron.
ion Atom that carries a charge because it has an unequal number of protons and electrons.
shell model Model of electron distribution in an atom.

Take-Home Message

Why do atoms interact?

» An atom's electrons are the basis of its chemical behavior.

» Shells represent all electron orbitals at one energy level in an atom. When the outermost shell is not full of electrons, the atom has a vacancy.

» Atoms with vacancies tend to interact with other atoms.

2.4 Chemical Bonds: From Atoms to Molecules

- Chemical bonds link atoms into molecules.
- The characteristics of a chemical bond arise from the properties of the atoms taking part in it.
- Link to building blocks of life 1.2

An atom can get rid of vacancies by participating in a **chemical bond**, which is an attractive force that arises between two atoms when their electrons interact. When atoms interact, they often form molecules. A **molecule** consists of atoms held together in a particular number and arrangement by chemical bonds.

Water is an example of a substance made of molecules. Each water molecule consists of three atoms: two hydrogen atoms bonded to the same oxygen atom (**Figure 2.7**). A water molecule consists of two or more elements, so it is a **compound**. Other molecules, includ-

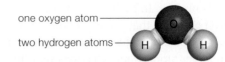

one oxygen atom —
two hydrogen atoms — H   H

Figure 2.7 The water molecule. Each water molecule has two hydrogen atoms bonded to the same oxygen atom.

ing molecular oxygen (a gas in air), have atoms of one element only.

The term "bond" applies to a continuous range of atomic interactions. However, we can categorize most bonds into distinct types based on their different properties. Which type forms depends on the atoms taking part in the molecule.

Ionic Bonds Two ions may stay together by the mutual attraction of their opposite charges, an association called an **ionic bond**. Ionic bonds can be quite strong. Ionically bonded sodium and chloride ions make sodium chloride (NaCl), which we know as common table salt. A crystal of this substance consists of a lattice of sodium and chloride ions interacting in ionic bonds (**Figure 2.8A**).

Ions retain their respective charges when participating in an ionic bond (**Figure 2.8B**). Thus, one "end" of an ionic bond has a positive charge, and the other "end" has a negative charge. Any such separation of charge into distinct positive and negative regions is called **polarity** (**Figure 2.8C**).

A sodium chloride molecule is polar because the chloride ion keeps a very strong hold on its extra electron. In other words, it is strongly electronegative. **Electronegativity** is a measure of an atom's ability to pull electrons away from another atom. Electronegativity is not the same as charge. Rather, an atom's electronegativity depends on its size, how many vacancies it has, and what other atoms it is interacting with.

An ionic bond is completely polar because the atoms participating in it have a very large difference in electronegativity. When atoms with a lower difference

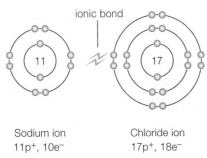

ionic bond

Sodium ion
11p+, 10e⁻

Chloride ion
17p+, 18e⁻

A The mutual attraction of opposite charges holds a sodium ion and a chloride ion together in an ionic bond.

Cl⁻ Na⁺

B Each crystal of table salt consists of many sodium and chloride ions locked together in a cubic lattice by ionic bonds.

C Ions taking part in an ionic bond retain their charge, so the molecule is polar. One side is positively charged (here represented by a *blue* overlay); the other side is negatively charged (*red* overlay).

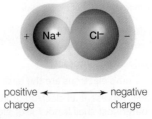

+ Na⁺   Cl⁻ –

positive ← → negative
charge   charge

Figure 2.8 Animated An example of ionic bonding: table salt, or NaCl.

chemical bond An attractive force that arises between two atoms when their electrons interact.
compound Molecule that has atoms of more than one element.
covalent bond Chemical bond in which two atoms share a pair of electrons.
electronegativity Measure of the ability of an atom to pull electrons away from other atoms.
ionic bond Chemical bond that consists of a strong mutual attraction between ions of opposite charge.
molecule Group of two or more atoms joined by chemical bonds.
polarity Separation of charge into positive and negative regions.

3.1 Fear of Frying

The human body requires only about a tablespoon of fat each day to stay healthy, but most people in developed countries eat far more than that. The average American eats about 70 pounds of fat per year, which may be part of the reason why the average American is overweight. Being overweight increases one's risk for many chronic illnesses. However, the total quantity of fat in the diet may be less important than the types of fats. Fats are more than inert molecules that accumulate in strategic areas of our bodies. They are major constituents of cell membranes, and as such they have powerful effects on cell function.

The typical fat molecule has three fatty acids, each with a long chain of carbon atoms that can vary a bit in structure. Fats with a certain arrangement of hydrogen atoms around those carbon chains are called *trans* fats (**Figure 3.1**). Small amounts of *trans* fats occur naturally in red meat and dairy products. However, most of the *trans* fats that humans eat come from partially hydrogenated vegetable oil, an artificial food product.

Hydrogenation is a manufacturing process that adds hydrogen atoms to vegetable oils in order to change them into solid fats. Procter & Gamble Co. developed partially hydrogenated soybean oil in 1908 as a substitute for the more expensive solid animal fats they had been using to make candles. However, the demand for candles began to wane as more households in the United States became wired for electricity, and P & G began to look for another way to sell its proprietary fat. Partially hydrogenated vegetable oil looks a lot like lard, so in 1911 the company began aggressively marketing it as a revolutionary new food: a solid cooking fat with a long shelf life, mild flavor, and lower cost than lard or butter.

By the mid-1950s, hydrogenated vegetable oil had become a major part of the American diet. It was (and still is) found in many manufactured and fast foods: french fries, butter substitutes, cookies, crackers, cakes and pancakes, peanut butter, pies, doughnuts, muffins, chips, granola bars, breakfast bars, chocolate, microwave popcorn, pizzas, burritos, chicken nuggets, fish sticks, and so on.

For decades, hydrogenated vegetable oil was considered more healthy than animal fats because it was made from plants, but we now know otherwise. The *trans* fats in hydrogenated vegetable oils raise the level of cholesterol in our blood more than any other fat, and they directly alter the function of our arteries and veins. The effects of such changes are quite serious. Eating as little as 2 grams a day of hydrogenated vegetable oils increases a person's risk of atherosclerosis (hardening of the arteries), heart attack, and diabetes. A small serving of french fries made with hydrogenated vegetable oil contains about 5 grams of *trans* fat.

All organisms consist of the same kinds of molecules, but small differences in the way those molecules are put together can have big effects in a living organism. With this concept, we introduce you to the chemistry of life. This is your chemistry. It makes you far more than the sum of your body's molecules.

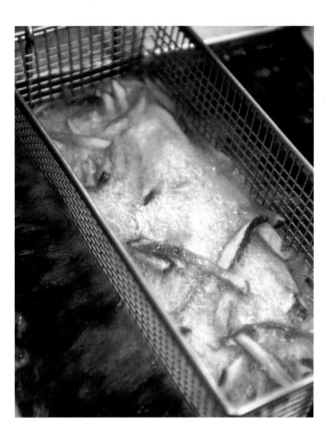

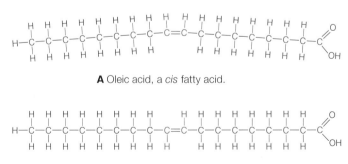

A Oleic acid, a *cis* fatty acid.

B Elaidic acid, a *trans* fatty acid.

Figure 3.1 *Trans* fats, an unhealthy food. A fat has three fatty acid tails, each a long carbon chain. Hydrogenation changes the arrangement of hydrogen atoms around double bonds in those chains from (**A**) a *cis* configuration to (**B**) a *trans* configuration. This small difference in structure makes a big difference in our bodies.

3.2 Organic Molecules

- All of the molecules of life are built with carbon atoms.
- We use different models to highlight different aspects of the same molecule.
- Links to Elements 2.2, Representing Molecules 2.4

Carbon—The Stuff of Life

Living things are mainly oxygen, hydrogen, and carbon. Most of the oxygen and hydrogen are in the form of water. Put water aside, and carbon makes up more than half of what is left.

The carbon in living organisms is part of the molecules of life—complex carbohydrates, lipids, proteins, and nucleic acids. These molecules consist primarily of hydrogen and carbon atoms, so they are said to be **organic**. The term is a holdover from a time when such molecules were thought to be made only by living things, as opposed to the "inorganic" molecules that formed by nonliving processes. The term persists, even though we now know that organic compounds were present on Earth long before organisms were, and we can also make them synthetically in laboratories.

Carbon's importance to life starts with its versatile bonding behavior. Each carbon atom can form covalent bonds with one, two, three, or four other atoms. Depending on the other elements in the resulting molecule, such bonds may be polar or nonpolar. Many organic compounds have a backbone—a chain of carbon atoms—to which other atoms attach. The ends of a backbone may join so that the carbon chain forms one or more ring structures (**Figure 3.2**). Such versatility means that carbon atoms can be assembled and remodeled into a variety of organic compounds.

Representing Structures of Organic Molecules

As you will see in the next few sections, the function of an organic molecule depends on its structure.

A Carbon's versatile bonding behavior allows it to form a variety of structures, including rings.

B Carbon rings form the framework of many sugars, starches, and fats, such as those found in doughnuts.

Figure 3.2 Carbon rings.

Researchers routinely make models of organic molecules such as proteins to study structure–function relationships, surface properties, changes during synthesis or biochemical processes, and molecular recognition. A molecule's structure can be modeled in various ways. The different models allow us to visualize different characteristics of the same molecule.

Structural formulas of organic molecules can be quite complex, even when the molecules are relatively small (**Figure 3.3A**). Thus, formulas of organic molecules are typically simplified. Some of the bonds may not be shown, but they are implied. Hydrogen atoms bonded to a carbon backbone may also be omitted, and other atoms as well. Carbon ring structures such as the ones that occur in glucose and other sugars are often represented as polygons. If no atom is shown at a corner or at the end of a bond, a carbon atom is implied there (**Figure 3.3B**).

Ball-and-stick models show the positions of individual atoms in three dimensions (**Figure 3.3C**). Single, double, and triple covalent bonds are all shown as one stick connecting two balls, which represent atoms. Ball size reflects relative sizes of the atoms, and ball color indicates the element according to a standard code. The most common color scheme shows carbon

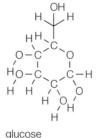

glucose

A A structural formula for an organic molecule can be very complicated.

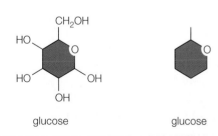

glucose

B Structural formulas of organic molecules are often simplified by using polygons as symbols for rings, and omitting the labels for some atoms.

glucose

C A ball-and-stick model shows the arrangement of atoms in three dimensions.

glucose

D A space-filling model shows a molecule's overall shape.

Figure 3.3 Modeling an organic molecule.

in black, oxygen in red, hydrogen in white, nitrogen in blue, and sulfur in yellow.

Space-filling models represent atomic volume most accurately (**Figure 3.3D**). This type of model shows the overall shape of an organic molecule. Atoms in space-filling models may be color-coded by element using the same scheme as ball-and-stick models.

Many organic molecules are so large that ball-and-stick or space-filling models of them may be incomprehensible. **Figure 3.4** shows three different ways to represent the same molecule, hemoglobin, a very large protein that colors your blood red. Hemoglobin transports oxygen to tissues throughout the body of all vertebrates (animals that have a backbone). A ball-and-stick or space-filling model of hemoglobin is so complicated that many interesting features of this molecule are obscured (**Figure 3.4A**).

To reduce visual complexity, other types of molecular models do not depict individual atoms. Surface models of large molecules can reveal large-scale features, such as folds or pockets, that can be difficult to see when individual atoms are shown. For example, in the surface model of hemoglobin in **Figure 3.4B**, you can see folds of the molecule that cradle hemes. Hemes are carbon ring structures that often have an iron atom at their center. They are part of many important proteins that you will encounter in this book.

Very large molecules such as hemoglobin are often shown as ribbon models. Such models highlight different features of the structure, such as coils or sheets. In a ribbon model of hemoglobin (**Figure 3.4C**), you can see that the protein consists of four coiled components, each of which folds around a heme.

Such structural details are clues to how a molecule functions. For example, hemoglobin, which is the main oxygen-carrier in vertebrate blood, has four hemes. Oxygen binds at the hemes, so each hemoglobin molecule can carry up to four molecules of oxygen.

organic Describes a compound that consists primarily of carbon and hydrogen atoms.

Take-Home Message

How are all of the molecules of life alike?

» The molecules of life (carbohydrates, lipids, proteins, and nucleic acids) are organic, which means they consist mainly of carbon and hydrogen atoms.

» The structure of an organic molecule starts with its carbon backbone, a chain of carbon atoms that may form a ring.

» We use different models to represent different characteristics of a molecule's structure. Considering a molecule's structural features gives us insight into how it functions.

A A space-filling model of hemoglobin.

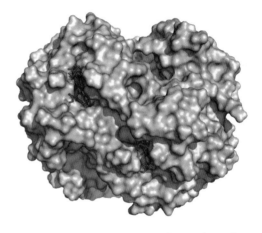

B A surface model of the same molecule reveals crevices and folds that are important for its function. Heme groups, in *red*, are cradled in pockets of the molecule.

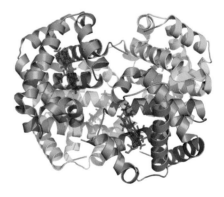

C A ribbon model of hemoglobin shows all four heme groups, also in *red*, held in place by the molecule's coils.

Figure 3.4 Visualizing the structure of hemoglobin, the oxygen-transporting molecule in red blood cells. Models that show individual atoms usually depict them color-coded by element. Other models may be shown in various colors, depending on which features are being studied.

3.3 Molecules of Life—From Structure to Function

- The function of organic molecules in biological systems begins with their structure.
- Links to Ions 2.3, Polarity 2.4, Acids and bases 2.6

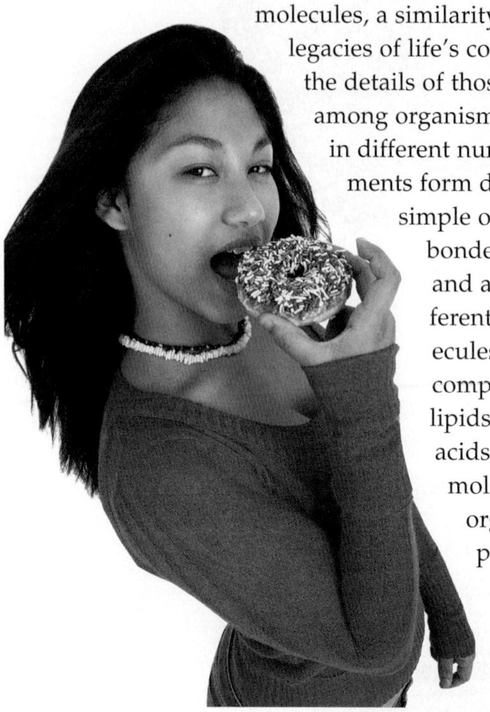

All biological systems are based on the same organic molecules, a similarity that is one of many legacies of life's common origin. However, the details of those molecules differ among organisms. Just as atoms bonded in different numbers and arrangements form different molecules, simple organic building blocks bonded in different numbers and arrangements form different versions of the molecules of life. Cells assemble complex carbohydrates, lipids, proteins, and nucleic acids from small organic molecules. These small organic molecules—simple sugars, fatty acids, amino acids, and nucleotides—are called **monomers** when they are used as subunits of larger molecules. Molecules that consist of multiple monomers are called **polymers**. The remainder of this chapter introduces the different types of biological molecules and the monomers from which they are built.

What Cells Do to Organic Compounds

Cells build polymers from monomers, and break down polymers to release monomers. Any process in which a molecule changes is called a **reaction**. Reactions that run constantly inside cells help them stay alive, grow, and reproduce. **Metabolism** refers to reactions and all other activities by which cells acquire and use energy as they make and break apart organic compounds (**Figure 3.5A**). Metabolism requires **enzymes**, which are organic molecules (usually proteins) that speed up reactions without being changed by them.

Many metabolic reactions build large organic molecules from smaller ones. With **condensation**, an enzyme covalently bonds two molecules together. Water (H—O—H) usually forms as a product of condensation when a hydroxyl group (—OH) from one of the molecules combines with a hydrogen atom (—H) from the other molecule (**Figure 3.5B**). **Hydrolysis**, which is the reverse of condensation, breaks apart large organic molecules into smaller ones (**Figure 3.5C**). Hydrolysis enzymes break a bond by attaching a hydroxyl group to one atom and a hydrogen atom to the other. The —OH and —H come from a water molecule, so this reaction requires water. We will revisit enzymes and metabolic reactions in Chapter 5.

Functional Groups

An organic molecule that consists only of hydrogen and carbon atoms is called a **hydrocarbon**. Methane, the simplest kind, is one carbon atom bonded to four hydrogen atoms. We use larger kinds for fuel. Hydrocarbons are generally nonpolar. **Functional groups**, which are clusters of atoms covalently bonded to a carbon atom of an organic molecule, impart additional chemical properties such as polarity or acidity.

All of the molecules of life have at least one functional group. The chemical behavior of these molecules arises mainly from the number, kind, and arrangement of their functional groups. **Table 3.1** lists a few functional groups that are common in biological molecules.

The hydroxyl group mentioned earlier is an example of a functional group. Hydroxyl groups add polar character to organic compounds such as alcohols and

A Metabolism refers to processes by which cells acquire and use energy as they make and break down molecules. Humans and other consumers break down the molecules in food. They use energy and raw materials from the breakdown to maintain themselves and to build new components.

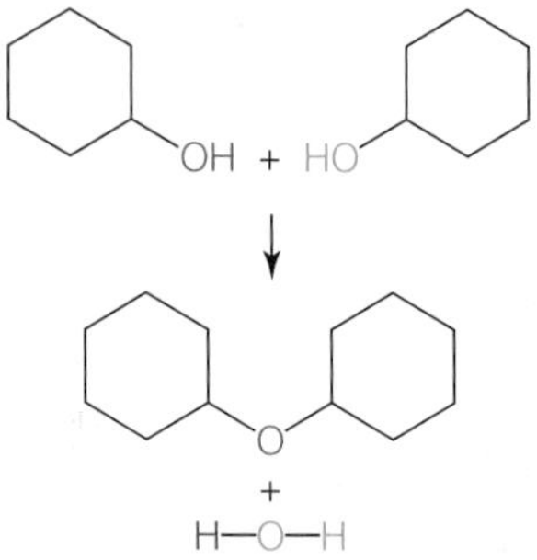

B Condensation. Cells build a large molecule from smaller ones by this reaction.

An enzyme removes a hydroxyl group from one molecule and a hydrogen atom from another. A covalent bond forms between the two molecules, and water also forms.

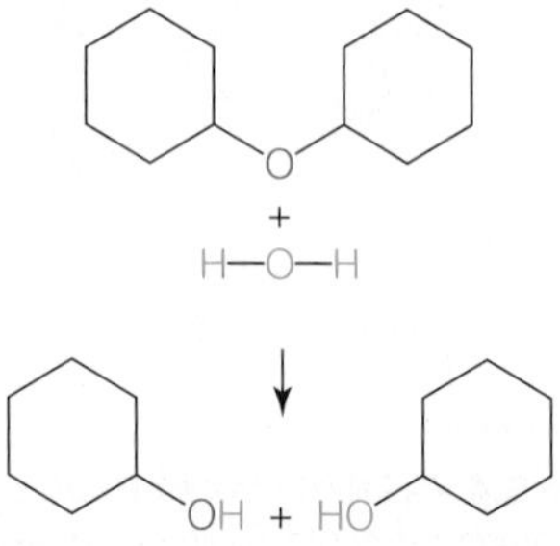

C Hydrolysis. Cells split a large molecule into smaller ones by this water-requiring reaction.

An enzyme attaches a hydroxyl group and a hydrogen atom (both from water) at the cleavage site.

Figure 3.5 Animated Metabolism. Two common reactions by which cells build and break down organic molecules are shown.

simple sugars. Methyl groups add nonpolar character. Adding a methyl group to a molecule reduces the solubility of a compound and may dampen the character of another functional group. Methyl groups added to DNA act like an "off" switch for this molecule; acetyl groups act like an "on" switch. Acetyl groups also carry two carbons from one molecule to another in many metabolic reactions.

Aldehyde and ketone groups are part of simple sugars. Some sugars convert to a ring form when the highly reactive aldehyde group on their terminal carbon reacts with a hydroxyl group on another carbon (Figure 3.6). Carboxyl groups make amino acids and fatty acids acidic; amine and amide groups make nucleotide bases basic.

Reactions that transfer phosphate groups from one molecule to another also transfer energy. Bonds between sulfhydryl groups stabilize the structure of many proteins, including those that make up human hair. Heat and some kinds of chemicals can temporarily break sulfhydryl bonds, which is why we can curl straight hair and straighten curly hair.

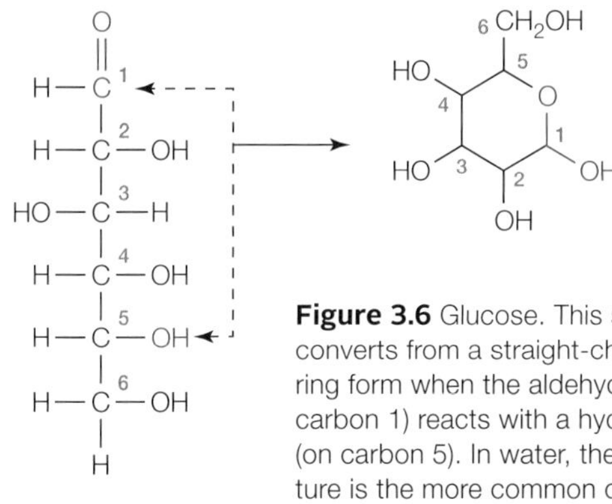

Figure 3.6 Glucose. This simple sugar converts from a straight-chain into a ring form when the aldehyde group (on carbon 1) reacts with a hydroxyl group (on carbon 5). In water, the cyclic structure is the more common one.

condensation Process by which enzymes build large molecules from smaller subunits; water also forms.

enzyme Compound (usually a protein) that speeds up a reaction without being changed by it.

functional group A group of atoms bonded to a carbon of an organic compound; imparts a specific chemical property to the molecule.

hydrocarbon Compound that consists only of carbon and hydrogen atoms.

hydrolysis Process by which an enzyme breaks a molecule into smaller subunits by attaching a hydroxyl group to one part and a hydrogen atom to the other.

metabolism All the enzyme-mediated chemical reactions by which cells acquire and use energy as they build and break down organic molecules.

monomers Molecules that are subunits of polymers.

polymer Molecule that consists of multiple monomers.

reaction Process of molecular change.

Table 3.1 Some Functional Groups in Biological Molecules

Group	Structure	Character	Found in:
acetyl	$-\!\!\overset{\displaystyle O}{\overset{\|}{C}}\!-CH_3$	polar, acidic	some proteins, coenzymes
aldehyde	$-\!\!\overset{\displaystyle O}{\overset{\|}{C}}\!-H$	polar, reactive	simple sugars
amide	$-\!\!\overset{\displaystyle O}{\overset{\|}{C}}\!-N$	weakly basic, stable, rigid	proteins nucleotide bases
amine	$-NH_2$	very basic	nucleotide bases amino acids
carboxyl	$-\!\!\overset{\displaystyle O}{\overset{\|}{C}}\!-OH$	very acidic	fatty acids amino acids
hydroxyl	$-OH$	polar	alcohols sugars
ketone	$\overset{\displaystyle O}{\overset{\|}{C}}$	polar, acidic	simple sugars nucleotide bases
methyl	$-CH_3$	nonpolar	fatty acids some amino acids
phosphate	$O-\!\!\overset{\displaystyle O}{\underset{\displaystyle OH}{\overset{\|}{P}}}\!-OH$	polar, reactive	nucleotides DNA RNA phospholipids proteins
sulfhydryl	$-SH$	reacts with metals, forms rigid disulfide bonds	cysteine many cofactors

---

Take-Home Message

How do organic molecules work in living systems?

» All life is based on the same organic compounds: complex carbohydrates, lipids, proteins, and nucleic acids.

» By processes of metabolism, cells assemble these molecules of life from monomers. They also break apart polymers into component monomers.

» Functional groups impart chemical characteristics to organic molecules. Such groups contribute to the function of biological molecules.

» An organic molecule's structure dictates its function in biological systems.

3.4 Carbohydrates

- Carbohydrates are the most plentiful biological molecules.
- Cells use some carbohydrates as structural materials; they use others for fuel, or to store or transport energy.
- Link to Hydrogen bonds 2.5

Carbohydrates are organic compounds that consist of carbon, hydrogen, and oxygen in a 1:2:1 ratio. Cells use different kinds as structural materials, for fuel, and for storing and transporting energy.

Carbohydrates in Biological Systems

Simple Sugars "Saccharide" is from *sacchar*, a Greek word that means sugar. **Monosaccharides** (one sugar unit) are the simplest type of carbohydrate, but they have extremely important roles as components of larger molecules. Common monosaccharides have a backbone of five or six carbon atoms, one carbonyl group, and two or more hydroxyl groups. Enzymes can easily break the bonds of monosaccharides to release energy (we will return to carbohydrate metabolism in Chapter 7). Monosaccharides are very soluble in water, so they move easily throughout the water-based internal environments of all organisms.

Monosaccharides that are components of the nucleic acids DNA and RNA have five carbon atoms. Glucose has six. Glucose can be used as a fuel to drive cellular processes, or as a structural material to build larger molecules. It can also be used as a precursor, or starting material, that is remodeled into other molecules. For example, cells of plants and many animals make vitamin C from glucose. Human cells are unable to make vitamin C, so we need to get it from our food.

Short-Chain Carbohydrates An oligosaccharide is a short chain of covalently bonded monosaccharides (*oligo*– means a few). **Disaccharides** consist of two sugar monomers. The lactose in milk is a disaccharide, with one glucose and one galactose unit. Sucrose, the most plentiful sugar in nature, has a glucose and a fructose

unit (**Figure 3.7**). Sucrose extracted from sugarcane or sugar beets is our table sugar. Oligosaccharides with three or more sugar units are often attached to lipids or proteins that have important functions in immunity.

Complex Carbohydrates The "complex" carbohydrates, or **polysaccharides**, are straight or branched chains of many sugar monomers—often hundreds or thousands of them. There may be one type or many types of monomers in a polysaccharide.

The most common polysaccharides are cellulose, glycogen, and starch. All consist of glucose monomers, but their chemical properties differ substantially. Why? The answer begins with differences in patterns of covalent bonding that link their glucose monomers.

Cellulose, the major structural material of plants, is the most abundant biological molecule in the biosphere. It consists of long, straight chains of glucose monomers. Hydrogen bonds lock the chains into tight, sturdy bundles (**Figure 3.8A**). In plants, these tough cellulose fibers act like reinforcing rods that help stems resist wind and other forms of mechanical stress.

Cellulose does not dissolve in water, and it is not easily broken down. Some bacteria and fungi make enzymes that break it apart into its component sugars, but humans and other mammals do not. Dietary fiber, or "roughage," usually refers to the cellulose and other indigestible polysaccharides in our vegetable foods. Bacteria that live in the guts of termites and grazers such as cattle and sheep help these animals digest the cellulose in plants.

In **starch**, a different covalent bonding pattern between glucose monomers makes a chain that coils up into a spiral (**Figure 3.8B**). Like cellulose, starch does not dissolve easily in water, but it is not as stable as cellulose. These properties make the molecule ideal for storing chemical energy in the watery, enzyme-filled interior of plant cells. Most plants make much more glucose than they can use. The excess is stored as starch inside cells that make up roots, stems, and

$$\text{glucose} \quad + \quad \text{fructose} \quad \longrightarrow \quad \text{sucrose} \quad + \quad \text{water}$$

Figure 3.7 The synthesis of a sucrose molecule is an example of a condensation reaction. You are already familiar with sucrose—it is common table sugar.

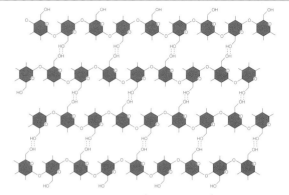

A Cellulose, a structural component of plants. Chains of glucose units stretch side by side and hydrogen-bond at many —OH groups. The hydrogen bonds stabilize the chains in tight bundles that form long fibers. Very few types of organisms can digest this tough, insoluble material.

B In amylose, one type of starch, a series of glucose units form a chain that coils. Starch is the main energy reserve in plants, which store it in their roots, stems, leaves, fruits, and seeds (such as coconuts).

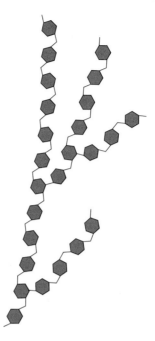

C Glycogen. In humans and other animals, this polysaccharide functions as an energy reservoir. It is stored in muscles and in the liver.

Figure 3.8 Structure of (**A**) cellulose, (**B**) starch, and (**C**) glycogen, and their typical locations in a few organisms. All three carbohydrates consist only of glucose units, but the different bonding patterns that link the subunits result in substances with very different properties.

leaves. However, because it is insoluble, starch cannot be transported out of the cells and distributed to other parts of the plant. When sugars are in short supply, hydrolysis enzymes break the bonds between starch's monomers to release glucose subunits. Humans also have enzymes that hydrolyze starch, so this carbohydrate is an important component of our food.

The covalent bonding pattern in **glycogen** forms highly branched chains of glucose monomers (**Figure 3.8C**). In animals, glycogen is the sugar-storage equivalent of starch in plants. Muscle and liver cells store it to meet a sudden need for glucose. These cells break down glycogen to release its glucose subunits.

carbohydrate Molecule that consists primarily of carbon, hydrogen, and oxygen atoms in a 1:2:1 ratio.
cellulose Polysaccharide; major structural material in plants.
disaccharide Polymer of two sugar subunits.
glycogen Polysaccharide; energy reservoir in animal cells.
monosaccharide Simple sugar; monomer of polysaccharides.
polysaccharide Polymer of many monosaccharides.
starch Polysaccharide; energy reservoir in plant cells.

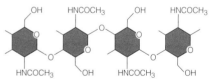

Figure 3.9 Chitin. This polysaccharide strengthens the hard parts of many small animals, such as crabs.

Chitin is a polysaccharide similar to cellulose. Its monomers are glucose with a nitrogen-containing carbonyl group (**Figure 3.9**). Long, unbranching chains of these monomers are linked by hydrogen bonds. As a structural material, chitin is durable, translucent, and flexible. It strengthens hard parts of many animals, including the outer cuticle of crabs, beetles, and ticks, and it reinforces the cell wall of many fungi.

Take-Home Message

What are carbohydrates?

» Simple carbohydrates (sugars), bonded together in different ways, form various types of complex carbohydrates.

» Cells use carbohydrates for energy or as structural materials.

3.5 Greasy, Oily—Must Be Lipids

■ Triglycerides, phospholipids, waxes, and steroids are lipids common in biological systems.

Lipids are fatty, oily, or waxy organic compounds. They vary in structure, but all are hydrophobic. Many lipids incorporate **fatty acids**, which are small organic molecules that consist of a hydrocarbon "tail" topped with a carboxyl group "head" (**Figure 3.10**). The tail of a fatty acid is hydrophobic (hence the name "fatty"), but the carboxyl group (the "acid" part of the name) makes the head hydrophilic. You are already familiar with the properties of fatty acids because these molecules are the main component of soap. The hydrophobic tails of the fatty acids in soap attract oily dirt, and the hydrophilic heads dissolve the dirt in water.

Saturated fatty acids have only single bonds in their tails. In other words, their carbon chains are fully saturated with hydrogen atoms (**Figure 3.10A**). Saturated fatty acid tails are flexible and they wiggle freely. The tails of **unsaturated fatty acids** have one or more double bonds that limit their flexibility (**Figure 3.10B,C**). **Figure 3.1** shows how these bonds are *cis* or *trans*, depending on the way the hydrogens are arranged around them.

Lipids in Biological Systems

Fats The carboxyl group of a fatty acid easily forms bonds with other molecules. **Fats** are lipids with one, two, or three fatty acids bonded to glycerol, which is an alcohol. A fatty acid attaches to a glycerol via its

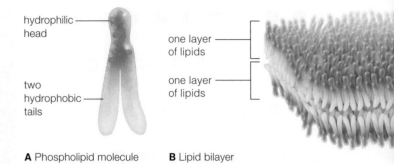

A Phospholipid molecule   **B** Lipid bilayer

Figure 3.12 Phospholipids as components of cell membranes. A double layer of phospholipids—the lipid bilayer—is the structural foundation of all cell membranes.

carboxyl group head. When it does, the fatty acid loses its hydrophilic character. When three fatty acids attach to a glycerol, the resulting molecule, which is called a **triglyceride**, is entirely hydrophobic (**Figure 3.11A**).

Because they are hydrophobic, triglycerides do not dissolve easily in water. Most "neutral" fats, such as butter and vegetable oils, are examples. Triglycerides are the most abundant and richest energy source in vertebrate bodies. They are concentrated in adipose tissue that insulates and cushions body parts.

Animal fats are saturated, which means they consist mainly of triglycerides with three saturated fatty acid tails. Saturated fats tend to remain solid at room temperature because their floppy saturated tails can pack tightly together. Most vegetable oils are unsaturated, which means these fats consist mainly of triglycerides with one or more unsaturated fatty acid tails. Each

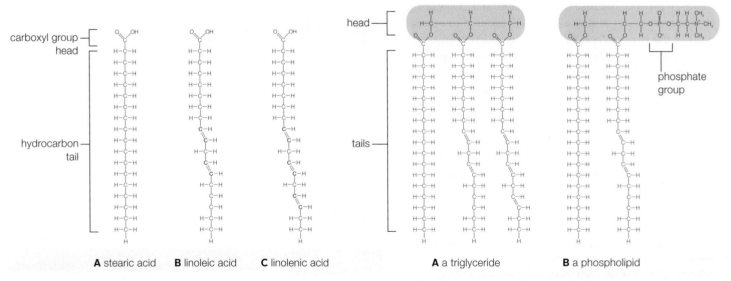

A stearic acid   **B** linoleic acid   **C** linolenic acid

Figure 3.10 Fatty acids. (**A**) The tail of stearic acid is fully saturated with hydrogen atoms. (**B**) Linoleic acid, with two double bonds, is unsaturated. The first double bond occurs at the sixth carbon from the end of the tail, so linoleic acid is called an omega-6 fatty acid. Omega-6 and (**C**) omega-3 fatty acids are "essential fatty acids": Your body does not make them, so they must come from food.

Figure 3.11 Animated Lipids with fatty acid tails. (**A**) Fatty acid tails of a triglyceride are attached to a glycerol head. (**B**) Fatty acid tails of a phospholipid are attached to a glycerol head with a phosphate group. **Figure It Out: Is the triglyceride saturated or unsaturated?**

Answer: Unsaturated

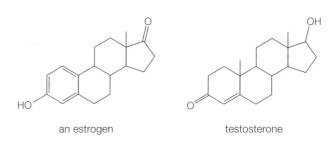

Figure 3.13 Estrogen and testosterone, steroid hormones that cause different traits to arise in males and females of many species such as wood ducks (*Aix sponsa*), pictured at *right*.

double bond in a fatty acid tail makes a rigid kink. Kinky tails do not pack tightly, so unsaturated fats are typically liquid at room temperature. The partially hydrogenated vegetable oils that you learned about in Section 3.1 are an exception. They are solid at room temperature. The special *trans* double bond keeps fatty acid tails straight, allowing them to pack tightly just like saturated fats do.

Phospholipids A **phospholipid** has two fatty acid tails and a head that contains a phosphate group (**Figure 3.11B**). The tails are hydrophobic, but the highly polar phosphate group makes the head very hydrophilic. The opposing properties of a phospholipid molecule give rise to cell membrane structure. Phospholipids are the most abundant lipids in cell membranes, which have two layers of lipids (**Figure 3.12**). The heads of one layer are dissolved in the cell's watery interior, and the heads of the other layer are dissolved in the cell's fluid surroundings. In such **lipid bilayers**, all of the hydrophobic tails are sandwiched between the hydrophilic heads. You will read more about the structure of cell membranes in Chapters 4 and 5.

Waxes A **wax** is a complex, varying mixture of lipids with long fatty acid tails bonded to long-chain alcohols or carbon rings. The molecules pack tightly, so the resulting substance is firm and water-repellent. Plant leaves secrete a layer of waxes that helps restrict water loss and keeps out parasites and other pests. Secreted waxes also protect, lubricate, and soften our skin and hair. Waxes, together with fats and fatty acids, make feathers waterproof. Bees store honey and raise new generations of bees inside honeycomb made from wax that they secrete.

Steroids All eukaryotic cell membranes contain **steroids**, lipids with a rigid backbone of four carbon rings and no fatty acid tails. Cholesterol, the most common steroid in animal cell membranes, also serves as a precursor for bile salts (which help digest fats), vitamin D (required to keep teeth and bones strong), and steroid hormones. Estrogens and testosterone, hormones that govern reproduction and the development of sexual traits, are steroid hormones (**Figure 3.13**).

fat Lipid that consists of a glycerol molecule with one, two, or three fatty acid tails.

fatty acid Organic compound that consists of a chain of carbon atoms with an acidic carboxyl group at one end. Carbon chain of saturated types has single bonds only; that of unsaturated types has one or more double bonds.

lipid Fatty, oily, or waxy organic compound.

lipid bilayer Double layer of lipids arranged tail-to-tail; structural foundation of all cell membranes.

phospholipid A lipid with a phosphate group in its hydrophilic head, and two nonpolar fatty acid tails; main constituent of eukaryotic cell membranes.

saturated fatty acid Fatty acid that contains no carbon–carbon double bonds.

steroid Type of lipid with four carbon rings and no fatty acid tails.

triglyceride A fat with three fatty acid tails.

unsaturated fatty acid Fatty acid that has one or more carbon–carbon double bonds in its tail.

wax Water-repellent mixture of lipids with long fatty acid tails bonded to long-chain alcohols or carbon rings.

Take-Home Message

What are lipids?

» Lipids are fatty, waxy, or oily organic compounds. Common types include fats, phospholipids, waxes, and steroids.

» Triglycerides are lipids that serve as energy reservoirs in vertebrate animals.

» Phospholipids are the main lipid component of cell membranes.

» Waxes are lipid components of water-repelling and lubricating secretions.

» Steroids are lipids that occur in cell membranes. Some are remodeled into other molecules.

3.6 Proteins—Diversity in Structure and Function

- All cellular processes involve proteins, the most diverse biological molecules.
- Cells build thousands of different proteins by stringing together amino acids in different orders.
- Link to Covalent bonding 2.4

Of all biological molecules, proteins are the most diverse in both structure and function. Structural proteins support cell parts and, as part of tissues, multicelled bodies. Feathers, hooves, and hair, as well as bones and other body parts, consist mainly of structural proteins. A tremendous number of different proteins, including some structural types, are active participants in all processes that sustain life. Most enzymes that drive metabolic reactions are proteins. Proteins move substances, help cells communicate, and defend the body.

From Structure to Function

Amino Acids Cells can make all of the thousands of different kinds of proteins they need from only twenty kinds of monomers called amino acids. **Proteins** are polymers of amino acids. An **amino acid** is a small organic compound with an amine group, a carboxyl group (the acid), and one or more atoms called an "R group." In most amino acids, all three groups are attached to the same carbon atom (**Figure 3.14**).

Peptide Bonds Protein synthesis involves covalently bonding amino acids into a chain. For each type of protein, instructions coded in DNA specify the order of the component amino acids.

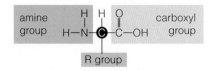

Figure 3.14 Generalized structure of amino acids. The complete structures of the twenty most common amino acids found in eukaryotic proteins are shown in Appendix I.

During protein synthesis, a condensation reaction links the amine group of one amino acid to the carboxyl group of the next. The bond that forms between the two amino acids is called a **peptide bond** (**Figure 3.15**). Enzymes repeat this bonding process hundreds or thousands of times, so a long chain of amino acids (a **polypeptide**) forms (**Figure 3.16**). The linear sequence of amino acids in the polypeptide is called the protein's primary structure ❶. You will learn more about protein synthesis in Chapter 9.

The Structure–Function Relationship One of the fundamental ideas in biology is that structure dictates function. This idea is particularly appropriate as applied to proteins, because a protein's biological activity arises from and depends on its shape.

There are several levels of protein structure beyond amino acid sequence. Even before a polypeptide is finished being synthesized, it begins to twist and fold as hydrogen bonds form among the amino acids of the chain. This hydrogen bonding may cause parts of the polypeptide to form flat sheets or coils (helices), patterns that constitute a protein's secondary structure ❷. The primary structure of each type of protein is unique, but most proteins have sheets and coils.

A Two amino acids (here, methionine and serine) are joined by condensation. A peptide bond forms between the carboxyl group of the methionine and the amine group of the serine.

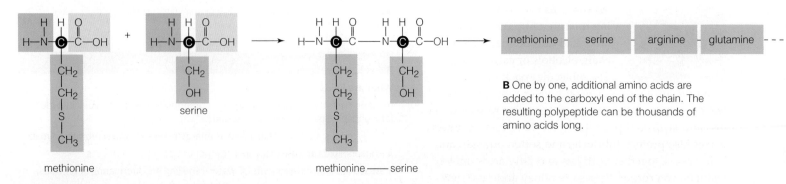

B One by one, additional amino acids are added to the carboxyl end of the chain. The resulting polypeptide can be thousands of amino acids long.

Figure 3.15 Animated Polypeptide formation. Chapter 9 offers a closer look at protein synthesis.
Figure It Out: What functional group forms with a peptide bond?   Answer: Amide

| lysine | glycine | glycine | arginine |

1 A protein's primary structure consists of a linear sequence of amino acids (a polypeptide chain). Each type of protein has a unique primary structure.

Figure 3.16 Animated Protein structure.

2 Secondary structure arises as a polypeptide chain twists into a coil (helix) or sheet held in place by hydrogen bonds between different parts of the molecule. The same patterns of secondary structure occur in many different proteins.

3 Tertiary structure occurs when a chain's coils and sheets fold up into a functional domain such as a barrel or pocket. In this example, the coils of a globin chain form a pocket.

4 Some proteins have quaternary structure, in which two or more polypeptide chains associate as one molecule. Hemoglobin, shown here, consists of four globin chains (*green* and *blue*). Each globin pocket now holds a heme group (*red*).

Much as an overly twisted rubber band coils back upon itself, hydrogen bonding between different parts of a protein make it fold up into even more compact domains. A domain is a part of a protein that is organized as a structurally stable unit. Such units are part of a protein's overall three-dimensional shape, or tertiary structure **3**. Tertiary structure is what makes a protein a working molecule. For example, sheets or coils of some proteins curl up into a barrel shape (*inset*). A barrel domain often functions as a tunnel for small molecules, allowing them to pass, for example, through a cell membrane. Globular domains of enzymes form chemically active pockets that can make or break bonds of other molecules.

5 Many proteins aggregate by the thousands into much larger structures, such as the keratin filaments that make up hair.

Many proteins also have quaternary structure, which means they consist of two or more polypeptide chains that are in close association or covalently bonded together. Most enzymes and many other proteins consist of two or more polypeptide chains that collectively form a roughly spherical shape **4**.

Some proteins aggregate by many thousands into much larger structures, with their polypeptide chains organized into strands or sheets. The keratin in your hair is an example **5**. Some fibrous proteins contribute to the structure and organization of cells and tissues. Others, such as the actin and myosin filaments in muscle cells, are part of the mechanisms that help cells, cell parts, and multicelled bodies move.

Sugars or lipids bonded to proteins make glycoproteins and lipoproteins, respectively. The molecules that allow a tissue or a body to recognize its own cells are glycoproteins, as are other molecules that help cells interact in immune responses. Some lipoproteins form when enzymes covalently bond lipids to a protein.

Other lipoproteins are aggregate structures that consist of variable amounts and types of proteins and lipids. These particles carry fats and cholesterol through the bloodstream. Low-density lipoprotein, or LDL, transports cholesterol out of the liver and into cells. High-density lipoprotein, or HDL (*right*), ferries the cholesterol that is released from dead cells back to the liver.

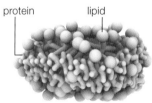
protein   lipid

an HDL particle

amino acid Small organic compound that is a subunit of proteins. Consists of a carboxyl group, an amine group, and a characteristic side group (R), all typically bonded to the same carbon atom.
peptide bond A bond between the amine group of one amino acid and the carboxyl group of another. Joins amino acids in proteins.
polypeptide Chain of amino acids linked by peptide bonds.
protein Organic compound that consists of one or more chains of amino acids (polypeptides).

Take-Home Message

What are proteins?

» Proteins are chains of amino acids. The order of amino acids in a polypeptide chain dictates the type of protein.

» Polypeptide chains twist and fold into coils, sheets, and loops, which fold and pack further into functional domains.

» A protein's shape is the source of its function.

3.7 Why Is Protein Structure So Important?

■ Changes in a protein's shape may have drastic health consequences.

Protein shape depends on hydrogen bonds and other interactions that heat, some salts, shifts in pH, or detergents can disrupt. Such disruption causes proteins to **denature**, which means they lose their shape. Once a protein's shape unravels, so does its function.

You can see denaturation in action when you cook an egg. A protein called albumin is a major component of egg white. Cooking does not disrupt the covalent bonds of albumin's primary structure, but it does destroy the hydrogen bonds that maintain the protein's shape. When a translucent egg white turns opaque, the albumin has been denatured. For a very few proteins, denaturation is reversible if normal conditions return, but albumin is not one of them. There is no way to uncook an egg.

Prion diseases such as mad cow disease (bovine spongiform encephalitis, or BSE) in cattle, Creutzfeldt–Jakob disease in humans, and scrapie in sheep, are the dire aftermath of a protein that changes shape. These infectious diseases may be inherited, but more often they arise spontaneously. All are characterized by relentless deterioration of mental and physical abilities that eventually causes death (**Figure 3.17A**).

All prion diseases begin with a protein that occurs normally in mammals. One such protein, PrPC, is found in cell membranes throughout the body. This copper-binding protein is especially abundant in brain cells, but we still know very little about what it does. Very rarely, a PrPC protein spontaneously misfolds so that it loses some of its coils. In itself, a single misfolded protein molecule would not pose much of a threat. However, when this particular protein misfolds, it becomes a **prion**, or infectious protein. The altered shape of a misfolded PrPC protein somehow causes normally folded PrPC proteins to misfold too. Because each protein that misfolds becomes infectious, the number of prions increases exponentially.

The shape of misfolded PrPC proteins allows them to align tightly into long fibers. Fibers of aggregated prion proteins begin to accumulate in the brain (**Figure 3.17B**). The water-repellent patches grow as more prions form, and they begin to disrupt brain cell function, causing symptoms such as confusion, memory loss, and lack of coordination. Tiny holes form in the brain as its cells die. Eventually, the brain becomes so riddled with holes that it looks like a sponge.

In the mid-1980s, an epidemic of mad cow disease in Britain was followed by an outbreak of a new variant of Creutzfeldt–Jakob disease (vCJD) in humans. Researchers isolated a prion similar to the one in scrapie-infected sheep from cows with BSE, and also from humans affected by the new type of Creutzfeldt–Jakob disease. How did the prion get from sheep to cattle to people? Prions are not denatured by cooking or typical treatments that inactivate other types of infectious agents. The cattle became infected by the prion after eating feed prepared from the remains of scrapie-infected sheep, and people became infected by eating beef from the infected cattle.

Two hundred people have died from vCJD since 1990. The use of animal parts in livestock feed is now banned in many countries, and the number of cases of BSE and vCJD has since declined. Cattle with BSE still turn up, but so rarely that they pose little threat to human populations.

denature To unravel the shape of a biological molecule.
prion Infectious protein.

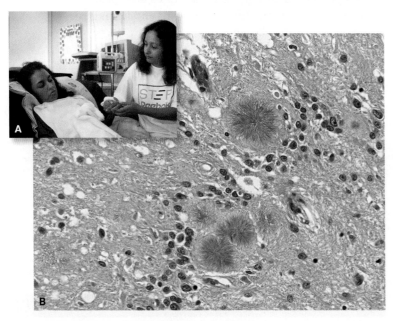

Figure 3.17 Variant Creutzfeldt–Jakob disease (vCJD).

(**A**) Charlene Singh, here being cared for by her mother, was one of the three people who developed symptoms of vCJD disease while living in the United States. Like the others, Singh most likely contracted the disease elsewhere; she spent her childhood in Britain. She was diagnosed in 2001, and she died in 2004.

(**B**) Slice of brain tissue from a person with vCJD. Characteristic holes and prion protein fibers radiating from several deposits are visible.

Take-Home Message

Why is protein structure important?

» A protein's function depends on its structure.

» Conditions that alter a protein's structure may also alter its function.

» Protein shape unravels if hydrogen bonds are disrupted.

3.8 Nucleic Acids

■ Nucleotides are subunits of DNA and RNA. Some have roles in metabolism.
■ Links to Inheritance 1.3, Diversity 1.4, Hydrogen bonds 2.5

Nucleotides are small organic molecules that function as energy carriers, enzyme helpers, chemical messengers, and subunits of DNA and RNA. Each consists of a sugar with a five-carbon ring, bonded to a nitrogen-containing base and one or more phosphate groups. The nucleotide **ATP** (adenosine triphosphate) has a row of three phosphate groups attached to its ribose sugar (**Figure 3.18A**). When the outer phosphate group of an ATP is transferred to another molecule, energy is transferred along with it. You will read about such phosphate-group transfers and their important metabolic role in Chapter 5.

Nucleic acids are polymers, chains of nucleotides in which the sugar of one nucleotide is joined to the phosphate group of the next (**Figure 3.18B**). An example

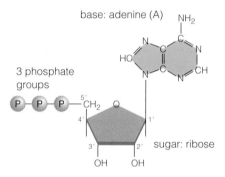

base: adenine (A)

3 phosphate groups

sugar: ribose

A ATP, a nucleotide monomer of RNA, and also an essential participant in many metabolic reactions.

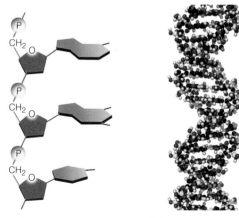

B A chain of nucleotides is a nucleic acid. The sugar of one nucleotide is covalently bonded to the phosphate group of the next, forming a sugar–phosphate backbone.

C DNA consists of two chains of nucleotides, twisted into a double helix. Hydrogen bonding maintains the three-dimensional structure of this nucleic acid.

Figure 3.18 Animated Nucleic acid structure.

Fear of Frying (revisited)

Trans fatty acids are relatively rare in unprocessed foods, so it makes sense from an evolutionary standpoint that our bodies may not have enzymes to deal with them efficiently. We do have enzymes that hydrolyze *cis* fatty acids, but these enzymes have difficulty breaking down *trans* fatty acids—a problem that may be a factor in the ill effects of *trans* fats on our health.

How would you vote? All prepackaged foods in the United States are now required to list *trans* fat content, but may be marked "zero grams of trans fats" even when a single serving contains up to half a gram. Should hydrogenated oils be banned from all food?

- -

is **RNA**, or ribonucleic acid, named after the ribose sugar of its component nucleotides. RNA consists of four kinds of nucleotide monomers, one of which is ATP. RNA molecules carry out protein synthesis, which we discuss in detail in Chapter 9.

DNA, or deoxyribonucleic acid, is a nucleic acid named after the deoxyribose sugar of its component nucleotides. A DNA molecule consists of two chains of nucleotides twisted into a double helix (**Figure 3.17C**). Hydrogen bonds between the nucleotides hold the two chains together.

Each cell starts life with DNA inherited from a parent cell. That DNA contains all information necessary to build a new cell and, in the case of multicelled organisms, an entire individual. The cell uses the order of nucleotide bases in DNA—the DNA sequence—to guide production of RNA and proteins. Parts of the sequence are identical or nearly so in all organisms, but most is unique to a species or an individual (Chapter 8 returns to DNA structure and function).

ATP Adenosine triphosphate. Nucleotide that consists of an adenine base, a ribose sugar, and three phosphate groups.
DNA Deoxyribonucleic acid. Nucleic acid that carries hereditary information about traits; consists of two nucleotide chains twisted in a double helix.
nucleic acid Single- or double-stranded chain of nucleotides joined by sugar–phosphate bonds; for example, DNA, RNA.
nucleotide Monomer of nucleic acids; has a five-carbon sugar, a nitrogen-containing base, and phosphate groups.
RNA Ribonucleic acid. Some types have roles in protein synthesis.

Take-Home Message

What are nucleotides and nucleic acids?

» Nucleotides are monomers of the nucleic acids DNA and RNA. Some have additional roles.

» DNA's nucleotide sequence encodes heritable information.

» RNA plays several important roles in the process by which a cell uses the instructions written in its DNA to build proteins.

Summary

Section 3.1 All organisms consist of the same kinds of molecules. Seemingly small differences in the way those molecules are put together can have big effects inside a living organism.

Section 3.2 Under present-day conditions in nature, only living things make complex carbohydrates and lipids, proteins, and nucleic acids. These molecules are **organic**—they consist mainly of carbon and hydrogen atoms.

Section 3.3 Hydrocarbons have only carbon and hydrogen atoms. Carbon chains or rings form the backbone of the molecules of life. **Functional groups** attached to the backbone influence the chemical character, and thus the function of these compounds. **Metabolism** includes all of the processes by which cells acquire and use energy as they make and break the bonds of organic compounds. By metabolic **reactions** such as **condensation**, **enzymes** build **polymers** from **monomers** of simple sugars, fatty acids, amino acids, and nucleotides. Reactions such as **hydrolysis** release the monomers by breaking apart the polymers.

Section 3.4 Enzymes assemble **disaccharides** and **polysaccharides** from simple **carbohydrate** (sugar) monomers. **Cellulose**, **glycogen**, and **starch** consist of **monosaccharides** bonded in different patterns. Cells use the different kinds for energy, and as structural materials.

Section 3.5 Lipids are fatty, oily, or waxy compounds that cells use for energy and as structural materials. All are nonpolar. **Fats** and some other lipids have **fatty acid** tails; **triglycerides** have three. **Unsaturated fatty acids** have carbon–carbon double bonds; only single bonds link carbons in **saturated fatty acids**. A **lipid bilayer** that consists mainly of **phospholipids** is the structural foundation of all cell membranes. **Waxes** are lipids that are part of water-repellent and lubricating secretions. **Steroids** occur in cell membranes; some are remodeled into other molecules.

Section 3.6 Structurally and functionally, **proteins** are the most diverse molecules of life. The shape of a protein is the source of its function. Protein structure begins as a linear sequence of **amino acids** linked by **peptide bonds** into a **polypeptide** (primary structure). Polypeptides twist into loops, sheets, and coils (secondary structure) that can pack further into functional domains (tertiary structure). Many proteins, including most enzymes, consist of two or more polypeptides (quaternary structure). Fibrous proteins aggregate by the thousands into much larger structures.

Section 3.7 A protein's structure dictates its function, so changes in a protein's structure may also alter its function. Hydrogen bonds and other molecular interactions that are responsible for a protein's shape may be disrupted by shifts in pH or temperature, or exposure to detergent or some salts. If that happens, the protein unravels, or **denatures**, and so loses its function. **Prion** diseases are a consequence of misfolded proteins.

Section 3.8 Nucleotides are small organic molecules consisting of a sugar, a phosphate group, and a nitrogen-containing base. Nucleotides are monomers of **DNA** and **RNA**, which are **nucleic acids**. Some nucleotides have additional functions. For example, **ATP** energizes many kinds of molecules by phosphate-group transfers. DNA encodes heritable information that guides the synthesis of RNA and proteins. RNA molecules interact with DNA to carry out protein synthesis.

Self-Quiz

Answers in Appendix III

1. Organic molecules consist mainly of _____ atoms.
   a. carbon
   b. carbon and oxygen
   c. carbon and hydrogen
   d. carbon and nitrogen

2. Each carbon atom can share pairs of electrons with as many as _____ other atom(s).

3. _____ groups impart polarity to alcohols.
   a. Hydroxyl ($-OH^-$)
   b. Phosphate ($-PO_4$)
   c. Methyl ($-CH_3$)
   d. Sulfhydryl ($-SH$)

4. _____ is a simple sugar (a monosaccharide).
   a. Glucose
   b. Sucrose
   c. Ribose
   d. Starch
   e. both a and c
   f. a, b, and c

5. Which three carbohydrates can be built using only glucose monomers?
   a. Starch, cellulose, and glycogen
   b. Glucose, sucrose, and ribose
   c. Cellulose, steroids, and polysaccharides
   d. Starch, chitin, and DNA
   e. Triglycerides, nucleic acids, and polypeptides

6. Unlike saturated fats, the fatty acid tails of unsaturated fats incorporate one or more _____ .
   a. phosphate groups
   b. glycerols
   c. double bonds
   d. single bonds

7. Is this statement true or false? Unlike saturated fats, all unsaturated fats are beneficial to health because their fatty acid tails kink and do not pack together.

8. Steroids are among the lipids with no _____ .
   a. double bonds
   b. fatty acid tails
   c. hydrogens
   d. carbons

9. Which of the following is a class of molecules that encompasses all of the other molecules listed?
   a. triglycerides
   b. fatty acids
   c. waxes
   d. steroids
   e. lipids
   f. phospholipids

10. _____ are to proteins as _____ are to nucleic acids.
    a. Sugars; lipids
    b. Sugars; proteins
    c. Amino acids; hydrogen bonds
    d. Amino acids; nucleotides

Effects of Dietary Fats on Lipoprotein Levels Cholesterol that is made by the liver or that enters the body from food does not dissolve in blood, so it is carried through the bloodstream by lipoproteins. Low-density lipoprotein (LDL) carries cholesterol to body tissues such as artery walls, where it can form deposits associated with cardio-vascular disease. Thus, LDL is often called "bad" choles-terol. High-density lipoprotein (HDL) carries cholesterol away from tissues to the liver for disposal, so HDL is often called "good" cholesterol.

In 1990, Ronald Mensink and Martijn Katan published a study that tested the effects of different dietary fats on blood lipoprotein levels. Their results are shown in **Figure 3.19**.

1. In which group was the level of LDL ("bad" cholesterol) highest?

2. In which group was the level of HDL ("good" cholesterol) lowest?

3. An elevated risk of heart disease has been correlated with increasing LDL-to-HDL ratios. Which group had the highest LDL-to-HDL ratio?

4. Rank the three diets from best to worst according to their potential effect on heart disease.

	Main Dietary Fats			
	cis fatty acids	*trans* fatty acids	saturated fats	optimal level
LDL	103	117	121	<100
HDL	55	48	55	>40
ratio	1.87	2.44	2.2	<2

Figure 3.19 Effect of diet on lipoprotein levels. Researchers placed 59 men and women on a diet in which 10 percent of their daily energy intake consisted of *cis* fatty acids, *trans* fatty acids, or saturated fats.

Blood LDL and HDL levels were measured after three weeks on the diet; averaged results are shown in mg/dL (milligrams per deciliter of blood). All subjects were tested on each of the diets. The ratio of LDL to HDL is also shown.

---

11. A denatured protein has lost its _____ .
    a. hydrogen bonds   c. function
    b. shape   d. all of the above

12. _____ consists of nucleotides.
    a. Sugars   c. DNA
    b. RNA   d. b and c

13. Which of the following is not found in DNA?
    a. amino acids   c. nucleotides
    b. sugars   d. phosphate groups

14. In the following list, identify the carbohydrate, the fatty acid, the amino acid, and the polypeptide:
    a. NH_2—CHR—COOH   c. (methionine)$_{20}$
    b. $C_6H_{12}O_6$ d. $CH_3(CH_2)_{16}COOH$

15. Match the molecules with the best description.
    ___ wax   a. protein primary structure
    ___ starch   b. an energy carrier
    ___ triglyceride   c. water-repellent secretions
    ___ DNA   d. carries heritable information
    ___ polypeptide   e. sugar storage in plants
    ___ ATP   f. richest energy source

16. Match each polymer with the most appropriate set of component monomers.
    ___ protein   a. glycerol, fatty acids, phosphate
    ___ phospholipid   b. amino acids, sugars
    ___ glycoprotein   c. glycerol, fatty acids
    ___ fat   d. nucleotides
    ___ nucleic acid   e. polysaccharide
    ___ carbohydrate   f. sugar, phosphate, base
    ___ nucleotide   g. amino acids
    ___ lipoprotein   h. glucose, fructose
    ___ sucrose   i. lipids, amino acids
    ___ glycogen   j. fatty acids, carbon rings
    ___ wax   k. glucose only

Critical Thinking

1. Lipoproteins are relatively large, spherical clumps of protein and lipid molecules that circulate in the blood of mammals. They are like suitcases that move cholesterol, fatty acid remnants, triglycerides, and phospholipids from one place to another in the body. Given what you know about the insolubility of lipids in water, which of the four kinds of lipids would you predict to be on the outside of a lipoprotein clump, bathed in the fluid portion of blood?

2. In 1976, a team of chemists in the United Kingdom was developing new insecticides by modifying sugars with chlorine (Cl_2), phosgene (Cl_2CO), and other toxic gases. One young mem-ber of the team misunderstood his verbal instruc-tions to "test" a new molecule. He thought he had been told to "taste" it. Luckily, the molecule was not toxic, but it was very sweet. It became the food additive sucralose.

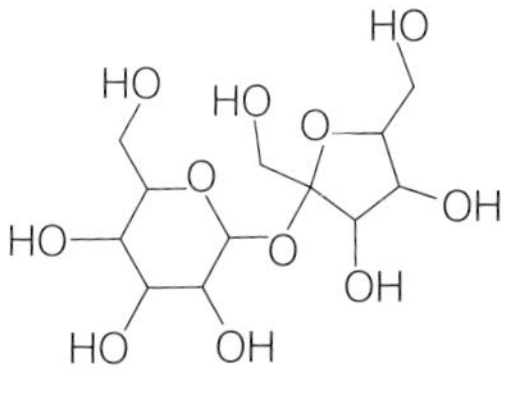

Sucralose has three chlorine atoms sub-stituted for three hydroxyl groups of sucrose (table sugar). It binds so strongly to sweet-taste receptors on the tongue that the human brain perceives it as 600 times sweeter than sucrose. Sucralose was originally marketed as an artificial sweetener called Splenda®, but it is now avail-able under several other brand names.

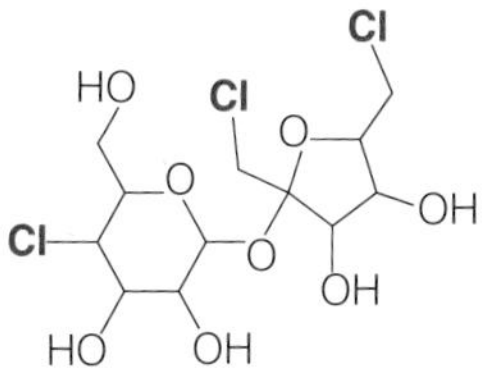

Researchers proved that the body does not recognize sucralose as a carbohydrate by feed-ing sucralose labeled with ^{14}C to volunteers. Analysis of the radioactive molecules in the volunteers' urine and feces showed that 92.8 percent of the sucralose passed through the body without being altered. Many people are worried that the chlorine atoms impart tox-icity to sucralose. How would you respond to that concern?

sucrose

sucralose

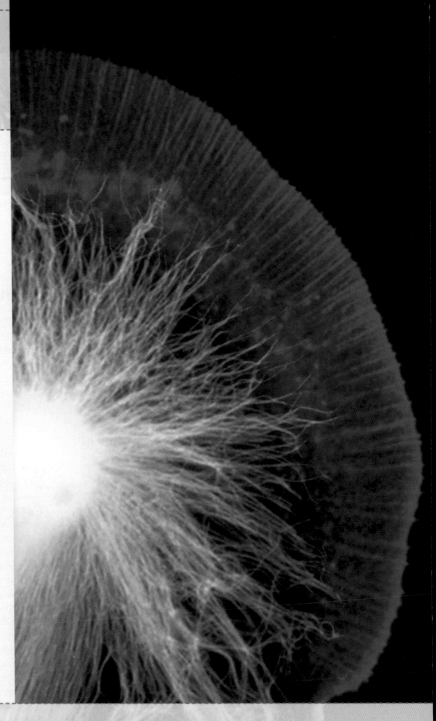

LEARNING ROADMAP

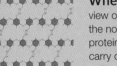

Where you have been Reflect on the Section 1.2 overview of life's levels of organization. You will now begin to see how the nonliving molecules of life—carbohydrates (3.4), lipids (3.5), proteins (3.6, 3.7), and nucleic acids (3.8)—form cells (1.2) and carry out functions that define life (1.3). You will also see an application of tracers (2.2), and revisit the philosophy of science (1.9).

Where you are now

What All Cells Have In Common
Every cell has a plasma membrane that separates its interior from the exterior environment. The interior contains cytoplasm, DNA, and other structures.

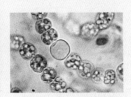

The "Prokaryotes"
Archaea and bacteria have no nucleus. In general, they are smaller and structurally simpler than eukaryotic cells, but they are by far the most numerous.

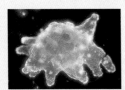

Eukaryotic Cells
Protists, plants, fungi, and animals are eukaryotes. Cells of these organisms differ in internal parts and surface specializations, but all start out life with a nucleus.

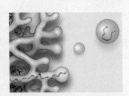

Organelles of Eukaryotes
Membranes around eukaryotic organelles maintain internal environments that allow these structures to carry out specialized functions within a cell.

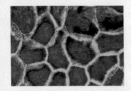

Other Cell Components
Cytoskeletal elements organize and move cells and cell components. Cells secrete protective and structural materials. Cell junctions connect cells to one another.

Where you are going Chapter 5 explores cell function; Chapter 6, photosynthesis; and Chapter 7, aerobic respiration. We revisit protein synthesis in Chapter 9, and control over it in Chapter 10. Some cellular structures are required for cell division (Chapter 11). Human genetic disorders return in Chapter 14. Chapter 19 details the structures, evolution, and metabolism of bacteria and archaea; Chapter 21, protists. Cell structures introduced in this chapter return in context of the physiology of animals (Chapters 31–42) and plants (Chapters 27–30). You will see some medical consequences of biofilms in Section 37.3.

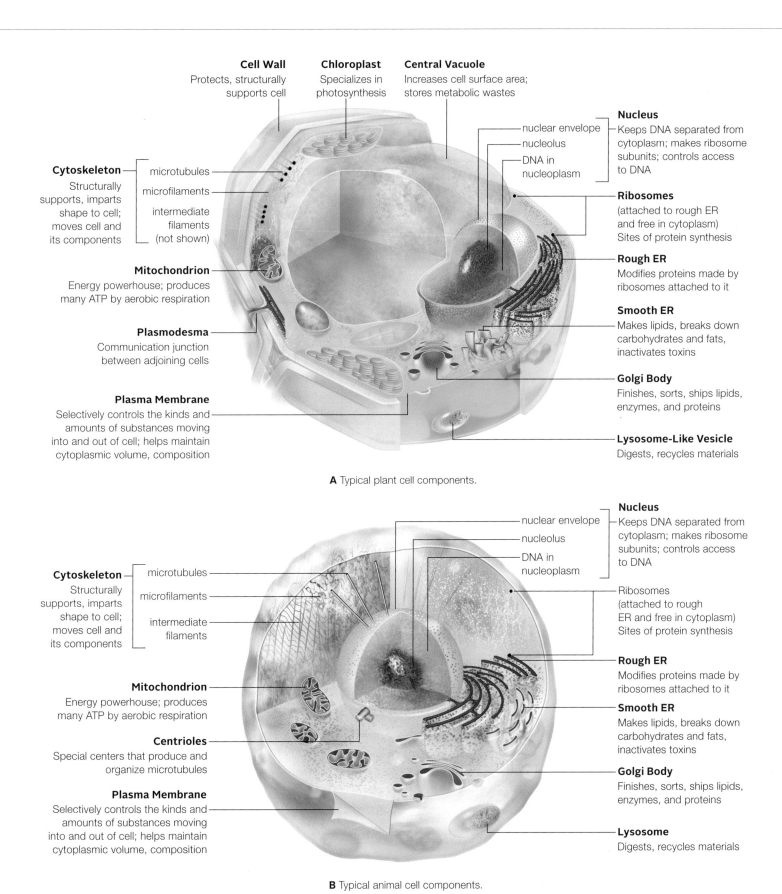

Cell Wall
Protects, structurally
supports cell

Chloroplast
Specializes in
photosynthesis

Central Vacuole
Increases cell surface area;
stores metabolic wastes

nuclear envelope

nucleolus

DNA in
nucleoplasm

Nucleus
Keeps DNA separated from
cytoplasm; makes ribosome
subunits; controls access
to DNA

Cytoskeleton
Structurally
supports, imparts
shape to cell;
moves cell and
its components

microtubules

microfilaments

intermediate
filaments
(not shown)

Ribosomes
(attached to rough ER
and free in cytoplasm)
Sites of protein synthesis

Rough ER
Modifies proteins made by
ribosomes attached to it

Mitochondrion
Energy powerhouse; produces
many ATP by aerobic respiration

Smooth ER
Makes lipids, breaks down
carbohydrates and fats,
inactivates toxins

Plasmodesma
Communication junction
between adjoining cells

Plasma Membrane
Selectively controls the kinds and
amounts of substances moving
into and out of cell; helps maintain
cytoplasmic volume, composition

Golgi Body
Finishes, sorts, ships lipids,
enzymes, and proteins

Lysosome-Like Vesicle
Digests, recycles materials

A Typical plant cell components.

Nucleus
Keeps DNA separated from
cytoplasm; makes ribosome
subunits; controls access
to DNA

nuclear envelope

nucleolus

DNA in
nucleoplasm

Cytoskeleton
Structurally
supports, imparts
shape to cell;
moves cell and
its components

microtubules

microfilaments

intermediate
filaments

Ribosomes
(attached to rough
ER and free in cytoplasm)
Sites of protein synthesis

Rough ER
Modifies proteins made by
ribosomes attached to it

Mitochondrion
Energy powerhouse; produces
many ATP by aerobic respiration

Smooth ER
Makes lipids, breaks down
carbohydrates and fats,
inactivates toxins

Centrioles
Special centers that produce and
organize microtubules

Golgi Body
Finishes, sorts, ships lipids,
enzymes, and proteins

Plasma Membrane
Selectively controls the kinds and
amounts of substances moving
into and out of cell; helps maintain
cytoplasmic volume, composition

Lysosome
Digests, recycles materials

B Typical animal cell components.

Figure 4.12 **Animated** Organelles and structures typical of (**A**) plant cells and (**B**) animal cells.

4.6 The Nucleus

- All of a eukaryotic cell's DNA is in its nucleus.
- The nucleus keeps eukaryotic DNA away from potentially damaging reactions in the cytoplasm.
- The nuclear envelope controls when DNA is accessed.
- Link to Fibrous proteins 3.6

As molecules go, DNA is gigantic. Unraveled and stretched out, the DNA in the nucleus of a single human cell would be about 2 meters (6–1/2 feet) long. All of that DNA fits into a nucleus about six microns in diameter.

A cell's nucleus serves two important functions. First, it keeps the cell's genetic material—its DNA—safe from metabolic processes that might damage it. Isolated in its own compartment, the cell's DNA stays separated from the bustling activity of the cytoplasm. Second, a nucleus controls the passage of certain molecules across its membrane.

Figure 4.13 shows the components of the nucleus. **Table 4.3** lists their functions. Let's zoom in on the individual components.

The Nuclear Envelope

The nuclear membrane, which is called the **nuclear envelope**, consists of two lipid bilayers folded together as a single membrane (**Figure 4.14**). Membrane proteins embedded in the two lipid bilayers aggregate into thousands of tiny pores that span the nuclear enve-

Table 4.3	Components of the Nucleus
Nuclear envelope	Pore-riddled double membrane that controls which substances enter and leave the nucleus
Nucleoplasm	Semifluid interior portion of the nucleus
Nucleolus	Rounded mass of proteins and copies of genes for ribosomal RNA used to construct ribosomal subunits
Chromatin	Total collection of all DNA molecules and associated proteins in the nucleus; all of the cell's chromosomes
Chromosome	One DNA molecule and many proteins associated with it

lope. The pores are anchored by the nuclear lamina, a dense mesh of fibrous proteins that supports the inner surface of the envelope. Some bacteria have membranes around their DNA, but we do not consider the bacteria to have nuclei because there are no pores in these membranes.

As you will see in Chapter 5, large molecules, including RNA and proteins, cannot cross a lipid bilayer on their own. Nuclear pores function as gateways for these molecules to enter and exit a nucleus. Protein synthesis offers an example of why this movement is important. Protein synthesis occurs in

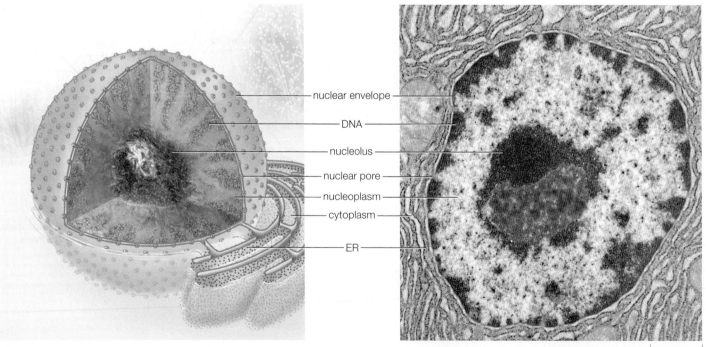

nuclear envelope

DNA

nucleolus

nuclear pore

nucleoplasm

cytoplasm

ER

1 μm

Figure 4.13 Animated The cell nucleus. TEM at *right*, nucleus of a mouse pancreas cell.

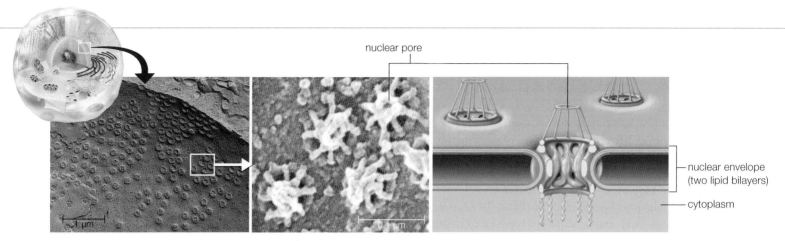

nuclear pore

nuclear envelope
(two lipid bilayers)

cytoplasm

A The outer surface of this nuclear envelope was split apart to reveal the pores that span the two lipid bilayers.

B Each nuclear pore is an organized cluster of membrane proteins that selectively allows certain substances to cross it on their way into and out of the nucleus.

Figure 4.14 **Animated** Structure of the nuclear envelope.

cytoplasm, and it requires the participation of many molecules of RNA. RNA is produced in the nucleus. Thus, RNA molecules must move from nucleus to cytoplasm, and they do so through nuclear pores. Proteins must move through the pores in the other direction, because RNA synthesis, which occurs in the nucleus, requires proteins produced in cytoplasm.

A cell can regulate the amounts and types of proteins it makes at a given time by selectively restricting the passage of certain molecules through nuclear pores. Later chapters return to details of protein synthesis and controls over it, as well as how molecules cross membranes.

The Nucleolus

The nucleus contains at least one **nucleolus** (plural, nucleoli), a dense, irregularly shaped region of proteins and nucleic acid where subunits of ribosomes are produced. The subunits pass through nuclear pores into the cytoplasm, where they join and become active in protein synthesis.

The DNA

The nuclear envelope encloses **nucleoplasm**, a viscous fluid, similar to cytoplasm, in which the cell's DNA is suspended. Most eukaryotic cells have a number of DNA molecules. That number is characteristic of the type of organism and the type of cell, but it varies widely among species. For instance, the nucleus of a normal oak tree cell contains 12 DNA molecules; a human body cell, 46; and a king crab cell, 208. Each molecule of DNA, together with its many attached

proteins, is called a **chromosome**. DNA molecules, together with their associated proteins, are collectively called **chromatin**.

Chromosomes are anchored to and organized by the nuclear lamina. They change in appearance over the lifetime of a cell. When a cell is not dividing, its chromatin is invisible in light micrographs. Just before a cell divides, the DNA in each chromosome is cop-

ied, or duplicated. As cell division begins, the duplicated chromosomes condense. As they do, they become visible in micrographs in their characteristic "X" forms (*inset*). In later chapters, we will look in more detail at the dynamic structure and function of eukaryotic chromosomes.

chromatin Collective term for DNA molecules together with their associated proteins.
chromosome A structure that consists of DNA and associated proteins; carries part or all of a cell's genetic information.
nuclear envelope A double membrane that constitutes the outer boundary of the nucleus. Pores in the membrane control which substances can cross.
nucleolus In a cell nucleus, a dense, irregularly shaped region where ribosomal subunits are assembled.
nucleoplasm Viscous fluid enclosed by the nuclear envelope.

Take-Home Message

What is the function of the cell nucleus?

» A nucleus protects and controls access to a eukaryotic cell's DNA.

» The nuclear envelope is a double lipid bilayer. Proteins embedded in it control the passage of molecules between the nucleus and the cytoplasm.

4.7 The Endomembrane System

■ The endomembrane system is a set of organelles that makes, modifies, and transports proteins and lipids.

■ Links to Lipids 3.5, Polypeptide formation 3.6

The **endomembrane system** is a series of interacting organelles between nucleus and plasma membrane. Its main function is to make lipids, enzymes, and other proteins destined for secretion, or for insertion into cell membranes. It also destroys toxins, recycles wastes, and has other specialized functions. The system's components vary among different types of cells, but here we present the most common ones (**Figure 4.15**).

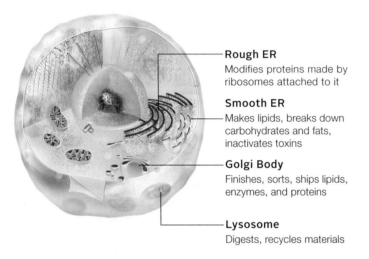

Rough ER
Modifies proteins made by ribosomes attached to it

Smooth ER
Makes lipids, breaks down carbohydrates and fats, inactivates toxins

Golgi Body
Finishes, sorts, ships lipids, enzymes, and proteins

Lysosome
Digests, recycles materials

Figure 4.15 Animated Endomembrane system, where many proteins are modified and lipids are built. These molecules are sorted and shipped to cellular destinations or to the plasma membrane for export.

A Series of Interacting Vessels

Endoplasmic Reticulum Part of the endomembrane system is an extension of the nuclear envelope called **endoplasmic reticulum**, or **ER**. ER forms a continuous compartment that folds into flattened sacs and tubes. Two kinds of ER, rough and smooth, are named for their appearance in electron micrographs. Thousands of ribosomes transiently attached to the outer surface of rough ER give this organelle its "rough" appearance. These ribosomes make polypeptides that thread into the interior of the ER as they are assembled ❶. Inside the ER, the polypeptides fold and take on their tertiary structure. Some of them become part of the ER membrane itself.

Cells that make, store, and secrete proteins have a lot of rough ER. For example, ER-rich gland cells in the pancreas make and secrete enzymes that help digest meals in the small intestine.

Smooth ER has no ribosomes on its surface ❷, so it does not make protein. Some of the polypeptides made in the rough ER end up as enzymes in the smooth ER. These enzymes assemble most of the lipids that form the cell's membranes. They also break down carbohydrates, fatty acids, and some drugs and poisons. In skeletal muscle cells, one type of smooth ER stores calcium ions and has a role in muscle contraction.

Vesicles Small, membrane-enclosed, saclike **vesicles** form in great numbers, in a variety of types, either on their own or by budding ❸. Many vesicles transport

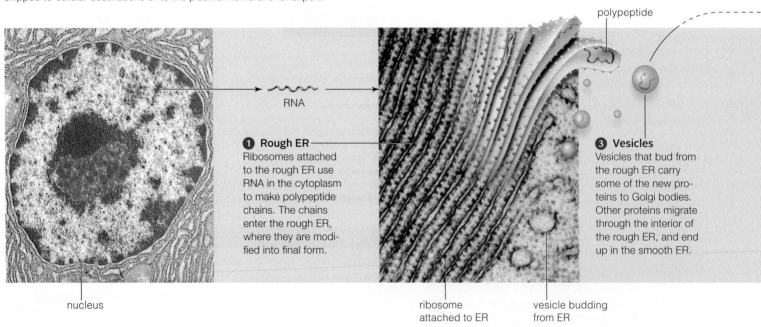

RNA

polypeptide

❶ Rough ER
Ribosomes attached to the rough ER use RNA in the cytoplasm to make polypeptide chains. The chains enter the rough ER, where they are modified into final form.

❸ Vesicles
Vesicles that bud from the rough ER carry some of the new proteins to Golgi bodies. Other proteins migrate through the interior of the rough ER, and end up in the smooth ER.

nucleus

ribosome attached to ER

vesicle budding from ER

substances from one organelle to another, or to and from the plasma membrane. Some form as a patch of plasma membrane sinks into the cytoplasm. Others bud from the ER or Golgi membranes and transport substances to the plasma membrane for export.

Vesicles that are a bit like trash cans collect and dispose of waste, debris, or toxins. Enzymes in ER-derived **peroxisomes** break down fatty acids, amino acids, and toxins such as alcohol. **Lysosomes** that bud from Golgi bodies take part in intracellular digestion. They contain powerful enzymes that can break down carbohydrates, proteins, nucleic acids, and lipids. Vesicles in white blood cells or amoebas deliver ingested bacteria, cell parts, and other debris to lysosomes for destruction.

central vacuole Fluid-filled vesicle in many plant cells.
endomembrane system Series of interacting organelles (endoplasmic reticulum, Golgi bodies, vesicles) between nucleus and plasma membrane; produces lipids, proteins.
endoplasmic reticulum (ER) Organelle that is a continuous system of sacs and tubes; extension of the nuclear envelope. Smooth ER makes lipids and breaks down carbohydrates and fatty acids; rough ER modifies polypeptides made by ribosomes on its surface.
Golgi body Organelle that modifies polypeptides and lipids; also sorts and packages the finished products into vesicles.
lysosome Enzyme-filled vesicle that functions in intracellular digestion.
peroxisome Enzyme-filled vesicle that breaks down amino acids, fatty acids, and toxic substances.
vacuole A fluid-filled organelle that isolates or disposes of waste, debris, or toxic materials.
vesicle Small, membrane-enclosed, saclike organelle; different kinds store, transport, or degrade their contents.

Vacuoles are vesicles that appear empty under a microscope. They form by the fusion of multiple vesicles, and have various functions depending on cell type. Many isolate or dispose of waste, debris, and toxins. Amino acids, sugars, ions, wastes, and toxins accumulate in the water-filled interior of a plant cell's large **central vacuole**. Fluid pressure in a central vacuole keeps plant cells plump, so stems, leaves, and other structures stay firm. As you can see in **Figure 4.11B**, a central vacuole typically takes up 50 to 90 percent of a plant cell's interior, with cytoplasm confined to a narrow zone between it and the plasma membrane.

Golgi Bodies Some vesicles fuse with and empty their contents into a **Golgi body**. This organelle has a folded membrane that typically looks like a stack of pancakes ❹. Enzymes in a Golgi body put finishing touches on proteins and lipids that have been delivered from the ER. They attach phosphate groups or oligosaccharides, and cleave certain polypeptides. The finished products are sorted and packaged into new vesicles that carry them to lysosomes or to the plasma membrane ❺.

Take-Home Message

What is the endomembrane system?

» The endomembrane system includes rough and smooth endoplasmic reticulum, vesicles, and Golgi bodies.

» This series of organelles works together mainly to synthesize and modify cell membrane proteins and lipids.

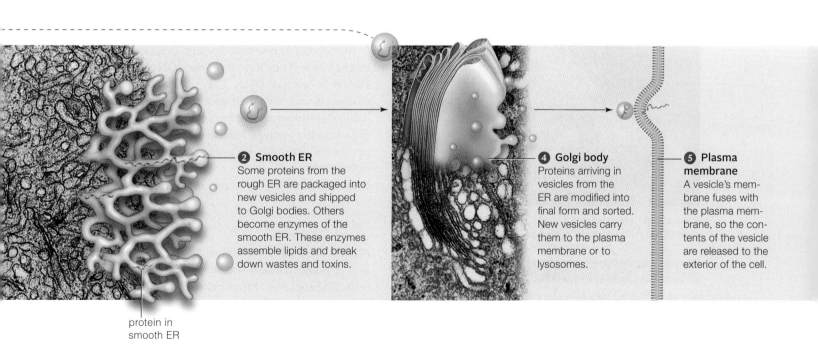

❷ **Smooth ER**
Some proteins from the rough ER are packaged into new vesicles and shipped to Golgi bodies. Others become enzymes of the smooth ER. These enzymes assemble lipids and break down wastes and toxins.

protein in smooth ER

❹ **Golgi body**
Proteins arriving in vesicles from the ER are modified into final form and sorted. New vesicles carry them to the plasma membrane or to lysosomes.

❺ **Plasma membrane**
A vesicle's membrane fuses with the plasma membrane, so the contents of the vesicle are released to the exterior of the cell.

4.8 Lysosome Malfunction

- When lysosomes do not work properly, some cellular materials are not properly recycled, with devastating results.

Lysosomes serve as vessels for waste disposal and recycling. Enzymes inside them break large molecules into smaller subunits that the cell can use as building material or eliminate. A deficiency or malfunction in one of these enzymes can cause molecules that would normally get broken down to accumulate instead.

Cells continually make and break down gangliosides, a kind of lipid. The turnover is especially brisk during early development. In Tay–Sachs disease, the lysosomal enzyme responsible for ganglioside breakdown misfolds and is destroyed. Affected infants usually seem normal for the first few months, but symptoms appear as gangliosides accumulate to higher and higher levels inside their nerve cells. Within three to six months the child becomes irritable, listless, and may have seizures. Blindness, deafness, and paralysis follow. Affected children usually die by age five (**Figure 4.16**).

Tay–Sachs is a genetic disease, so it is heritable. It occurs in all populations, but it is most common in Jews of eastern European descent. Researchers continue to explore potential therapies such as blocking ganglioside synthesis, using gene therapy to deliver a normal version of the missing enzyme to the brain, or infusing normal blood cells from umbilical cords. All treatments are still considered experimental, and Tay–Sachs is still incurable.

Figure 4.16 Conner Hopf was diagnosed with Tay–Sachs disease at age 7–1/2 months. He died at 22 months.

Take-Home Message

Are all organelle types necessary for life?

» Defects in the function of an organelle can have devastating consequences to health.

4.9 Other Organelles

- Eukaryotic cells make most of their ATP in mitochondria.
- Plastids function in storage and photosynthesis in plants and some types of algae.
- Links to Metabolism 3.3, ATP 3.8

Mitochondria

A **mitochondrion** (plural, mitochondria) is a type of eukaryotic organelle that specializes in making ATP (**Figure 4.17**). Aerobic respiration, an oxygen-requiring series of reactions that proceeds in mitochondria, extracts more energy from organic compounds than any other metabolic pathway. With each breath, you are taking in oxygen mainly for the mitochondria in your trillions of aerobically respiring cells.

Typical mitochondria are between 1 and 4 micrometers in length. Some are branched. These organelles can change shape, split in two, and fuse together. Each has two membranes, one highly folded inside the other. This arrangement creates two compartments. During aerobic respiration, hydrogen ions accumulate between the two membranes. The buildup causes the ions to flow across the inner mitochondrial membrane, and this flow drives ATP formation (Chapter 7 returns to the details of aerobic respiration).

Nearly all eukaryotic cells have mitochondria, but the number varies by the type of cell and by the organism. For example, a yeast cell might have only one mitochondrion, but a human skeletal muscle cell may have a thousand or more. Cells that have a very high demand for energy tend to have many mitochondria. Some eukaryotes that live in oxygen-free environments have no mitochondria; these organisms produce ATP in special organelles called hydrogenosomes.

Mitochondria resemble bacteria in size, form, and biochemistry. They have their own DNA, which is similar to bacterial DNA. They divide independently of the cell, and have their own ribosomes. Such clues led to the endosymbiont hypothesis, which states that mitochondria evolved from aerobic bacteria that took up permanent residence inside a host cell, a process called endosymbiosis. We will explore evidence for the endosymbiont hypothesis in Section 19.6.

Plastids

Plastids are double-membraned organelles that function in photosynthesis or storage in plant and algal cells. Chloroplasts, chromoplasts, and amyloplasts are common types of plastids.

Photosynthetic cells of plants and many protists contain **chloroplasts**, which are plastids specialized for photosynthesis. Most chloroplasts are oval or disk-shaped (Figure 4.18A). Each has two outer membranes enclosing a semifluid interior, the stroma, that contains enzymes and the chloroplast's own DNA. In the stroma, a third, highly folded membrane forms a single, continuous compartment. The folded membrane, which resembles stacks of flattened disks, is called the thylakoid membrane (Figure 4.18B). Photosynthesis occurs at this membrane.

The thylakoid membrane incorporates many pigments, the most abundant of which is chlorophyll. The abundance of green chlorophylls in plant cell chloroplasts is the reason most plants are green. During photosynthesis, chlorophylls and other molecules in the thylakoid membrane harness the energy in sunlight to drive the synthesis of ATP. The ATP is then used inside the stroma to build carbohydrates from carbon dioxide and water. (Chapter 6 describes the process of photosynthesis in more detail.) In many ways, chloroplasts resemble photosynthetic bacteria, and like mitochondria they may have evolved by endosymbiosis.

Chromoplasts make and store pigments other than chlorophylls. They have an abundance of carotenoids,  a pigment that colors many flowers, leaves, fruits, and roots red or orange. For example, as a tomato ripens, its green chloroplasts are converted to red chromoplasts, and the color of the fruit changes. The *inset* shows carotenoid-containing chromoplasts in cells of a red bell pepper.

Amyloplasts are unpigmented plastids. Typical amyloplasts store starch grains, and are notably abundant in starch-storing cells of stems, tubers (underground stems), and seeds. Starch-packed amyloplasts are dense and heavy compared to cytoplasm. In some plant cells, they function as gravity-sensing organelles.

chloroplast Organelle of photosynthesis in the cells of plants and many protists.
mitochondrion Organelle that produces ATP by aerobic respiration in eukaryotes.
plastid Category of double-membraned organelle in plants and algal cells. Different types specialize in storage or photosynthesis; e.g., chloroplast, amyloplast.

Take-Home Message

What eukaryotic organelles are specialized for producing ATP?

» Mitochondria are eukaryotic organelles that produce ATP from organic compounds in reactions that require oxygen.

» Chloroplasts are plastids that carry out photosynthesis in cells of plants and many protists.

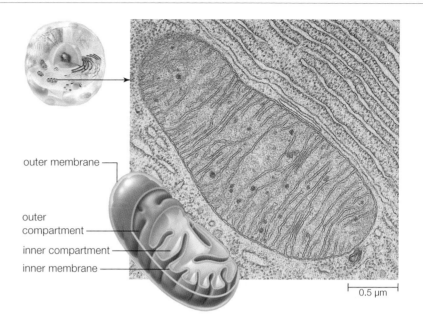

outer membrane
outer compartment
inner compartment
inner membrane

0.5 µm

Figure 4.17 Animated Sketch and transmission electron micrograph of a mitochondrion. This organelle specializes in producing large quantities of ATP.

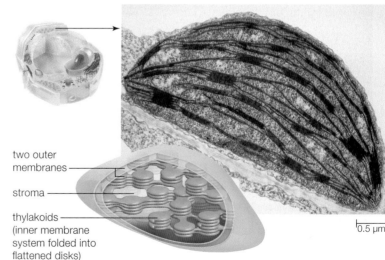

A Photosynthetic cells in a leaf of *Plagiomnium ellipticum*, a moss.

50 µm

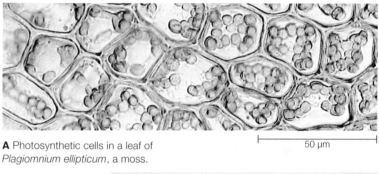

two outer membranes
stroma
thylakoids (inner membrane system folded into flattened disks)

0.5 µm

B Each chloroplast is enclosed by two outer membranes. Photosynthesis occurs at a much-folded inner membrane. The transmission electron micrograph shows a chloroplast from a tobacco leaf (*Nicotiana tabacum*). The lighter patches are nucleoids where DNA is stored.

Figure 4.18 The chloroplast.

4.10 The Dynamic Cytoskeleton

- Eukaryotic cells have an extensive and dynamic internal framework called a cytoskeleton.
- Links to Protein structure and function 3.6, 3.7

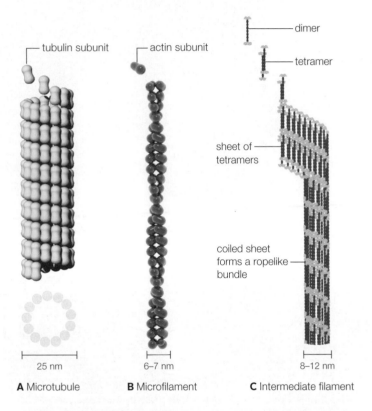

A Microtubule — tubulin subunit — 25 nm

B Microfilament — actin subunit — 6–7 nm

C Intermediate filament — dimer — tetramer — sheet of tetramers — coiled sheet forms a ropelike bundle — 8–12 nm

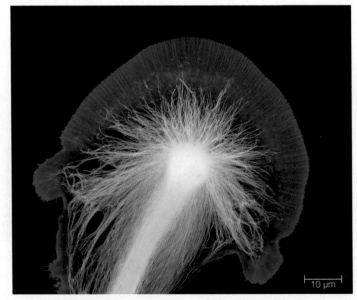

D A fluorescence micrograph shows microtubules (*yellow*) and actin microfilaments (*blue*) in the growing end of a nerve cell. These cytoskeletal elements support and guide the cell's lengthening.

10 μm

Figure 4.19 Animated Cytoskeletal elements.

Between the nucleus and plasma membrane of all eukaryotic cells is a system of interconnected protein filaments collectively called the **cytoskeleton**. Elements of the cytoskeleton reinforce, organize, and move cell structures, and often the whole cell. Some are permanent; others form only at certain times.

Microtubules are long, hollow cylinders that consist of subunits of the protein tubulin (**Figure 4.19A**). They form a dynamic scaffolding for many cellular processes, rapidly assembling when they are needed, disassembling when they are not. For example, before a eukaryotic cell divides, microtubules assemble, separate the cell's duplicated chromosomes, then disassemble. As another example, microtubules that form in the growing end of a young nerve cell support and guide its lengthening in a particular direction.

Microfilaments are fibers that consist primarily of subunits of the globular protein actin (**Figure 4.19B**). They strengthen or change the shape of eukaryotic cells. Crosslinked, bundled, or gel-like arrays of them make up the **cell cortex**, a reinforcing mesh under the plasma membrane. Actin microfilaments that form at the edge of a cell drag or extend it in a certain direction (**Figure 4.19D**). Myosin and actin microfilaments interact to bring about muscle cell contraction.

Intermediate filaments that support cells and tissues are the most stable elements of the cytoskeleton (**Figure 4.19C**). These filaments form a framework that lends structure and resilience to cells and tissues. Some kinds underlie and reinforce membranes. The nuclear lamina, for example, consists of intermediate filaments of fibrous proteins called lamins.

Among many accessory molecules associated with cytoskeletal elements are **motor proteins**, which move cell parts when energized by a phosphate-group transfer from ATP. A cell is like a bustling train station, with molecules and structures being moved continuously throughout its interior. Motor proteins are like freight trains, dragging their cellular cargo along tracks of dynamically assembled microtubules and microfilaments (**Figure 4.20**). Motor proteins also move external structures such as flagella and cilia. Eukaryotic flagella are structures that whip back and forth to propel cells such as sperm (*right*) through fluid. They have a different structure and type of motion than flagella of bacteria and archaea. **Cilia** (singular, cilium) are short, hairlike structures that project from the

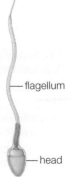

flagellum

head

68 UNIT I PRINCIPLES OF CELLULAR LIFE

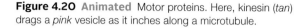

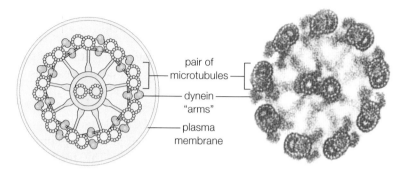

Figure 4.20 **Animated** Motor proteins. Here, kinesin (*tan*) drags a *pink* vesicle as it inches along a microtubule.

surface of some cells. Cilia are usually more profuse than flagella. The coordinated waving of many cilia propels cells through fluid, and stirs fluid around stationary cells. A motor protein called dynein interacts with organized arrays of microtubules to bring about movement of eukaryotic flagella and cilia. A special 9+2 array of microtubules extends lengthwise through these motile structures (**Figure 4.21**). The array consists of nine pairs of microtubules ringing another pair in the center. The microtubules grow from a barrel-shaped organelle called the **centriole**, which remains below the finished array as a **basal body**.

Amoebas and other types of eukaryotic cells form **pseudopods**, or "false feet." As these temporary, irregular lobes bulge outward, they move the cell and engulf a target such as prey. Elongating microfilaments force the lobe to advance in a steady direction. Motor proteins that are attached to the microfilaments drag the plasma membrane along with them.

basal body Organelle that develops from a centriole.
cell cortex Reinforcing mesh of cytoskeletal elements under a plasma membrane.
centriole Barrel-shaped organelle from which microtubules grow.
cilium Short, movable structure that projects from the plasma membrane of some eukaryotic cells.
cytoskeleton Dynamic framework of protein filaments that support, organize, and move eukaryotic cells and their internal structures.
intermediate filament Stable cytoskeletal element that structurally supports cells and tissues.
microfilament Reinforcing cytoskeletal element; a fiber of actin subunits.
microtubule Cytoskeletal element involved in cellular movement; hollow filament of tubulin subunits.
motor protein Type of energy-using protein that interacts with cytoskeletal elements to move the cell's parts or the whole cell.
pseudopod A temporary protrusion that helps some eukaryotic cells move and engulf prey.

Take-Home Message

What is a cytoskeleton?

» A cytoskeleton of protein filaments is the basis of eukaryotic cell shape, internal structure, and movement.

» Microtubules organize eukaryotic cells and help move their parts. Networks of microfilaments reinforce their surfaces. Intermediate filaments strengthen and maintain the shape of animal cells and tissues.

» When energized by ATP, motor proteins move along tracks of microtubules and microfilaments. As part of cilia, flagella, and pseudopods, they can move the whole cell.

A Sketch and micrograph of a eukaryotic flagellum, cross-section. Like a cilium, it contains a 9+2 array: a ring of nine pairs of microtubules plus one pair at its core. Stabilizing spokes and linking elements that connect to the microtubules keep them aligned in a radial pattern. (Plasma membrane not visible in the micrograph.)

pair of microtubules

dynein "arms"

plasma membrane

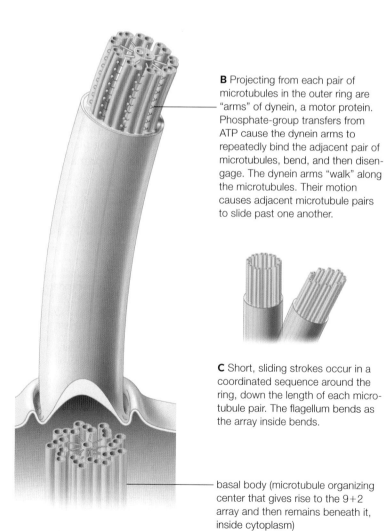

B Projecting from each pair of microtubules in the outer ring are "arms" of dynein, a motor protein. Phosphate-group transfers from ATP cause the dynein arms to repeatedly bind the adjacent pair of microtubules, bend, and then disengage. The dynein arms "walk" along the microtubules. Their motion causes adjacent microtubule pairs to slide past one another.

C Short, sliding strokes occur in a coordinated sequence around the ring, down the length of each microtubule pair. The flagellum bends as the array inside bends.

basal body (microtubule organizing center that gives rise to the 9+2 array and then remains beneath it, inside cytoplasm)

Figure 4.21 **Animated** Mechanism of movement of eukaryotic flagella and cilia.

4.11 Cell Surface Specializations

■ Many cells secrete materials that form a covering or matrix outside their plasma membrane.
■ Links to Tissue 1.2, Chitin 3.4

Matrixes Between and Around Cells

Many cells secrete an **extracellular matrix (ECM)**, a complex mixture of fibrous proteins and polysaccharides. The composition and function of ECM varies by the type of cell that secretes it.

A cell wall is an example of ECM. Bacteria and archaea secrete a wall around their plasma membrane, as do fungi, plants, and some protists. The structure of the wall differs among these groups, but in all cases it protects, supports, and imparts shape to the cell. The cell wall is also porous: Water and solutes easily cross it on the way to and from the plasma membrane. Cells could not live without exchanging these substances with their environment.

A plant cell wall forms as a young cell secretes pectin and other polysaccharides onto the outer surface of its plasma membrane. The sticky coating is shared between adjacent cells, and it cements them together. Each cell then forms a **primary wall** by secreting strands of cellulose into the coating. Some of the coating remains as the middle lamella, a sticky layer in between the primary walls of abutting plant cells (**Figure 4.22A**). Being thin and pliable, a primary wall

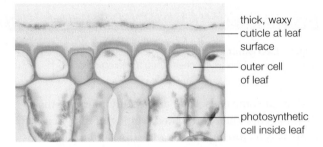

Figure 4.23 A plant ECM. Section through a plant leaf showing cuticle, a protective covering of deposits secreted by living cells.

allows a growing plant cell to enlarge and change shape. At maturity, cells in some plant tissues stop enlarging and begin to secrete material onto the primary wall's inner surface. These deposits form a firm **secondary wall** (**Figure 4.22B**). One of the materials deposited is **lignin**, an organic compound that makes up as much as 25 percent of the secondary wall of cells in older stems and roots. Lignified plant parts are stronger, more waterproof, and less susceptible to plant-attacking organisms than younger tissues.

A **cuticle** is a type of ECM secreted by cells at a body surface. The cuticle secreted by cells at the surface of a leaf or stem consists of waxes and proteins (**Figure 4.23**). Plant cuticle helps a plant retain water and fend off insects. The cuticle of crabs, spiders, and other arthropods is mainly chitin, a tough polysaccharide.

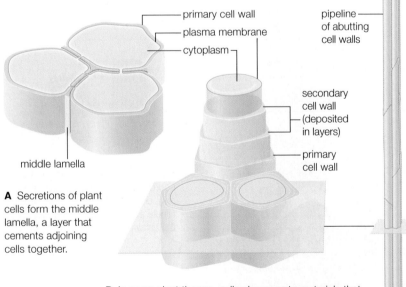

A Secretions of plant cells form the middle lamella, a layer that cements adjoining cells together.

B In many plant tissues, cells also secrete materials that are deposited in layers on the inner surface of their primary wall. These layers strengthen the wall and maintain its shape. The walls remain after the cells die, and become part of the pipelines that carry water through the plant.

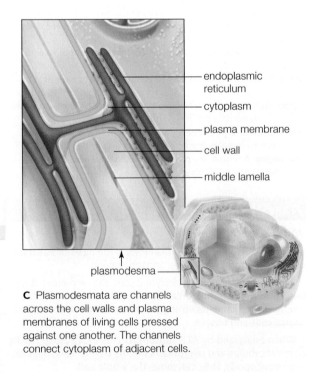

C Plasmodesmata are channels across the cell walls and plasma membranes of living cells pressed against one another. The channels connect cytoplasm of adjacent cells.

Figure 4.22 Animated Some characteristics of plant cell walls.

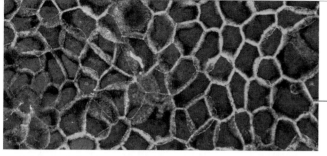

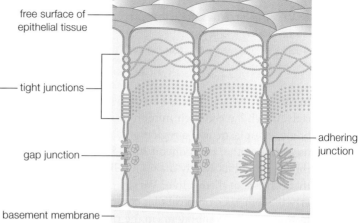

Figure 4.24 Animated Three types of cell junctions in animal tissues: tight junctions, gap junctions, and adhering junctions.

In the micrograph *above*, a profusion of tight junctions (*green*) seals abutting surfaces of kidney cell membranes and forms a waterproof tissue. The DNA in each cell nucleus appears *red*.

Inside animal bodies, other types of ECM support and organize tissues, and some have roles in cell signaling. For example, bone is an ECM composed mostly of the fibrous protein collagen, and hardened by deposits of calcium and phosphorus.

Cell Junctions

ECM does not prevent a cell from interacting with other cells or the surroundings. In multicelled species, such interaction occurs by way of **cell junctions**, which are structures that connect a cell to other cells and to the environment. Cells send and receive ions, molecules, or signals through some junctions. Other kinds help cells recognize and stick to each other and to extracellular matrix.

In plants, open channels called **plasmodesmata** (singular, plasmodesma) extend across cell walls, connecting the cytoplasm of adjoining cells (**Figure 4.22C**). Substances such as water, nutrients, and signaling molecules can flow quickly from cell to cell through these cell junctions.

Three types of cell junctions are common in animal bodies (**Figure 4.24**). In tissues that line body surfaces and internal cavities, rows of proteins that form **tight junctions** between plasma membranes prevent body fluids from seeping between adjacent cells. To cross these tissues, fluid must pass directly through the cells. For example, an abundance of tight junctions in the lining of the stomach normally keeps acidic fluid from leaking out. If a bacterial infection damages this lining, acid and enzymes can erode the underlying layers. The result is a painful peptic ulcer.

Strong **adhering junctions** composed of adhesion proteins snap cells to one another. They also connect microfilaments and intermediate filaments inside the cell to ECM outside the cell. Skin and other tissues

that are subject to abrasion or stretching have a lot of adhering junctions. These cell junctions also strengthen contractile tissues such as heart muscle.

Gap junctions form channels that connect the cytoplasm of adjoining cells, thus permitting ions and small molecules to pass directly from the cytoplasm of one cell to another. By opening or closing, these channels allow entire regions of cells to respond to a single stimulus. Heart muscle and other tissues in which the cells perform a coordinated action have many of these communication channels.

adhering junction Cell junction composed of adhesion proteins; anchors cells to each other and extracellular matrix.
cell junction Structure that connects a cell to another cell or to extracellular matrix.
cuticle Secreted covering at a body surface.
extracellular matrix (ECM) Complex mixture of cell secretions; supports cells and tissues; has roles in cell signaling.
gap junction Cell junction that forms a channel across the plasma membranes of adjoining animal cells.
lignin Material that stiffens cell walls of vascular plants.
plasmodesmata Cell junctions that connect the cytoplasm of adjacent plant cells.
primary wall The first cell wall of young plant cells.
secondary wall Lignin-reinforced wall that forms inside the primary wall of a plant cell.
tight junctions Arrays of fibrous proteins; join epithelial cells and collectively prevent fluids from leaking between them.

Take-Home Message

What structures form on the outside of eukaryotic cells?

» Cells of many protists, nearly all fungi, and all plants have a porous wall around the plasma membrane. Animal cells do not have walls.

» Plant cell secretions form a waxy cuticle that helps protect the exposed surfaces of soft plant parts.

» Cell secretions form extracellular matrixes between cells in many tissues.

» Cells make structural and functional connections with one another and with extracellular matrix in tissues.

4.12 The Nature of Life

■ We define life by describing the set of properties that is unique to living things.
■ Links to Life's levels of organization 1.2, Homeostasis 1.3, Philosophy of science 1.9

In this chapter, you learned about the structure of cells, which have at their minimum a plasma membrane, cytoplasm, and a region of DNA. Most cells have many other components in addition to these things. But what exactly makes a cell, or an organism that consists of them, alive? We say that life is a property that emerges from cellular components, but a collection of those components in the right amounts and proportions is not necessarily alive.

We know intuitively what "life" is, but defining it unambiguously is challenging, if not impossible. We can more easily describe what sets the living apart from the nonliving, but even that can be tricky. For example, living things have a high proportion of the organic molecules of life, but so do the remains of dead organisms in seams of coal. Living things use energy to reproduce themselves, but computer viruses, which are arguably not alive, can do that too. So how do biologists, who study life as a profession, describe life? The short answer is that their best description is very long. It consists of a list of properties associated with things we know to be alive. You already know about two of these properties of organisms:

1. They make and use the organic molecules of life.
2. They consist of one or more cells.

The remainder of this book details the other properties of living things:

3. They engage in self-sustaining biological processes such as metabolism and homeostasis.
4. They change over their lifetime, for example by growing, maturing, and aging.
5. They use DNA as their hereditary material.
6. They have the collective capacity to change over successive generations, for example by adapting to environmental pressures.

Collectively, these properties characterize living things as different from nonliving things.

Carbon, hydrogen, oxygen, and other atoms of organic molecules are the stuff of you, and us, and all of life. Yet it takes far more than organic molecules to complete the picture. Life continues only as long as a continuous flow of energy sustains its organization, because it takes energy to assemble molecules into cells. Life is no more and no less than a marvelously complex system for prolonging order. Sustained by

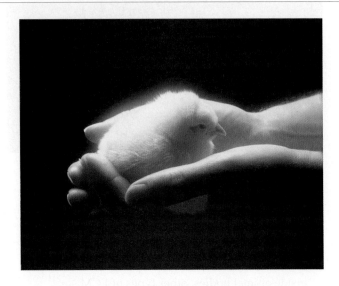

energy inputs, it continues by a defining capacity for self-reproduction. With energy and the hereditary codes of DNA, matter becomes organized, generation after generation. Even with the death of individuals, life elsewhere is prolonged. With each death, molecules are released and may be cycled as raw materials for new generations.

Take-Home Message

What is life?

» We describe the characteristic of "life" in terms of a set of properties unique to living things.

» In living things, the molecules of life are organized as one or more cells that engage in self-sustaining biological processes.

» Organisms use DNA as their hereditary material.

» Living things change over lifetimes, and over generations.

Summary

Section 4.1 Bacteria are found in all parts of the biosphere, including the human body. Huge numbers inhabit our intestines, but most of these are beneficial. A few can cause disease. Contamination of food with disease-causing bacteria can result in food poisoning that is sometimes fatal.

Section 4.2 Cells differ in size, shape, and function, but all start out life with a **plasma membrane**, **cytoplasm**, and a region of DNA. Most cells have additional components (**Table 4.4**).

In eukaryotic cells, DNA is contained within a **nucleus**, which is a membrane-enclosed **organelle**. All cell membranes, including the plasma membrane and

(continued on page 74)

Food for Thought (revisited)

One food safety measure involves sterilization, which kills *E. coli* O157:H7 and other bacteria. Recalled, contaminated ground beef is typically cooked or otherwise sterilized, then processed into ready-to-eat products. Raw beef trimmings are effectively sterilized when sprayed with ammonia and ground to a paste. The resulting meat product is routinely added as a filler to hamburger patties, fresh ground beef, hot dogs, lunch meats, sausages, frozen entrees, canned foods, and other items sold to quick service restaurants, hotel and restaurant chains, institutions, and school lunch programs.

Meat, poultry, milk, and fruits sterilized by exposure to radiation are available in supermarkets. By law, irradiated foods must be marked with the symbol on the *right*, but foods sterilized with chemicals are not currently required to carry any disclosure. Some worry that sterilization may alter food or leave harmful chemicals in it. Whether any health risks are associated with consuming sterilized foods is unknown.

How would you vote? Some think the safest way to protect consumers from food poisoning is by sterilizing food with chemicals or radiation to kill any bacteria that may be in it. Others think we should prevent food from getting contaminated in the first place by enacting stricter laws governing farming practices and food preparation. Both approaches add cost to our food. Which approach would you choose?

Table 4.4 Summary of Typical Components of Prokaryotic and Eukaryotic Cells

Cell Component	Main Functions	Bacteria, Archaea	Eukaryotes			
			Protists	Fungi	Plants	Animals
Cell wall	Protection, structural support	✓*	✓*	✓	✓	–
Plasma membrane	Control of substances moving into and out of cell	✓	✓	✓	✓	✓
Nucleus	Physical separation of DNA from cytoplasm	–**	✓	✓	✓	✓
DNA	Encoding of hereditary information	✓	✓	✓	✓	✓
Nucleolus	Assembly of ribosome subunits	–	✓	✓	✓	✓
Ribosome	Protein synthesis	✓	✓	✓	✓	✓
Endoplasmic reticulum (ER)	Initial modification of polypeptide chains; lipid synthesis	–	✓	✓	✓	✓
Golgi body	Final modification of proteins, lipid assembly, and packaging of both for use inside cell or export	–	✓	✓	✓	✓
Lysosome	Intracellular digestion	–	✓	✓*	✓*	✓
Mitochondrion	Aerobic production of ATP	–	✓	✓	✓	✓
Hydrogenosome	Anaerobic production of ATP	–	✓*	✓*	–	✓*
Photosynthetic pigments	Light-to-energy conversion	✓*	✓*	–	✓	–
Chloroplast	Photosynthesis; some starch storage	–	✓*	–	✓	–
Central vacuole	Increasing cell surface area; storage	–	–	✓*	✓	–
Bacterial flagellum	Locomotion through fluid surroundings	✓*	–	–	–	–
Eukaryotic flagellum or cilium	Locomotion through or motion within fluid surroundings	–	✓*	✓*	✓*	✓
Cytoskeleton	Cell shape; internal organization; basis of cell movement and, in many cells, locomotion	✓	✓*	✓*	✓*	✓

*Known to be present in some species.
** One or two lipid bilayers surround the DNA of some species

organelle membranes, are selectively permeable and consist mainly of phospholipids organized as a lipid bilayer. The **surface-to-volume ratio** limits cell size.

By the **cell theory**, all organisms consist of one or more cells; the cell is the smallest unit of life; each new cell arises from another, preexisting cell; and a cell passes hereditary material to its offspring.

 Section 4.3 Most cells are far too small to see with the naked eye, so we use microscopes to observe them. Different types of microscopes and techniques reveal different internal and external details of cells.

 Section 4.4 Bacteria and archaea, informally grouped as "prokaryotes," are the most diverse forms of life. These single-celled organisms have no nucleus, but they have **nucleoids** and **ribosomes**. Many have a permeable but protective **cell wall** and a sticky capsule, as well as motile structures (**flagella**) and other projections (**pili**). Some have **plasmids** in addition to their single chromosome. Bacteria and other microbial organisms often share living arrangements in **biofilms**.

 Section 4.5 Protists, fungi, plants, and animals are eukaryotic. Cells of these organisms start out life with membrane-enclosed organelles, including a nucleus. Organelles compartmentalize tasks and substances that are sensitive or dangerous to the rest of the cell.

 Section 4.6 A nucleus protects and controls access to a eukaryotic cell's DNA, which is typically distributed among a characteristic number of **chromosomes**. A **nuclear envelope** surrounds **nucleoplasm**. In the nucleus, ribosome subunits are produced in dense, irregularly shaped **nucleoli**.

 Sections 4.7, 4.8 The **endomembrane system** is a system of interacting organelles that includes ER, Golgi bodies, and vesicles. **Endoplasmic reticulum (ER)** is a continuous system of sacs and tubes extending from the nuclear envelope. Ribosome-studded rough ER makes polypeptides; smooth ER assembles lipids and degrades carbohydrates and fatty acids. **Golgi bodies** modify peptides and lipids before sorting them into **vesicles**. Different types of vesicles store, degrade, or transport substances through the cell. Enzymes in **peroxisomes** break down substances such as amino acids, fatty acids, and toxins. **Lysosomes** contain enzymes that break down cellular debris for recycling. Fluid-filled **vacuoles** store and dispose of waste, debris, and toxins. Fluid pressure inside a **central vacuole** keeps plant cells plump, thus keeping plant parts firm.

 Section 4.9 Double-membraned **mitochondria** specialize in making ATP by breaking down organic compounds in the oxygen-requiring pathway of aerobic respiration. Different types of **plastids** are specialized for photosynthesis or storage. In eukaryotes, photosynthesis takes place inside **chloroplasts**. Pigment-filled chromoplasts and starch-filled amyloplasts are used for storage, and also serve additional roles.

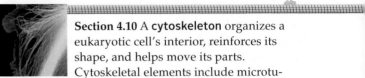

 Section 4.10 A **cytoskeleton** organizes a eukaryotic cell's interior, reinforces its shape, and helps move its parts. Cytoskeletal elements include microtubules, microfilaments, and intermediate filaments. Interactions between ATP-driven **motor proteins** and hollow, dynamically assembled **microtubules** bring about cellular movement. A **microfilament** mesh called the **cell cortex** reinforces plasma membranes. Elongating microfilaments bring about movement of **pseudopods**. **Intermediate filaments** lend structural support to cells and tissues.

Centrioles give rise to a special 9+2 array of microtubules inside **cilia** and eukaryotic flagella, then remain beneath these motile structures as **basal bodies**.

 Section 4.11 Many cells secrete a complex mixture of fibrous proteins and polysaccharides onto their surfaces. The secretions form an **extracellular matrix (ECM)** that supports cells and tissues, and also functions in cell-to-cell signaling. Most prokaryotes, protists, fungi, and all plant cells secrete a wall around the plasma membrane. Older plant cells secrete a rigid, **lignin**-containing **secondary wall** inside their pliable **primary wall**. Many eukaryotic cell types also secrete a waxy, protective **cuticle**. **Cell junctions** connect animal cells to one another and to ECM. **Plasmodesmata** connect the cytoplasm of adjacent plant cells. In animals, **gap junctions** form open channels between adjacent cells; **adhering junctions** anchor cells to one another and to ECM; and **tight junctions** form a waterproof seal between cells in some tissues.

Section 4.12 We describe the quality of "life" as a set of properties that are collectively unique to living things. Living things consist of cells that engage in self-sustaining biological processes, pass their hereditary material (DNA) to offspring by mechanisms of reproduction, and have the capacity to change over successive generations.

Self-Quiz

Answers in Appendix III

1. Despite the diversity of cell type and function, all cells have these three things in common:
    a. cytoplasm, DNA, and organelles
    b. a plasma membrane, DNA, and proteins
    c. cytoplasm, DNA, and a plasma membrane
    d. carbohydrates, nucleic acids, and proteins

2. Every cell is descended from another cell. This idea is part of _____ .
    a. evolution
    b. the theory of heredity
    c. the cell theory
    d. cell biology

DATA ANALYSIS ACTIVITIES

Abnormal Motor Proteins Cause Kartagener Syndrome

An abnormal form of the motor protein dynein causes Kartagener syndrome, a genetic disorder characterized by chronic sinus and lung infections. Biofilms form in the thick mucus that collects in the airways, and the resulting bacterial activities and inflammation damage tissues.

Affected men can produce sperm but are infertile (**Figure 4.25**). Some have become fathers after a doctor injects their sperm cells directly into eggs. Review **Figure 4.25**, then explain how abnormal dynein could cause the observed effects.

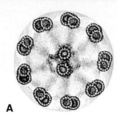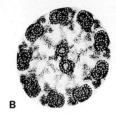

A **B**

Figure 4.25 Cross-section of the flagellum of a sperm cell from (**A**) a human male affected by Kartagener syndrome and (**B**) an unaffected male.

3. The surface-to-volume ratio _____ .
   a. does not apply to prokaryotic cells
   b. constrains cell size
   c. is part of the cell theory
   d. b and c

4. True or false? Ribosomes are only found in bacteria and archaea.

5. Unlike eukaryotic cells, bacterial cells _____ .
   a. have no plasma membrane   c. have no nucleus
   b. have RNA but not DNA   d. a and c

6. True or false? Some protists start out life with no nucleus.

7. Cell membranes consist mainly of a _____ .
   a. carbohydrate bilayer and proteins
   b. protein bilayer and phospholipids
   c. lipid bilayer and proteins

8. Enzymes contained in _____ break down worn-out organelles, bacteria, and other particles.
   a. lysosomes   c. endoplasmic reticulum
   b. amyloplasts   d. peroxisomes

9. Put the following structures in order according to the pathway of a secreted protein:
   a. plasma membrane   c. endoplasmic reticulum
   b. Golgi bodies   d. post-Golgi vesicles

10. The main function of the endomembrane system is building and modifying _____ and _____ .

11. Is this statement true or false? The plasma membrane is the outermost component of all cells. Explain.

12. Which of the following organelles contains no DNA?
   a. nucleus   c. mitochondrion
   b. Golgi body   d. chloroplast

13. Cytoskeletal elements called _____ form a reinforcing mesh under the nuclear envelope.
   a. intermediate filaments   c. actin filaments
   b. microtubules   d. microfilaments

14. No animal cell has a _____ .
   a. plasma membrane   c. lysosome
   b. flagellum   d. cell wall

15. _____ connect the cytoplasm of plant cells.
   a. Plasmodesmata   c. Tight junctions
   b. Adhering junctions   d. a and b

16. Intermediate filaments are a feature of _____ cells.
   a. eukaryotic   c. animal
   b. all   d. algal

17. Match each term with the best description.
   ___ centriole   a. shows surface details
   ___ ECM   b. feature of secondary walls
   ___ cuticle   c. basal body
   ___ lignin   d. connectivity
   ___ SEM   e. protective covering

18. Match each cell component with its specialization.
   ___ mitochondrion   a. protein synthesis
   ___ chloroplast   b. associates with
   ___ ribosome   ribosomes
   ___ smooth ER   c. fatty acid breakdown
   ___ Golgi body   d. sorts and ships
   ___ rough ER   e. assembles lipids
   ___ peroxisome   f. photosynthesis
   ___ amyloplast   g. ATP production
   ___ flagellum   h. movement
                        i. stores starch

Critical Thinking

1. In a classic episode of *Star Trek*, a gigantic amoeba engulfs an entire starship. Spock blows the cell to bits before it has a chance to reproduce. Think of at least one problem a biologist would have with this particular scenario.

2. Many plant cells form a secondary wall on the inner surface of their primary wall. Speculate on the reason why the secondary wall does not form on the outer surface of the primary wall.

3. A student is examining different samples with a microscope. She discovers a single-celled organism swimming in a freshwater pond (*right*). What kind of microscope is she using?

4. Which structures can you identify in the organism on the *right*? Is it a prokaryotic or eukaryotic cell? Can you be more specific about the type of cell based on what you know about cell structure? Look ahead to Section 21.3 to check your answers.

5 Ground Rules of Metabolism

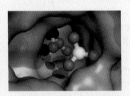

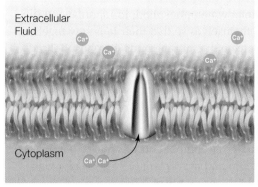
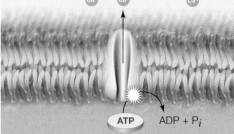
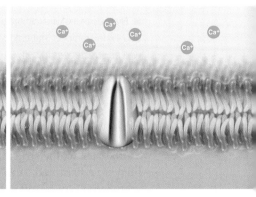

A Two calcium ions bind to the transport protein.

B Energy in the form of a phosphate group is transferred from ATP to the protein. The transfer causes the protein to change shape so that it ejects the calcium ions to the opposite side of the membrane.

C After it loses the calcium ions, the transport protein resumes its original shape.

Figure 5.25 Active transport. This is a model of a plasma membrane calcium pump.

its gradient across a cell membrane. After a solute binds to an active transporter, an energy input (often in the form of a phosphate-group transfer from ATP) changes the shape of the protein. The change causes the transporter to release the solute to the other side of the membrane.

A **calcium pump** is an example of an active transporter. This protein moves calcium ions across cell membranes (**Figure 5.25**). Calcium ions act as potent messengers inside cells, and many enzymes have allosteric sites that bind these ions. Thus, their presence in cytoplasm is tightly regulated. Calcium pumps in the plasma membrane of all eukaryotic cells can keep the concentration of calcium in cytoplasm thousands of times lower than it is in extracellular fluid.

Cotransporters are active transport proteins that move two substances at the same time, in the same or opposite directions across a membrane. Nearly all of the cells in your body have cotransporters called sodium–potassium pumps (**Figure 5.26**). Sodium ions (Na^+) in the cytoplasm diffuse into the pump's open channel and bind to its interior. A phosphate-group transfer from ATP causes the pump to change shape. Its channel opens to extracellular fluid, where it releases the Na^+. Then, potassium ions (K^+) from extracellular fluid diffuse into the channel and bind to its interior. The transporter releases the phosphate

group and reverts to its original shape. The channel opens to the cytoplasm, where it releases the K^+.

Bear in mind, the membranes of all cells, not just those of animals, have active transporters. For example, active transporters in plant leaf cells pump sugars into tubes that distribute them throughout the plant body.

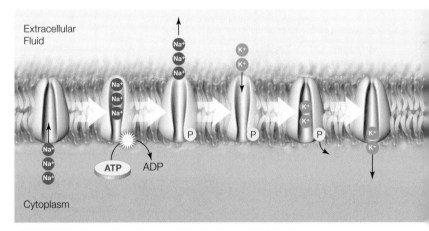

Figure 5.26 Cotransport. This model shows how a sodium–potassium pump transports sodium ions (Na^+, *purple*) from cytoplasm to extracellular fluid, and potassium ions (K^+, *green*) in the other direction across the plasma membrane. A phosphate-group transfer from ATP provides energy for the transport.

active transport Energy-requiring mechanism in which a transport protein pumps a solute across a cell membrane against its concentration gradient.
calcium pump Active transport protein; pumps calcium ions across a cell membrane against their concentration gradient.
passive transport Mechanism by which a concentration gradient drives the movement of a solute across a cell membrane through a transport protein. Requires no energy input.

Take-Home Message

How do molecules or ions that cannot diffuse through a lipid bilayer cross a cell membrane?

» Transport proteins help specific molecules or ions to cross cell membranes.

» In passive transport, a solute binds to a protein that releases it on the opposite side of the membrane. The movement is driven by the solute's concentration gradient.

» In active transport, a transport protein pumps a solute across a membrane against its concentration gradient. The movement is driven by an energy input, as from ATP.

5.10 Membrane Trafficking

- By processes of exocytosis and endocytosis, cells take in and expel particles that are too big for transport proteins, as well as substances in bulk.
- Links to Lipoproteins 3.6, Endomembrane system 4.7, Cytoskeleton 4.10

Endocytosis and Exocytosis

Think back on the structure of a lipid bilayer. When a bilayer is disrupted, such as when part of the plasma membrane pinches off as a vesicle, it seals itself. Why? The disruption exposes the nonpolar fatty acid tails of the phospholipids to their watery surroundings.

Remember, in water, phospholipids spontaneously rearrange themselves so that their tails stay together. When a patch of membrane buds, its phospholipid tails are repelled by water on both sides. The water "pushes" the phospholipid tails together, which helps round off the bud as a vesicle, and also seals the rupture in the membrane.

Patches of membrane constantly move to and from the cell surface as components of vesicles (**Figure 5.27**). The formation and movement of vesicles involves motor proteins and requires ATP.

By **exocytosis**, a vesicle moves to the cell's surface, and the protein-studded lipid bilayer of its membrane fuses with the plasma membrane. As the exocytic vesicle loses its identity, its contents are released to the surroundings.

There are three pathways of **endocytosis**, but they all take up substances near the cell's surface. A small patch of plasma membrane balloons inward, and then it pinches off after sinking farther into the cytoplasm. The membrane patch becomes the outer boundary of an endocytic vesicle, which delivers its contents to an organelle or stores them in a cytoplasmic region.

With receptor-mediated endocytosis, molecules of a hormone, vitamin, mineral, or another substance bind to receptors on the plasma membrane. A shallow pit forms in the membrane patch under the receptors. The pit sinks into the cytoplasm and closes back on itself, and in this way it becomes a vesicle (**Figure 5.28**).

The term "receptor-mediated endocytosis" is a bit misleading, because receptors also function in phagocytosis. **Phagocytosis** ("cell eating") is an endocytic

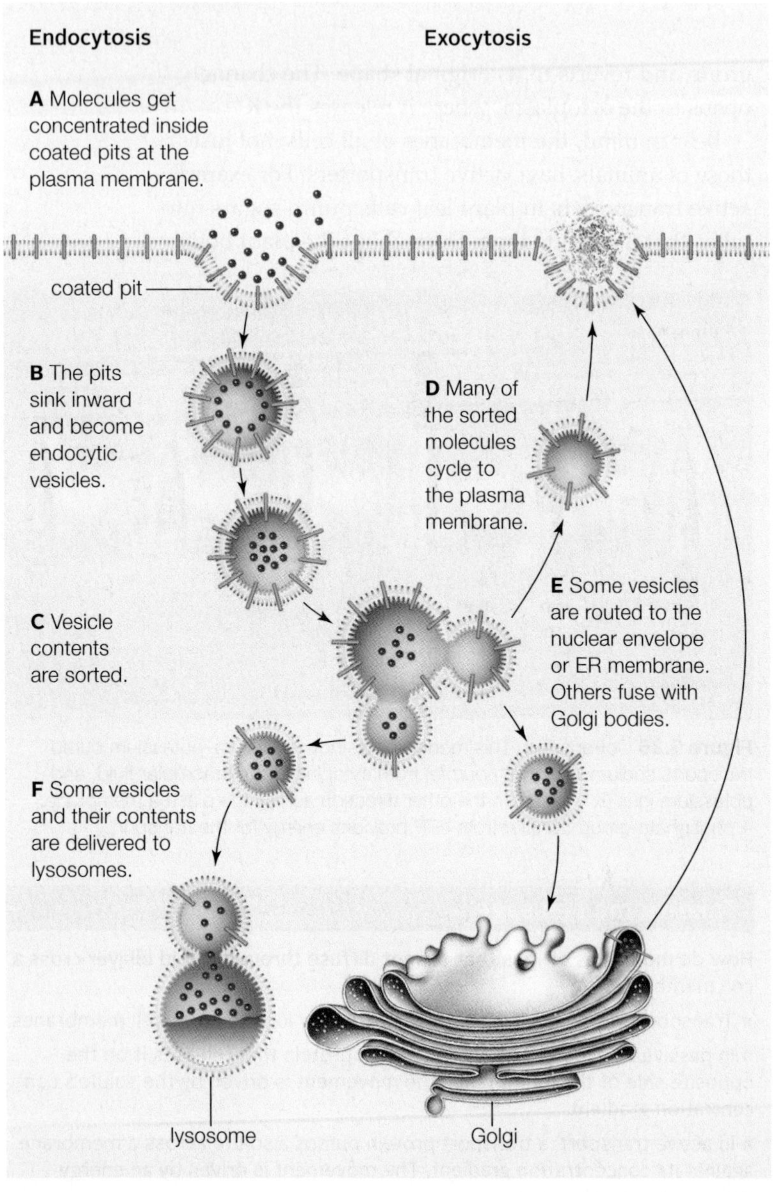

Endocytosis

A Molecules get concentrated inside coated pits at the plasma membrane.

coated pit

B The pits sink inward and become endocytic vesicles.

C Vesicle contents are sorted.

F Some vesicles and their contents are delivered to lysosomes.

lysosome

Exocytosis

D Many of the sorted molecules cycle to the plasma membrane.

E Some vesicles are routed to the nuclear envelope or ER membrane. Others fuse with Golgi bodies.

Golgi

Figure 5.27 Animated Membrane trafficking: endocytosis and exocytosis.

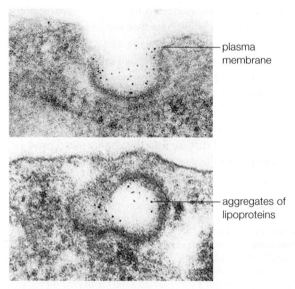

plasma membrane

aggregates of lipoproteins

Figure 5.28 Endocytosis of lipoproteins.

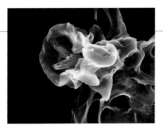

A Pseudopods of a white blood cell surround *Tuberculosis* bacteria (*red*).

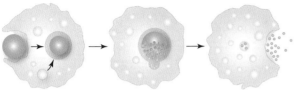

B Endocytic vesicle forms.

C Lysosome fuses with vesicle; enzymes digest pathogen.

D Cell uses the digested material or expels it.

Figure 5.29 Animated Phagocytosis. A phagocytic cell's pseudopods (extending lobes of cytoplasm) surround bacteria. The plasma membrane above the bulging lobes fuses and forms an endocytic vesicle. Once inside the cytoplasm, the vesicle fuses with a lysosome, which digests its contents.

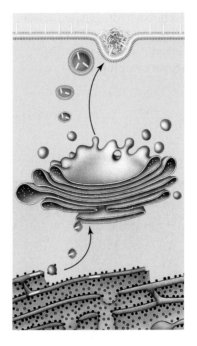

Figure 5.30 How membrane proteins become oriented to the inside or the outside of the cell.

Proteins of the plasma membrane are assembled in the ER, and finished inside Golgi bodies. The proteins (shown in *white*) become part of vesicle membranes that bud from the Golgi. The membrane proteins automatically become oriented in the proper direction when the vesicles fuse with the plasma membrane.

Figure It Out: What process does the upper arrow represent?

Answer: Exocytosis

pathway in which phagocytic cells such as amoebas engulf microorganisms, cellular debris, or other particles. In animals, macrophages and other white blood cells engulf and digest pathogenic viruses and bacteria, cancerous body cells, and other threats (**Figure 5.29A**). Phagocytosis begins when receptors bind to a particular target. The binding causes microfilaments to assemble in a mesh under the plasma membrane. The microfilaments contract, forcing some cytoplasm and plasma membrane above it to bulge outward as a lobe, or pseudopod. Pseudopods engulf a target and merge as a vesicle, which sinks into the cytoplasm and fuses with a lysosome (**Figure 5.29B,C**). Enzymes in the lysosome break down the vesicle's contents. The resulting molecular bits may be recycled by the cell, or expelled by exocytosis (**Figure 5.29D**).

Pinocytosis is an endocytic pathway that brings materials in bulk into the cell. It is not as selective as receptor-mediated endocytosis. An endocytic vesicle forms around a small volume of the extracellular fluid regardless of the kinds of substances dissolved in it.

Membrane Cycling

The composition of a plasma membrane begins in the ER. There, membrane proteins and lipids are made and modified, and both become part of vesicles that transport them to Golgi bodies for final modification. The finished proteins and lipids are repackaged as new vesicles that travel to the plasma membrane and fuse with it. The lipids and proteins of the vesicle membrane become part of the plasma membrane. This is the process by which new plasma membrane forms.

Figure 5.30 shows what happens when an exocytic vesicle fuses with the plasma membrane. Golgi bodies package membrane proteins facing the inside of a vesicle, so after the vesicle fuses with the plasma membrane, the proteins face the extracellular environment.

As long as a cell is alive, exocytosis and endocytosis continually replace and withdraw patches of its plasma membrane. If the cell is not enlarging, the total area of the plasma membrane remains more or less constant. Membrane lost as a result of endocytosis is replaced by membrane arriving as exocytic vesicles.

endocytosis Process by which a cell takes in a small amount of extracellular fluid by the ballooning inward of its plasma membrane.
exocytosis Process by which a cell expels a vesicle's contents to extracellular fluid.
phagocytosis "Cell eating"; an endocytic pathway by which a cell engulfs particles such as microbes or cellular debris.
pinocytosis Endocytosis of bulk materials.

A Toast to Alcohol Dehydrogenase (revisited)

In the human body, alcohol dehydrogenase (ADH) converts ethanol to acetaldehyde, an organic molecule even more toxic than ethanol and the most likely source of various hangover symptoms:

$$\text{ethanol} + \text{NAD}^+ \xrightarrow{\text{ADH}} \text{acetaldehyde} + \text{NADH}$$

A different enzyme, aldehyde dehydrogenase (ALDH), very quickly converts the toxic acetaldehyde to non-toxic acetate:

$$\text{acetaldehyde} + \text{NAD}^+ \xrightarrow{\text{ALDH}} \text{acetate} + \text{NADH} + \text{H}^+$$

Thus, the overall pathway of ethanol metabolism in humans is:

$$\text{ethanol} \xrightarrow[\text{NAD}^+ \quad \text{NADH}]{\text{ADH}} \text{acetaldehyde} \xrightarrow[\text{NAD}^+ \quad \text{NADH}]{\text{ALDH}} \text{acetate}$$

In the average adult human body, this metabolic pathway can detoxify between 7 and 14 grams of ethanol per hour. The average alcoholic beverage contains between 10 and 20 grams of ethanol, which is why having more than one drink in any two-hour interval may result in a hangover.

In most organisms, the main function of alcohol dehydrogenase is to detoxify the tiny quantities of alcohols that form in some metabolic pathways. In animals, the enzyme also detoxifies alcohols made by gut-inhabiting bacteria, and those in foods such as ripe fruit. Despite the small amounts of alcohol that humans encounter naturally, our bodies make at least nine different kinds of alcohol dehydrogenase. It is interesting to speculate about why so many of them have evolved.

We do understand how defects in the ADH enzyme affect our alcohol metabolism. For example, if a person's ADH is overactive, acetaldehyde accumulates faster than ALDH can detoxify it:

$$\text{ethanol} \xrightarrow{\text{ADH}} \begin{array}{c}\text{acetaldehyde}\\\text{acetaldehyde}\\\text{acetaldehyde}\end{array} \xrightarrow{\text{ALDH}} \text{acetate}$$

People with an overactive form of ADH become flushed and feel ill after drinking even a small amount of alcohol. The unpleasant experience may be part of the reason that these people are less likely to become alcoholic than other people.

Having underactive ALDH also causes acetaldehyde to accumulate:

$$\text{ethanol} \xrightarrow{\text{ADH}} \begin{array}{c}\text{acetaldehyde}\\\text{acetaldehyde}\\\text{acetaldehyde}\end{array} \xrightarrow{\text{X}} \text{acetate}$$

Underactive ALDH is associated with the same effect—and the same protection from alcoholism—as

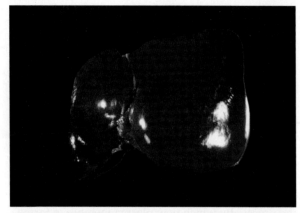

A Normal, healthy human liver.

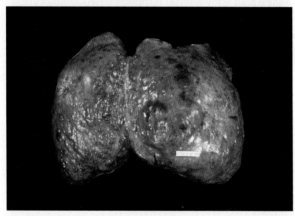

B Cirrhotic liver.

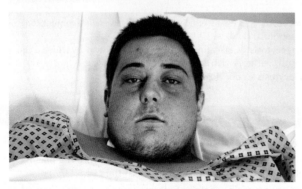

C Gary Reinbach, who died in 2009 from alcoholic liver disease shortly after this photograph was taken. His orange skin is a symptom of cirrhosis. He was 22 years old.

Figure 5.31 Alcoholic liver disease.

overactive ADH. Both types of variant enzymes are common in people of Asian descent. For this reason, the alcohol flushing reaction is informally called "Asian flush."

Having an underactive ADH enzyme has the opposite effect. It slows alcohol metabolism, so people with low ADH activity may not feel the ill effects of drinking alcoholic beverages as much as other people. When these people drink alcohol, they have a tendency to become alcoholics. Compulsive, uncontrolled drinking damages an alcoholic's health and social relationships. The study mentioned in Section 5.1 showed that one-quarter of the undergraduate students who binged also had other signs of alcoholism.

Alcoholics will continue to drink despite the knowledge that doing so has tremendous negative consequences. In the United States, alcohol abuse is the leading cause of cirrhosis of the liver. The liver becomes so scarred, hardened, and filled with fat that it loses its function (**Figure 5.31A,B**). It stops making the protein albumin, so the solute balance of body fluids is disrupted, and the legs and abdomen swell with watery fluid. It cannot remove drugs and other toxins from the blood, so they accumulate in the brain—which impairs mental functioning and alters personality. Restricted blood flow through the liver causes veins to enlarge and rupture, so internal bleeding is a risk. The damage to the body results in a heightened susceptibility to diabetes and liver cancer. Once cirrhosis has been diagnosed, a person has about a 50 percent chance of dying within 10 years (**Figure 5.31C**).

How would you vote? Transplantation is a last-resort treatment for a failed liver, but even so there are not nearly enough donors for everyone who needs a liver transplant. Liver failure can be a result of factors that are generally  beyond an individual's control, such as inherited disease, cancer, illness, or infection. The damage to an alcoholic's liver is self-inflicted. Should lifestyle be a factor in deciding who gets a donated liver for transplant?

Summary

 Section 5.1 Currently the most serious drug problem on college campuses is binge drinking, which is a symptom of alcoholism. Drinking more alcohol than the body's enzymes can detoxify can be lethal in both the short term and the long term.

 Section 5.2 Kinetic energy, **potential energy**, and other forms of **energy** cannot be created or destroyed (**first law of thermodynamics**). Energy can be converted from one form to another and transferred between objects or systems. Energy tends to disperse spontaneously (**second law of thermodynamics**). Some energy disperses with every transfer, usually as heat. **Entropy** is a measure of how much the energy of a system is dispersed.

Living things maintain their organization only as long as they harvest energy from someplace else. Energy flows in one direction through the biosphere, starting mainly from the sun, then into and out of ecosystems. Producers and then consumers use the captured energy to assemble, rearrange, and break down organic molecules that cycle among organisms throughout ecosystems.

 Section 5.3 Cells store and retrieve free energy by making and breaking chemical bonds in metabolic reactions, in which **reactants** are converted to **products**. **Endergonic** reactions require a net energy input. **Exergonic** reactions end with a net energy release. **Activation energy** is the minimum energy required to start a reaction.

 Section 5.4 Enzymes greatly enhance the rate of reactions without being changed by them, a process called **catalysis**. Enzymes lower a reaction's activation energy by boosting local concentrations of **substrates**, orienting substrates in positions that favor reaction, inducing the fit between a substrate and the enzyme's **active site** (**induced-fit model**), and sometimes excluding water, all of which bring on a substrate's **transition state**. Each type of enzyme works best within a characteristic range of temperature, salt concentration, and pH.

 Section 5.5 Cells build, convert, and dispose of substances in enzyme-mediated reaction sequences called **metabolic pathways**. Controls over enzymes allow cells to conserve energy and resources by producing only what they require. **Allosteric** sites are points of control by which a cell adjusts the types and amounts of substances it makes. **Feedback inhibition** is an example of enzyme control. **Redox** (oxidation–reduction) **reactions** in **electron transfer chains** allow cells to harvest energy in manageable increments.

 Section 5.6 Most enzymes require **cofactors**, which are metal ions or organic **coenzymes**. Cofactors in some **antioxidants** help them stop reactions with oxy-

gen that produce free radicals. **ATP** functions as an energy carrier between reaction sites in cells. It has three phosphate bonds; when a phosphate group is transferred to another molecule, the energy of the bond is transferred along with it. Phosphate-group transfers (**phosphorylations**) to and from ATP couple reactions that release energy with reactions that require energy. Cells regenerate ATP in the **ATP/ADP cycle**.

Section 5.7 A cell membrane is a mosaic of proteins and lipids (mainly phospholipids) organized as a bilayer. Membranes of bacteria and eukaryotic cells can be described as a **fluid mosaic**; those of archaea are not fluid. Proteins transiently or permanently associated with a membrane carry out most membrane functions. All cell membranes have **transport proteins**. Plasma membranes also incorporate **receptor proteins**, **adhesion proteins**, enzymes, and **recognition proteins**.

Section 5.8 Molecules or ions tend to spread spontaneously (**diffuse**), with the eventual result being a gradual and complete mixing. A **concentration gradient** is a difference in the **concentration** of a substance between adjoining regions of fluid. The steepness of the gradient, temperature, solute size, charge, and pressure influence the diffusion rate.

Osmosis is the diffusion of water across a selectively permeable membrane, from the region with a lower solute concentration (**hypotonic**) toward the region with a higher solute concentration (**hypertonic**). There is no net movement of water between **isotonic** solutions. **Osmotic pressure** is the amount of **turgor** (fluid pressure against a cell membrane or wall) that stops osmosis.

Section 5.9 Gases, water, and small nonpolar molecules can diffuse across a lipid bilayer. Most other molecules, and ions in particular, cross only with the help of transport proteins, which allow a cell or membrane-enclosed organelle to control which substances enter and exit.

The types of transport proteins in a membrane determine which substances can cross it. **Active transport** proteins such as **calcium pumps** use energy, such as a phosphate transfer from ATP, to pump a solute against its concentration gradient. **Passive transport** proteins work without an energy input; a solute's movement is driven by its concentration gradient.

Section 5.10 Substances in bulk and large particles are moved across plasma membranes by processes of exocytosis and endocytosis. With **exocytosis**, a cytoplasmic vesicle fuses with the plasma membrane, and its contents are released to the outside of the cell. The vesicle's membrane lipids and proteins become part of the plasma membrane. With **endocytosis**, a patch of plasma membrane balloons into the cell, and forms a vesicle that sinks into the cytoplasm. **Pinocytosis** is not as specific as **phagocytosis**, a receptor-mediated path-

way by which cells engulf particles or cell debris. Plasma membrane lost by endocytosis is replaced by exocytosis.

Self-Quiz

Answers in Appendix III

1. _____ is life's primary source of energy.
   a. Food   b. Water   c. Sunlight   d. ATP

2. Which of the following statements is not correct?
   a. Energy cannot be created or destroyed.
   b. Energy cannot change from one form to another.
   c. Energy tends to disperse spontaneously.

3. Entropy _____ .
   a. disperses   c. always increases, overall
   b. is a measure of disorder   d. b and c

4. If we liken a chemical reaction to an energy hill, then a(n) _____ reaction is an uphill run.
   a. endergonic   c. catalytic
   b. exergonic   d. both a and c

5. If we liken a chemical reaction to an energy hill, then activation energy is like _____ .
   a. a burst of speed
   b. coasting downhill
   c. a bump at the top of the hill

6. Enzymes _____ .
   a. are proteins, except for a few RNAs
   b. lower the activation energy of a reaction
   c. are changed by the reactions they catalyze
   d. a and b

7. _____ are always changed by participating in a reaction. (Choose all that are correct.)
   a. Enzymes   c. Reactants
   b. Cofactors   d. Coenzymes

8. One environmental factor that influences enzyme function is _____ .
   a. temperature   c. light
   b. wind   d. radioactivity

9. A metabolic pathway _____ .
   a. may build or break down molecules
   b. generates heat
   c. can include an electron transfer chain
   d. all of the above

10. A molecule that donates electrons becomes _____ , and the one that accepts electrons becomes _____ .
    a. reduced; oxidized   c. oxidized; reduced
    b. ionic; electrified   d. electrified; ionic

11. An antioxidant _____ .
    a. prevents other molecules from being oxidized
    b. is necessary in the human diet
    c. balances charge
    d. oxidizes free radicals

12. Ions or molecules tend to diffuse from a region where they are _____ (more/less) concentrated to another where they are _____ (more/less) concentrated.

13. _____ cannot easily diffuse across a lipid bilayer.
    a. Water   c. Ions
    b. Gases   d. all of the above

14. Transporters that require an energy boost help sodium ions across a cell membrane. This is a case of _____ .
    a. passive transport   c. facilitated diffusion
    b. active transport   d. b and c

One Tough Bug The genus *Ferroplasma* consists of a few species of acid-loving archaea. One species, *F. acidarmanus*, was discovered to be the main constituent of slime streamers (a type of biofilm) deep inside an abandoned California copper mine (**Figure 5.32**). *F. acidarmanus* cells living on the surfaces of the streamers use an energy-harvesting pathway that combines oxygen with iron-sulfur compounds in minerals such as pyrite. Oxidizing these minerals dissolves them, so groundwater that seeps into the mine ends up accumulat-

Figure 5.32 A one-meter (3 feet) wide section of a stream deep inside one of the most toxic sites in the United States: Iron Mountain Mine, in California. The water is hot (around 40°C, or 104°F), has a pH of zero, and is heavily laden with arsenic and other toxic metals. The slime streamers growing in it are a biofilm dominated by a species of archaea, *Ferroplasma acidarmanus*.

ing extremely high concentrations of metal ions such as copper, zinc, cadmium, and arsenic. Another oxidation product of sulfide minerals, sulfuric acid, lowers the pH of the resulting solution to zero.

  F. acidarmanus cells are unwalled, yet they are able to maintain their internal pH at a cozy 5 despite living in an environment with a composition essentially the same as hot battery acid. Thus, researchers investigating *Ferroplasma* metabolic enzymes were surprised to discover that most of the cells' enzymes function best at low pH (**Figure 5.33**).

1. What does the dashed line signify?

2. Of the four enzymes profiled in the graph, how many function optimally at a pH lower than 5? How many retain significant function at pH 5?

3. What is the optimal pH for *Ferroplasma* carboxylesterase?

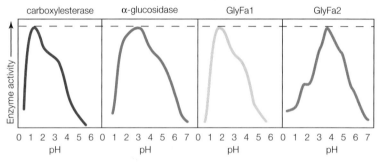

Figure 5.33 pH anomaly of *Ferroplasma* enzymes. The graphs show the pH profiles of four enzymes isolated from *Ferroplasma*. Researchers had expected these enzymes to function best at the cells' cytoplasmic pH (5). *Source:* Golyshina et al., *Environmental Microbiology* (2006) 8(3): 416–425.

15. Immerse a human red blood cell in a hypotonic solution, and water _____ .
    a. diffuses into the cell
    b. diffuses out of the cell
    c. shows no net movement
    d. moves in by endocytosis

16. Vesicles form during _____ .
    a. endocytosis
    b. exocytosis
    c. phagocytosis
    d. symbiosis
    e. a through c
    f. all of the above

17. Match each term with its most suitable description.
    ___ reactant
    ___ phagocytosis
    ___ first law of thermodynamics
    ___ product
    ___ cofactor
    ___ diffusion
    ___ passive transport
    ___ active transport

    a. assists enzymes
    b. forms at reaction's end
    c. enters a reaction
    d. requires energy boost
    e. one cell engulfs another
    f. energy cannot be created or destroyed
    g. faster with a gradient
    h. no energy boost required

Critical Thinking

1. Beginning physics students are often taught the basic concepts of thermodynamics with two phrases: First, you can't win. Second, you can't break even. Explain.

2. How is diffusion like entropy?

3. Water molecules tend to diffuse in response to their own concentration gradient. How can water be more or less concentrated?

4. Dixie Bee wanted to make JELL-O shots for her next party, but felt guilty about encouraging her guests to consume alcohol. She tried to compensate for the toxicity of the alcohol by adding pieces of healthy fresh pineapple to the shots, but when she did, the JELL-O never solidified. What happened? Hint: JELL-O is mainly sugar and a gelatinous mixture of proteins.

5. The enzyme trypsin is sold as a dietary enzyme supplement. Explain what happens to trypsin taken with food.

6. One molecule of catalase can inactivate about 6 million hydrogen peroxide molecules per minute by combining them two at a time. Catalase also inactivates other toxins, including ethanol. Given that its active site specifically binds to hydrogen peroxide, how do you think this enzyme acts on other molecules?

7. Catalase combines two hydrogen peroxide molecules ($H_2O_2 + H_2O_2$) to make two molecules of water. A gas also forms. What is the gas?

8. Hydrogen peroxide bubbles if dribbled on an open cut but does not bubble on unbroken skin. Explain why.

6 Where It Starts—Photosynthesis

LEARNING ROADMAP

Where you have been This chapter explores the main metabolic pathways (Sections 5.5, 5.6) by which organisms harvest energy from the sun (5.2, 5.3). We revisit experimental design (1.6), electrons and energy levels (2.3), bonds (2.4), carbohydrates (3.4), plastids (4.9), plant cell specializations (4.7, 4.11), membrane proteins (5.7), and gradients (5.8).

Where you are now

The Rainbow Catchers
The flow of energy through the biosphere starts when photosynthetic pigments absorb the energy of light. In photoautotrophs, that energy drives carbohydrate synthesis.

What Is Photosynthesis?
Photosynthesis has two stages in chloroplasts. In the first stage, cells harvest light energy. Molecules that form in this stage power the formation of sugars in the second stage.

Making ATP and NADPH
ATP forms in the light-dependent reactions. NADPH forms in a noncyclic pathway that releases oxygen. ATP also forms in a cyclic pathway that does not release oxygen.

Making Sugars
In the second stage of photosynthesis, sugars are assembled from CO_2. The reactions run on ATP and NADPH—molecules that formed in the first stage of photosynthesis.

Adaptations
Details of light-independent reactions that vary among organisms are evolutionary adaptations to environmental conditions that prevail in different habitats.

Where you are going You will see in Chapter 7 how molecules originally assembled by photosynthesizers are disassembled to harvest energy stored in their bonds. Chapter 22 returns to evolutionary adaptations of plants, and Chapters 27—30 return to plant structure and function. In Chapter 46, you will see how photosynthetic organisms sustain almost all life on Earth and how carbon cycles through the biosphere. Chapter 48 returns to human impacts on the biosphere.

6.1 Biofuels

Today, the expression "food is fuel" is not just about eating. With fossil fuel prices soaring, there is an increasing demand for biofuels, which are oils, gases, or alcohols made from organic matter that is not fossilized. Much of the material currently used for biofuel production in the United States consists of food crops—mainly corn, soybeans, and sugarcane. Growing these crops in large quantities is typically expensive and damaging to the environment, and using them to make biofuel competes with our food supply. The diversion of food crops to biofuel production may be contributing to the worldwide increases in food prices we are seeing now.

How did we end up competing with our vehicles for food? We both run on the same fuel: energy that plants have stored in chemical bonds. Fossil fuels such as petroleum, coal, and natural gas formed from the remains of ancient swamp forests that decayed and compacted over millions of years. These fuels consist mainly of molecules originally assembled by ancient plants. Biofuels—and foods—consist mainly of molecules originally assembled by modern plants.

Autotrophs harvest energy directly from the environment, and obtain carbon from inorganic molecules (*auto*– means self; –*troph* refers to nourishment). Plants and most other autotrophs make their own food by photosynthesis, a process in which they use the energy of sunlight to assemble carbohydrates—sugars—from carbon dioxide and water. Directly or indirectly, photosynthesis also feeds most other life on Earth. Animals and other **heterotrophs** get energy and carbon by breaking down organic molecules assembled by other organisms (*hetero*– means other). We and almost all other organisms sustain ourselves by extracting energy from organic molecules that photosynthesizers make.

A lot of energy is locked up in the chemical bonds of molecules made by plants. That energy can fuel heterotrophs, as when an animal cell powers ATP synthesis by breaking the bonds of sugars. It can also fuel our cars, which run on energy released by burning biofuels or fossil fuels. Both processes are fundamentally the same: They release energy by breaking the bonds of organic molecules. Both use oxygen to break those bonds, and both produce carbon dioxide.

Corn and other food crops are rich in oils, starches, or sugars that can be easily converted to biofuels. The starch in corn kernels, for example, can be enzymatically broken down to glucose, which is converted to ethanol by heterotrophic bacteria or yeast. Making biofuels from other types of plant matter requires additional steps, because these materials contain a higher proportion of cellulose. Breaking down this tough, insoluble carbohydrate to its glucose monomers adds a lot of cost to the biofuel product. Researchers are currently trying to find cost-effective ways to break down the abundant cellulose in fast-growing weeds such as switchgrass (**Figure 6.1**), and agricultural wastes such as wood chips, wheat straw, cotton stalks, and rice hulls.

autotroph Organism that makes its own food using carbon from inorganic molecules such as CO_2, and energy from the environment.
heterotroph Organism that obtains energy and carbon from organic compounds assembled by other organisms.
photosynthesis Metabolic pathway by which most autotrophs capture light energy and use it to make sugars from CO_2 and water.

Figure 6.1 Biofuels. *Above*, Ratna Sharma and Mari Chinn are researching ways to reduce the cost of producing biofuel from renewable sources such as wild grasses and agricultural wastes. *Right*, switchgrass (*Panicum virgatum*) grows wild in North American prairies.

6.2 Sunlight as an Energy Source

■ Photosynthetic organisms use pigments to capture the energy of sunlight.

■ Links to Electrons 2.3, Bonding 2.4, Carbohydrates 3.4, Chromoplasts 4.9, Energy 5.2, Metabolism 5.5

Energy flow through nearly all ecosystems on Earth begins when photosynthesizers intercept energy from the sun. Harnessing the energy of sunlight for work is a complicated business, or we would have been able to do it in an economically sustainable way by now. Plants do it by converting light energy to chemical energy, which they and most other organisms use to drive cellular work. The first step involves capturing light. In order to understand how that happens, you have to understand a little about the nature of light.

Most of the energy that reaches Earth's surface is in the form of visible light, so it should not be surprising that visible light is the energy that drives photosynthesis. Visible light is a very small part of a large spectrum of electromagnetic energy radiating from the sun. Like all other forms of electromagnetic energy, light travels in waves, moving through space a bit like waves moving across an ocean. The distance between the crests of two successive waves is a **wavelength**. Light wavelength is measured in nanometers (nm).

Visible light travels in wavelengths between 380 and 750 nm (**Figure 6.2A**). Our eyes perceive all of these wavelengths combined as white light, and particular wavelengths in this range as different colors. White light separates into its component colors when it passes through a prism, or raindrops that act as tiny prisms. A prism bends longer wavelengths more than it bends shorter ones, so a rainbow of colors forms.

Light travels in waves, but it is also organized in packets of energy called photons. A photon's energy and its wavelength are related, so all photons traveling at the same wavelength carry the same amount of energy. Photons that carry the least amount of energy travel in longer wavelengths; those that carry the most energy travel in shorter wavelengths (**Figure 6.2B**). Photons of wavelengths shorter than about 380 nanometers carry enough energy to alter or break the chemical bonds of DNA and other biological molecules. That is why UV (ultraviolet) light, x-rays, and gamma rays are a threat to life.

Pigments: The Rainbow Catchers

Photosynthesizers use pigments to capture light. A **pigment** is an organic molecule that selectively absorbs light of specific wavelengths. Wavelengths of light that are not absorbed are reflected, and that reflected light gives each pigment its characteristic color.

Chlorophyll a is the most common photosynthetic pigment in plants, and also in photosynthetic protists and bacteria. Chlorophyll *a* absorbs violet and red

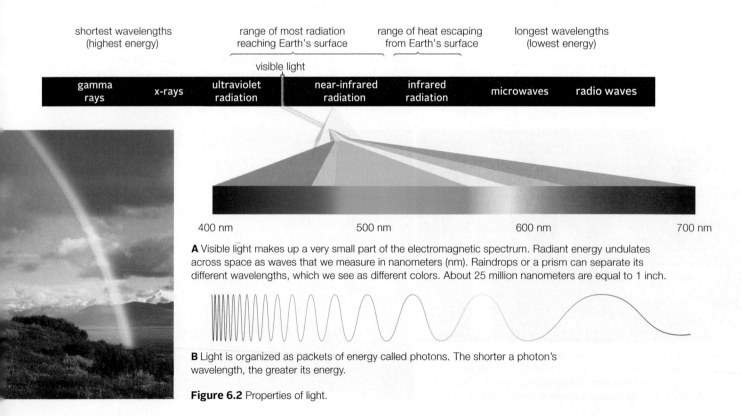

shortest wavelengths (highest energy)		range of most radiation reaching Earth's surface	range of heat escaping from Earth's surface	longest wavelengths (lowest energy)	

visible light

| gamma rays | x-rays | ultraviolet radiation | near-infrared radiation | infrared radiation | microwaves | radio waves |

400 nm   500 nm   600 nm   700 nm

A Visible light makes up a very small part of the electromagnetic spectrum. Radiant energy undulates across space as waves that we measure in nanometers (nm). Raindrops or a prism can separate its different wavelengths, which we see as different colors. About 25 million nanometers are equal to 1 inch.

B Light is organized as packets of energy called photons. The shorter a photon's wavelength, the greater its energy.

Figure 6.2 Properties of light.

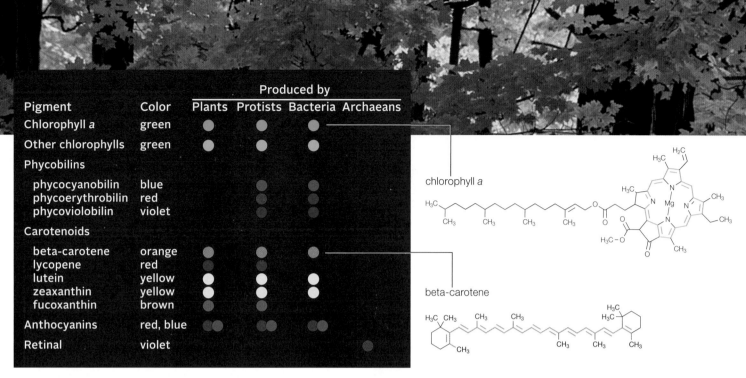

Pigment	Color	Produced by			
		Plants	Protists	Bacteria	Archaeans
Chlorophyll a	green	●	●	●	
Other chlorophylls	green	●	●	●	
Phycobilins					
phycocyanobilin	blue		●	●	
phycoerythrobilin	red		●	●	
phycoviolobilin	violet		●	●	
Carotenoids					
beta-carotene	orange	●	●	●	
lycopene	red	●	●		
lutein	yellow	●	●	●	
zeaxanthin	yellow	●	●	●	
fucoxanthin	brown	●	●		
Anthocyanins	red, blue	●●	●●	●●	
Retinal	violet				●

chlorophyll a

beta-carotene

Figure 6.3 Examples of photosynthetic pigments. *Left*, photosynthetic pigments can collectively absorb almost all visible light wavelengths.

Right, the light-catching part of a pigment (shown in color) is the region in which single bonds alternate with double bonds. These and many other pigments are derived from evolutionary remodeling of the same compound, as is heme (Section 5.6). Heme is a red pigment.

light, so it appears green to us. Accessory pigments, including other chlorophylls, work together with chlorophyll *a* to harvest a wide range of light wavelengths for photosynthesis. The colors of a few of the 600 or so known accessory pigments are shown in **Figure 6.3**.

Accessory pigments are multipurpose molecules. Antioxidant properties help protect plants and other organisms from the damaging effects of UV light in the sun's rays; appealing colors attract animals to ripening fruit or pollinators to flowers. You may already be familiar with some of these molecules. For example, carrots are orange because they contain beta-carotene (β-carotene). A tomato's color changes from green to red as it ripens because its chlorophyll-containing chloroplasts develop into lycopene-containing chromoplasts (Section 4.9). Roses are red and violets are blue because of their anthocyanin content.

Most photosynthetic organisms use a combination of pigments for photosynthesis. In plants, chlorophylls are usually so abundant that they mask the colors of other pigments, so leaves typically appear green. The green leaves of many plants change color during autumn because they stop making pigments in preparation for a period of dormancy. Chlorophyll breaks down faster than the other pigments, so the leaves turn red, orange, yellow, or violet as their chlorophyll content declines and their accessory pigments become visible.

The light-trapping part of a pigment is an array of atoms in which single bonds alternate with double bonds. Electrons in such arrays easily absorb photons, so pigment molecules function a bit like antennas that are specialized for receiving light energy of only certain wavelengths.

Absorbing a photon excites electrons. Remember, an energy input can boost an electron to a higher energy level (Section 2.3). The excited electron returns quickly to a lower energy level by emitting the extra energy. As you will see, photosynthetic cells can capture energy emitted from an electron returning to a lower energy level. Arrays of chlorophylls and other photosynthetic pigments in these cells hold on to the energy by passing it back and forth. When the energy reaches a special pair of chlorophylls, the reactions of photosynthesis begin.

chlorophyll a Main photosynthetic pigment in plants.
pigment An organic molecule that can absorb light of certain wavelengths.
wavelength Distance between the crests of two successive waves.

Take-Home Message

How do photosynthesizers absorb light?

» Energy radiating from the sun travels through space in waves and is organized as packets called photons.

» The spectrum of radiant energy from the sun includes visible light. Humans perceive different wavelengths of visible light as different colors. The shorter the wavelength, the greater the energy.

» Pigments absorb light at specific wavelengths. Photosynthetic species use pigments such as chlorophyll *a* to harvest the energy of light for photosynthesis.

6.3 Exploring the Rainbow

- Photosynthetic pigments work together to harvest light of different wavelengths.
- Link to Experiments 1.6

A Light micrograph of cells in a strand of *Chladophora*. Theodor Engelmann used this green alga in an early experiment to show that certain colors of light are best for photosynthesis.

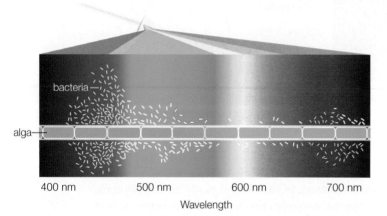

B Engelmann directed light through a prism so that bands of colors crossed a water droplet on a microscope slide. The water held a strand of *Chladophora* and oxygen-requiring bacteria. The bacteria clustered around the algal cells that were releasing the most oxygen—the ones that were most actively engaged in photosynthesis. Those cells were under red and violet light. These results constituted one of the first absorption spectra.

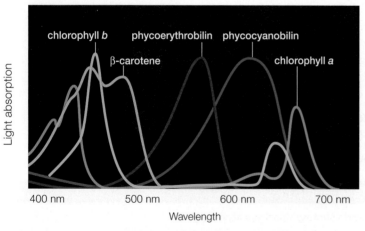

C Absorption spectra of chlorophylls *a* and *b*, β-carotene, and two phycobilins reveal the efficiency with which these pigments absorb different wavelengths of visible light. Line color indicates the characteristic color of each pigment.

Figure 6.4 Animated Discovery that photosynthesis is driven by particular wavelengths of light.

Figure It Out: Of the five pigments represented in C, which three are the main photosynthetic pigments in *Chladophora*?

Answer: Chlorophyll *a*, chlorophyll *b*, and β-carotene

In 1882, botanist Theodor Engelmann designed an experiment to test his hypothesis that the color of light affects the rate of photosynthesis. It had long been known that photosynthesis releases oxygen, so Engelmann used the amount of oxygen released by photosynthetic cells as a measure of how much photosynthesis was occurring in them. He used a prism to divide a ray of light into its component colors, then directed the resulting spectrum across a single strand of *Chladophora*, a photosynthetic alga (**Figure 6.4A**), suspended in a drop of water.

Oxygen-sensing equipment had not yet been invented, so Engelmann used oxygen-requiring bacteria to show him where the oxygen concentration in the water was highest. The bacteria moved through the water and gathered mainly where violet or red light fell across the strand of algae (**Figure 6.4B**). Engelmann concluded that the algal cells illuminated by light of these colors were releasing the most oxygen—a sign that violet and red light are the best for driving photosynthesis in these cells.

Engelmann's experiment allowed him to correctly identify the colors of light (red and violet) that are most efficient at driving photosynthesis in *Chladophora*. His results constituted an absorption spectrum, a graph that shows how efficiently a substance absorbs the different wavelengths of light. Peaks in the graph indicate wavelengths of light that the substance absorbs best (**Figure 6.4C**). Engelmann's results represent the combined absorption spectra of all the photosynthetic pigments in *Chladophora*.

Most photosynthetic organisms use a combination of pigments to drive photosynthesis, and the combination differs by species. Why? Different proportions of wavelengths in sunlight reach different parts of Earth. The particular set of pigments a species makes is an adaptation to the particular wavelengths of light available in its native habitat. For example, water absorbs light between wavelengths of 500 and 600 nm less efficiently than other wavelengths. Pigments in algae that live deep underwater absorb light in the range of 500–600 nm, which is the range that is most abundant in deep water. Phycobilins are the most common pigments in these algae.

Take-Home Message

Why do cells use more than one photosynthetic pigment?

» A combination of pigments allows a photosynthetic organism to most efficiently capture the particular range of light wavelengths that reaches the habitat in which it evolved.

6.4 Overview of Photosynthesis

- In plants and other photosynthetic eukaryotes, photosynthesis occurs in chloroplasts.
- Photosynthesis occurs in two stages.
- Links to Plastids 4.9, Metabolic Pathways 5.5

The **chloroplast** is an organelle that specializes in photosynthesis in plants and many protists (**Figure 6.5**). Plant chloroplasts have two outer membranes, and are filled with a semifluid matrix called **stroma**. Stroma contains the chloroplast's own DNA, some ribosomes, and an inner, much-folded **thylakoid membrane**. The folds of a thylakoid membrane typically form stacks of disks (thylakoids) that are connected by channels. The space inside all of the disks and channels is one continuous compartment.

Photosynthesis is often summarized by the following equation:

$$6\,CO_2 + 6H_2O \xrightarrow{\text{light energy}} C_6H_{12}O_6 + 6O_2$$

carbon dioxide   water   glucose   oxygen

However, photosynthesis is not a single reaction. It is a metabolic pathway, a series of many reactions that occur in two stages. The first stage is carried out by molecules embedded in the thylakoid membrane. It is driven by light, so the collective reactions of this stage are called the **light-dependent reactions**.

Two alternative sets of light-dependent reactions, a noncyclic and a cyclic pathway, both convert light energy to chemical bond energy of ATP. The noncyclic pathway, which is the main one in chloroplasts, yields NADPH and O_2 in addition to ATP:

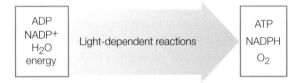

The reactions of the second stage of photosynthesis, which run in the stroma, break apart molecules of carbon dioxide and water, and reassemble their atoms as glucose. Light energy does not power these synthesis reactions, so they are collectively called the **light-independent reactions**. They run on energy delivered by coenzymes produced in the first stage:

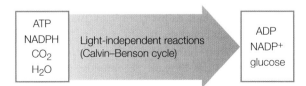

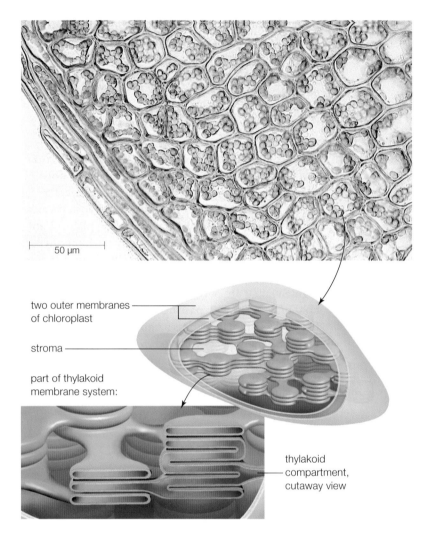

two outer membranes of chloroplast

stroma

part of thylakoid membrane system:

thylakoid compartment, cutaway view

Figure 6.5 **Animated** The chloroplast: site of photosynthesis in the cells of typical leafy plants. The micrograph shows chloroplast-stuffed cells in a leaf of *Plagiomnium ellipticum*, a type of moss.

chloroplast Organelle specialized for photosynthesis in plants and some protists.
light-dependent reactions First stage of photosynthesis; convert light energy to chemical energy of ATP and NADPH.
light-independent reactions Second stage of photosynthesis; use ATP and NADPH to assemble sugars from water and CO_2.
stroma Semifluid matrix between the thylakoid membrane and the two outer membranes of a chloroplast.
thylakoid membrane A chloroplast's highly folded inner membrane system; forms a continuous compartment in the stroma.

Take-Home Message

Where in a eukaryotic cell do the reactions of photosynthesis take place?

» In the first stage of photosynthesis, light energy drives the formation of ATP and NADPH, and oxygen is released. In eukaryotic cells, these light-dependent reactions occur at the thylakoid membrane of chloroplasts.

» The second stage of photosynthesis, the light-independent reactions, occur in the stroma of chloroplasts. ATP and NADPH drive the synthesis of carbohydrates from water and carbon dioxide.

6.5 Light-Dependent Reactions

■ The reactions of the first stage of photosynthesis convert the energy of light to the energy of chemical bonds.
■ Links to Electrons and energy levels 2.3, Chloroplasts 4.9, Energy 5.2, Electron transfer chains 5.5, Membrane properties 5.7, Gradients 5.8

When a pigment in a thylakoid membrane absorbs a photon, the photon's energy boosts one of the pigment's electrons to a higher energy level. The electron quickly emits the extra energy and drops back down to its unexcited state. In a thylakoid membrane, the energy emitted by excited electrons is not lost, because light-harvesting complexes can keep it in play. A light-harvesting complex is a circular array of chlorophylls, accessory pigments, and proteins. Millions of them are embedded in each thylakoid membrane (**Figure 6.6**). Pigments that are part of a light-harvesting complex hold on to energy by passing it back and forth, a bit like volleyball players pass a ball among team members. The energy gets volleyed from complex to complex until a photosystem absorbs it. **Photosystems** are groups of hundreds of chlorophylls, accessory pigments, and other molecules that work as a unit to begin the chemical reactions of photosynthesis.

The Noncyclic Pathway

Thylakoid membranes contain two kinds of photosystems, type I and type II, which were named in the order of their discovery. They work together in a set of reactions called the noncyclic pathway of photosynthesis (**Figure 6.7A**). These reactions begin when energy being passed among light-harvesting complexes reaches a photosystem II (**Figure 6.8**). At the center of each photosystem are two very closely associated chlorophyll *a* molecules (a "special pair"). When a

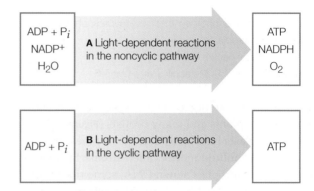

Figure 6.7 Summary of the inputs and outputs of photosynthesis.

photosystem absorbs energy, electrons are ejected from its special pair ❶. These electrons immediately enter an electron transfer chain in the thylakoid membrane.

A photosystem can lose only a few electrons before it must be restocked with more. Where do replacements come from? Photosystem II gets more electrons by removing them from water molecules in the thylakoid compartment. This reaction causes the water molecules to dissociate into hydrogen ions and oxygen ❷. The released oxygen diffuses out of the cell as O_2 gas. This and any other process by which a molecule is broken apart by light energy is called **photolysis**.

The actual conversion of light energy to chemical energy occurs when a photosystem donates electrons to an electron transfer chain ❸. Light does not take part in chemical reactions, but electrons do. The electrons pass from one molecule of the chain to the next in a series of redox reactions. With each reaction, the electrons release a bit of their extra energy.

Molecules of the electron transfer chain use the released energy to move hydrogen ions (H^+) across the membrane, from the stroma to the thylakoid compartment ❹. Thus, the flow of electrons through electron transfer chains sets up and maintains a hydrogen ion gradient across the thylakoid membrane. This gradient motivates hydrogen ions in the thylakoid compartment to move back into the stroma. However, ions cannot diffuse through lipid bilayers (Section 5.8). H^+ leaves the thylakoid compartment only by flowing through membrane transport proteins called ATP synthases ❼.

Hydrogen ion flow through an ATP synthase causes this protein to attach a phosphate group to ADP, so ATP forms in the stroma ❽. The process by which the flow of electrons through electron transfer chains drives ATP formation is called chemiosmosis, or **electron transfer phosphorylation**.

After the electrons have moved through the first electron transfer chain, they are accepted by a photosystem I. When this photosystem absorbs light energy,

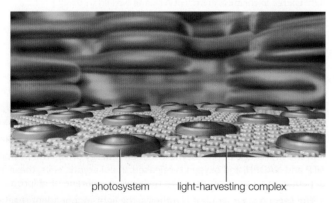

photosystem   light-harvesting complex

Figure 6.6 Artist's view of some of the components of the thylakoid membrane as seen from the stroma. Molecules of electron transfer chains and ATP synthases are also present, but not shown for clarity.

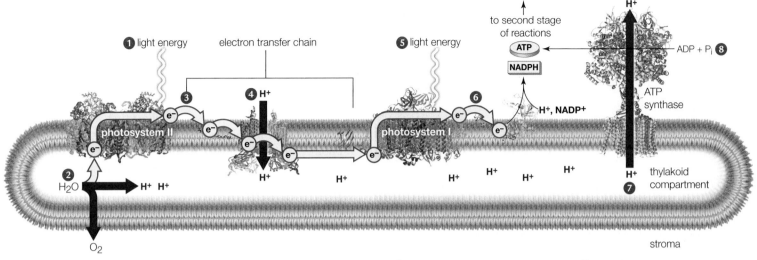

1 Light energy ejects electrons from a photosystem II.

2 The photosystem pulls replacement electrons from water molecules, which break apart into oxygen and hydrogen ions. The oxygen leaves the cell as O_2.

3 The electrons enter an electron transfer chain in the thylakoid membrane.

4 Energy lost by the electrons as they move through the electron transfer chain is used to pump hydrogen ions from the stroma into the thylakoid compartment. A hydrogen ion gradient forms across the thylakoid membrane.

5 Light energy ejects electrons from a photosystem I. Replacement electrons come from an electron transfer chain in the thylakoid membrane.

6 The electrons move through a second electron transfer chain, then combine with $NADP^+$ and H^+, so NADPH forms.

7 Hydrogen ions in the thylakoid compartment are propelled through the interior of ATP synthases by their gradient across the thylakoid membrane.

8 Hydrogen ion flow causes ATP synthases to attach phosphate to ADP, so ATP forms in the stroma.

Figure 6.8 Animated Noncyclic pathway of photosynthesis. Electrons that travel through two different electron transfer chains end up in NADPH, which delivers them to sugar-building reactions in the stroma. The cyclic pathway (not shown) uses a third type of electron transfer chain.

electrons are ejected from its special pair of chlorophylls **5**. These electrons enter a different electron transfer chain. At the end of this second chain, the coenzyme $NADP^+$ accepts the electrons along with H^+, so NADPH forms **6**:

$$NADP^+ + 2e^- + H^+ \longrightarrow NADPH$$

ATP and NADPH continue to form as long as electrons continue to flow through transfer chains in the thylakoid membrane, and electrons flow through the chains as long as water, $NADP^+$, and light are plentiful. The flow of electrons slows—and so does ATP and NADPH production—at night, or when water or $NADP^+$ is scarce.

The Cyclic Pathway

At high oxygen levels, or when NADPH accumulates in the stroma, the noncyclic pathway backs up and stalls. Even when the noncyclic pathway is not running, cells continue producing ATP by photosynthesis with the cyclic pathway (**Figure 6.7B**). This pathway involves photosystem I and an electron transfer chain that cycles electrons back to it. The chain that acts in the cyclic pathway uses electron energy to move hydrogen

ions into the thylakoid compartment. The resulting hydrogen ion gradient drives ATP formation, just as it does in the noncyclic pathway. However, NADPH does not form, because electrons at the end of this chain are accepted by a photosystem I, not $NADP^+$. Oxygen (O_2) does not form either, because photosystem I does not rely on photolysis to resupply itself with electrons.

electron transfer phosphorylation Process in which electron flow through electron transfer chains sets up a hydrogen ion gradient that drives ATP formation. Also called chemiosmosis.
photolysis Process by which light energy breaks down a molecule.
photosystem Cluster of pigments and proteins that converts light energy to chemical energy in photosynthesis.

Take-Home Message

What happens during the light-dependent reactions of photosynthesis?

» In the light-dependent reactions, chlorophylls and other pigments in the thylakoid membrane transfer the energy of light to photosystems.

» Absorbing energy causes photosystems to eject electrons that enter electron transfer chains in the membrane. The flow of electrons through the transfer chains sets up hydrogen ion gradients that drive ATP formation.

» In the noncyclic pathway, oxygen is released and electrons end up in NADPH.

» A cyclic pathway involving only photosystem I allows the cell to continue making ATP even when the noncyclic pathway is not running. NADPH does not form, and oxygen is not released.

6.6 Energy Flow in Photosynthesis

- Energy flow in the light-dependent reactions is an example of how organisms harvest energy from their environment.
- Links to Energy in metabolism 5.3, Redox reactions 5.5

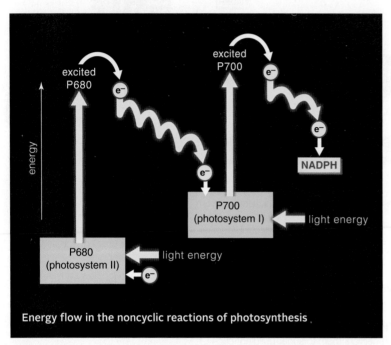

Energy flow in the noncyclic reactions of photosynthesis

A The noncyclic pathway involves a one-way flow of electrons from water, to photosystem II, to photosystem I, to NADPH. As long as electrons continue to flow through the two electron transfer chains, H+ continues to be carried across the thylakoid membrane, and ATP and NADPH keep forming. Light provides the energy boosts that keep the pathway going.

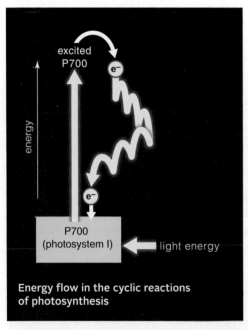

Energy flow in the cyclic reactions of photosynthesis

B In the cyclic pathway, electrons ejected from photosystem I are returned to it. As long as electrons continue to pass through its electron transfer chain, H+ continues to be carried across the thylakoid membrane, and ATP continues to form. Light provides the energy boost that keeps the cycle going.

Figure 6.9 Energy flow in the light-dependent reactions of photosynthesis. The P700 in photosystem I absorbs photons of a 700-nanometer wavelength. The P680 of photosystem II absorbs photons at 680 nanometers. Energy inputs boost P700 and P680 to an excited state in which they lose electrons.

A recurring theme in biology is that organisms use energy harvested from the environment to drive cellular processes. Energy flow in the light-dependent reactions of photosynthesis is a classic example of how that happens. **Figure 6.9** compares energy flow in the two pathways of light-dependent reactions.

The simpler cyclic pathway evolved first, and still operates in nearly all photosynthesizers. Later, the photosynthetic machinery in some organisms evolved so that photosystem II became part of it. That modification was the beginning of a combined sequence of reactions that pulls electrons away from water molecules, with the release of hydrogen ions and oxygen. Photosystem II is the only biological system strong enough to oxidize water.

In the noncyclic reactions, electrons that leave photosystem II do not return to it. Instead, they end up in NADPH, a powerful reducing agent (electron donor). In the cyclic reactions, electrons lost from photosystem I are cycled back to it. No NADPH forms, and no oxygen is released. In both the cyclic and noncyclic reactions, molecules in electron transfer chains use electron energy to shuttle H+ across the thylakoid membrane. Hydrogen ions accumulate in the thylakoid compartment, forming a gradient that powers ATP synthesis.

Today, the plasma membrane of different species of photosynthetic bacteria incorporates either type I or type II photosystems. Cyanobacteria use both types, as do plants and all photosynthetic protists. Which of the two pathways predominates at any given time depends on the organism's immediate metabolic demands for ATP and NADPH.

Having the alternate pathways is efficient, because cells can direct energy to producing NADPH and ATP or to producing ATP alone. NADPH accumulates when it is not being used, such as when sugar production declines during cold winters. The excess NADPH backs up the noncyclic pathway, so the cyclic pathway predominates. The cell still makes ATP, but not NADPH. When sugar production is in high gear, NADPH is being used quickly. It does not accumulate, and the noncyclic pathway is the predominant one.

Take-Home Message

How does energy flow during the light-dependent reactions of photosynthesis?

» Light provides energy inputs that keep electrons moving through electron transfer chains.

» Energy lost by electrons as they move through the chains sets up a hydrogen ion gradient that drives the synthesis of ATP alone, or ATP and NADPH.

6.7 Light-Independent Reactions

■ The chloroplast is a sugar factory operated by enzymes of the Calvin–Benson cycle. The cyclic, light-independent reactions are the "synthesis" part of photosynthesis.
■ Links to Carbohydrates 3.4, Phosphorylation 5.6

The enzyme-mediated reactions of the **Calvin–Benson cycle** build sugars in the stroma of chloroplasts (**Figure 6.10**). These reactions are light-independent because light energy does not power them. Instead, they run on ATP and NADPH that formed in the light-dependent reactions.

Light-independent reactions use carbon atoms from CO_2 to make sugars. Extracting carbon atoms from an inorganic source and incorporating them into an organic molecule is a process called **carbon fixation**. In most plants, photosynthetic protists, and some bacteria, the enzyme **rubisco** fixes carbon by attaching CO_2 to five-carbon RuBP (ribulose bisphosphate) ❶.

The six-carbon intermediate that forms by this reaction is unstable, so it splits right away into two three-carbon molecules of PGA (phosphoglycerate). Each PGA receives a phosphate group from ATP, and hydrogen and electrons from NADPH. Thus, ATP energy and the reducing power of NADPH convert each molecule of PGA into a molecule of PGAL (phosphoglyceraldehyde), a phosphorylated sugar ❷.

In later reactions, two or more of the three-carbon PGAL molecules can be combined and rearranged to form larger carbohydrates. Glucose, remember, has six carbon atoms (*left*). To make one glucose molecule, six CO_2 must be attached to six RuBP molecules, so twelve PGAL form. Two PGAL combine to form one glucose molecule ❸. The ten remaining PGAL regenerate the starting compound of the cycle, RuBP ❹.

Plants can use the glucose they make in the light-independent reactions as building blocks for other organic molecules, or they can break it down to access the energy held in its bonds. However, most of the glucose is converted at once to sucrose or starch by other pathways that conclude the light-independent reactions. Excess glucose is stored in the form of starch grains inside the stroma of chloroplasts. When sugars

Calvin–Benson cycle Light-independent reactions of photosynthesis; cyclic carbon-fixing pathway that forms sugars from CO_2.
carbon fixation Process by which carbon from an inorganic source such as carbon dioxide gets incorporated into an organic molecule.
rubisco Ribulose bisphosphate carboxylase. Carbon-fixing enzyme of the Calvin–Benson cycle.

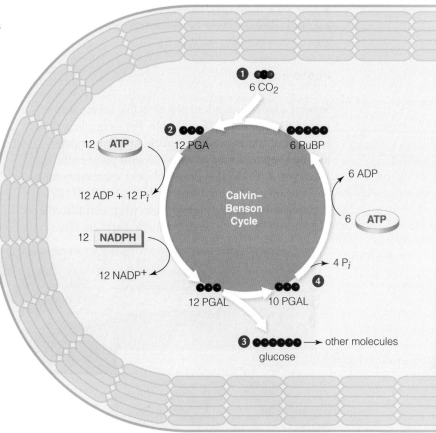

Figure 6.10 Animated Light-independent reactions of photosynthesis. The sketch shows a cross-section of a chloroplast with the light-independent reactions cycling in the stroma.

The steps shown are a summary of six cycles of the Calvin–Benson reactions. Black balls signify carbon atoms. Appendix V details the reaction steps.

❶ Six CO_2 diffuse into a photosynthetic cell, and then into a chloroplast. Rubisco attaches each to a RuBP molecule. The resulting intermediates split, so twelve molecules of PGA form.

❷ Each PGA molecule gets a phosphate group from ATP, plus hydrogen and electrons from NADPH. Twelve PGAL form.

❸ Two PGAL combine to form one glucose molecule.

❹ The remaining ten PGAL receive phosphate groups from ATP. The transfer primes them for endergonic reactions that regenerate the 6 RuBP.

are needed in other parts of the plant, the starch is broken down to sugar monomers and exported from the cell.

Take-Home Message

What happens during the light-independent reactions of photosynthesis?

» The light-independent reactions of photosynthesis run on the bond energy of ATP and the energy of electrons donated by NADPH. Both molecules formed in the light-dependent reactions.

» Collectively called the Calvin–Benson cycle, these carbon-fixing reaction use hydrogen (from NADPH), and carbon and oxygen (from CO_2) to build sugars.

6.8 Adaptations: Different Carbon-Fixing Pathways

- Environments differ, and so do details of photosynthesis.
- Links to Central vacuole 4.7, Surface specializations 4.11, Controls over metabolic reactions 5.5

Photorespiration

Several adaptations allow plants to live where water is scarce or sporadically available. For example, a thin, waterproof coating called a cuticle prevents water loss by evaporation from aboveground plant parts. However, a cuticle also prevents gases from entering and exiting a plant by diffusing through cells at the surfaces of leaves and stems. Gases play a critical role in photosynthesis, so photosynthetic parts are often studded with tiny, closable gaps called **stomata** (singular, stoma). When stomata are open, carbon dioxide for the light-independent reactions can diffuse from air into the plant's photosynthetic tissues, and oxygen produced by the light-dependent reactions can diffuse from photosynthetic cells into the air.

Plants that use only the Calvin–Benson cycle to fix carbon are called **C3 plants**, because three-carbon PGA is the first stable intermediate to form in the light-independent reactions. C3 plants typically conserve water on dry days by closing their stomata. However, when stomata are closed, oxygen produced by the light-dependent reactions cannot escape from the plant. Oxygen that accumulates in photosynthetic tissues lim-

A *Hordeum vulgare* (barley) and other C3 plants have two kinds of mesophyll cells: palisade and spongy. The light-dependent and light-independent reactions occur in both cell types.

A In *Eleusine coracana* (a type of millet) and other C4 plants, carbon is fixed the first time in mesophyll cells, which are near air spaces in the leaf. Carbon fixation occurs for the second time in bundle-sheath cells, which ring the leaf veins.

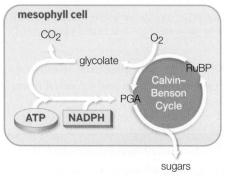

B On dry days, stomata close and oxygen accumulates inside leaves. The excess causes rubisco to attach oxygen instead of carbon to RuBP. This is photorespiration, and it makes sugar production inefficient in C3 plants.

Figure 6.11 Animated Light-independent reactions in C3 plants.

B C4 plants. Oxygen also builds up inside leaves when stomata close during photosynthesis.

An additional pathway in these plants keeps the CO_2 concentration high enough in bundle-sheath cells to prevent photorespiration.

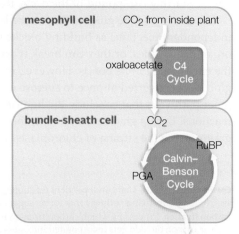

Figure 6.12 Animated Light-independent reactions in C4 plants.

7.1 Mighty Mitochondria

Leah Chalcraft (**Figure 7.1**) started to lose her sense of balance and coordination when she was five years old; six years later she was in a wheelchair. Her little brother Joshua could not walk by the time he was eleven, and became blind soon afterward. Both had heart problems; both had spinal fusion surgery.

Leah and Josh's symptoms are those of a genetic disorder called Friedreich's ataxia, the origins of which began billions of years ago. Before the noncyclic pathway of photosynthesis evolved, molecular oxygen had been a very small component of Earth's early atmosphere. Afterward, the new abundance of atmospheric oxygen exerted tremendous selection pressure on organisms that existed at the time. Oxygen reacts with metals, including enzyme cofactors, and free radicals form during those reactions (Section 2.3). The ancient cells had no way to detoxify the radicals, so most of them quickly died out. Only a few types persisted in deep water, muddy sediments, and other **anaerobic** (oxygen-free) habitats. New metabolic pathways that evolved in the survivors detoxified oxygen radicals. Cells with such pathways were the first **aerobic** organisms—they could live in the presence of oxygen.

One of the new pathways, aerobic respiration, put the reactive properties of oxygen to use. In modern eukaryotic cells, most of the aerobic respiration pathway takes place inside mitochondria (Section 4.9). Like chloroplasts, mitochondria have an internal folded membrane system that allows them to make ATP very efficiently. Electron transfer chains in this membrane set up hydrogen ion gradients that power ATP synthesis. At the end of these chains, electrons are transferred to oxygen molecules.

ATP participates in almost all cellular reactions, so a cell benefits from making a lot of it. However, aerobic respiration is a risky business. When an oxygen molecule (O_2) accepts electrons from an electron transfer chain, it dissociates into oxygen atoms. Most of the atoms immediately combine with hydrogen ions and end up in water molecules. Occasionally, however, an oxygen atom escapes this final reaction. The atom has an unpaired electron, so it is a free radical.

Mitochondria cannot detoxify free radicals, so they rely on antioxidant enzymes and vitamins in the cell's cytoplasm to do it for them. The system works well, at least most of the time. However, a genetic disorder or an unfortunate encounter with a toxin or pathogen can result in a missing antioxidant, or a defective component of the mitochondrial electron transfer chain. In either case, free radicals accumulate and destroy first the function of mitochondria, then the cell.

Hundreds of disorders are associated with free radical damage in mitochondria, and more are being discovered all the time. Nerve and brain cells, which require a lot of ATP, are particularly affected. Symptoms can range from mild to major neurological deficits, blindness, strokes, seizures, and disabling muscle weakness. In Friedreich's ataxia, a protein called frataxin is not properly transported into mitochondria. This protein regulates the synthesis of iron-containing proteins in mitochondrial electron transfer chains. Iron atoms that were supposed to be incorporated into the proteins accumulate. When too much iron accumulates in mitochondria, too many free radicals form, and these destroy the molecules of life faster than they can be repaired or replaced. Eventually, the mitochondria stop working, and the cell dies. The resulting symptoms cause many of those affected, including Josh Chalcraft, to die as young adults.

aerobic Involving or occurring in the presence of oxygen.
anaerobic Occurring in the absence of oxygen.

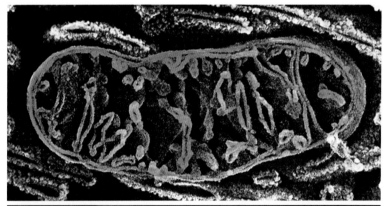

Figure 7.1 Sister, brother, and mitochondria. *Top*, a mitochondrion's folded internal membrane is the source of its function. *Bottom*, Leah and Joshua Chalcraft. Both developed Friedreich's ataxia, a genetic disorder in which excessive free radical formation in mitochondria causes a progressive loss of motor and sensory function. Assistive equipment has allowed them to go to school and to work in productive jobs. Josh died in 2009, at age 24.

7.2 Overview of Carbohydrate Breakdown Pathways

■ Photoautotrophs use the ATP they produce by photosynthesis to make sugars.

■ Most organisms, including photoautotrophs, make ATP by breaking down sugars and other organic compounds.

■ Links to Energy flow 5.2, Metabolic pathways and redox reactions 5.5, Coenzymes 5.6, Photosynthesis 6.4, Electron transfer phosphorylation 6.5, Reducing agents 6.6

In Chapter 6, you learned how photoautotrophs capture energy from the sun, and store it in the form of carbohydrates. Photoautotrophs and most other organisms use energy stored in carbohydrates to run the diverse reactions that sustain life. However, carbohydrates rarely participate in such reactions, so how do cells harness their energy? In order to use the energy stored in carbohydrates, cells must first transfer it to molecules such as ATP, which does participate in many of the energy-requiring reactions that a cell runs. The energy transfer occurs when cells break the bonds that hold a carbohydrate together. Free energy released as those bonds are broken drives ATP synthesis.

Aerobic Respiration

A few different pathways yield ATP by breaking down carbohydrates, but a typical eukaryotic cell uses **aerobic respiration** most of the time. This equation summarizes the overall pathway of aerobic respiration:

$$C_6H_{12}O_6 \; + \; 6O_2 \longrightarrow 6CO_2 \; + \; 6H_2O$$

glucose oxygen carbon dioxide water

Note that aerobic respiration requires oxygen (a by-product of photosynthesis), and it produces carbon dioxide and water (the same raw materials that photosynthesizers use to make sugars). With this connection, the cycling of carbon, hydrogen, and oxygen through the biosphere comes full circle (**Figure 7.2**).

Aerobic respiration begins with a set of reactions in the cytoplasm (**Figure 7.3A**). These reactions, which are collectively called **glycolysis**, convert one six-carbon molecule of glucose into two molecules of **pyruvate**, an organic compound with a three-carbon backbone.

Aerobic respiration continues with two more stages that occur inside mitochondria. During the second stage, the pyruvate is converted to acetyl–CoA, which enters the Krebs cycle. Carbon dioxide that forms in the second-stage reactions leaves the cell (**Figure 7.3B**).

Electrons and hydrogen ions released by the reactions of the first two stages are picked up by two coenzymes, NAD$^+$ (nicotinamide adenine dinucleotide) and FAD (flavin adenine dinucleotide). When these

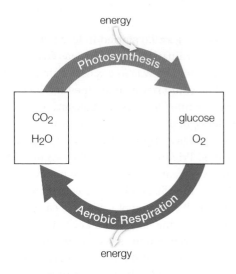

Figure 7.2 The global connection between photosynthesis and aerobic respiration. Note the cycling of materials, and the one-way flow of energy (compare **Figure 5.5**).

two coenzymes are carrying electrons and hydrogen, they are reduced (Section 5.6), and we refer to them as NADH and FADH$_2$.

Few ATP form during the first two stages of aerobic respiration. The big payoff occurs in the third stage, after coenzymes give up electrons and hydrogen to electron transfer chains. These chains power electron transfer phosphorylation, the process in which hydrogen ion gradients drive ATP formation (**Figure 7.3C**). Oxygen in mitochondria accepts electrons and H$^+$ at the end of the transfer chains, so water forms.

Anaerobic Fermentation Pathways

Most types of eukaryotic cells use aerobic respiration exclusively, or they use it most of the time. Many bacteria, archaea, and protists harvest energy from carbohydrates by anaerobic pathways of **fermentation**. Fermentation and aerobic respiration begin with the same reactions—glycolysis—in the cytoplasm. After glycolysis, the pathways diverge (**Figure 7.4**). Reactions that conclude fermentation pathways occur in cytoplasm, and do not include electron transfer chains. Electrons are accepted by organic molecules (not

aerobic respiration Oxygen-requiring metabolic pathway that breaks down carbohydrates to produce ATP.
fermentation Metabolic pathway that breaks down carbohydrates to produce ATP; does not require oxygen.
glycolysis Set of reactions in which glucose or another sugar is broken down to two pyruvate for a net yield of two ATP.
pyruvate Three-carbon end product of glycolysis.

Aerobic Respiration

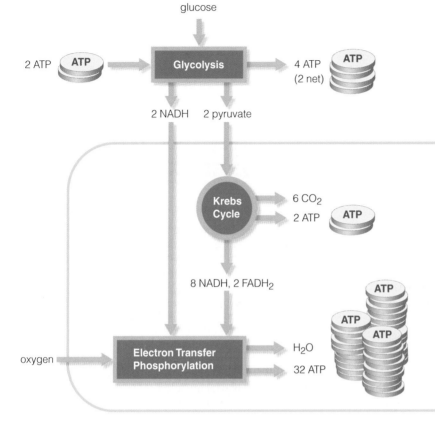

In the Cytoplasm

A The first stage, glycolysis, occurs in the cell's cytoplasm. Enzymes convert a glucose molecule to 2 pyruvate for a net yield of 2 ATP. 2 NAD^+ combine with electrons and hydrogen ions during the reactions, so 2 NADH also form.

In the Mitochondrion

B The second stage occurs in mitochondria. The 2 pyruvate are converted to a molecule that enters the Krebs cycle. CO_2 forms and leaves the cell. 2 ATP, 8 NADH, and 2 $FADH_2$ form during the reactions.

C The third and final stage, electron transfer phosphorylation, occurs inside mitochondria. 10 NADH and 2 $FADH_2$ donate electrons and hydrogen ions to electron transfer chains. Electron flow through the chains sets up hydrogen ion gradients that drive the formation of ATP. Oxygen is the final electron acceptor at the end of the chains.

Figure 7.3 Animated Overview of aerobic respiration. The reactions start in the cytoplasm and end in mitochondria.

Figure It Out: What is aerobic respiration's typical net yield of ATP?

Answer: 38 − 2 = 36 ATP per glucose

oxygen) at the end of the reactions, so fermentation pathways do not require oxygen to proceed.

Fermentation provides enough energy to sustain many anaerobic species, including bacteria, fungi, and single-celled protists that inhabit sea sediments, ani-

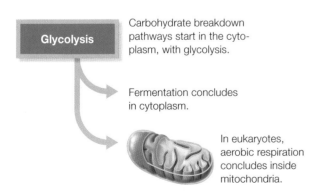

Carbohydrate breakdown pathways start in the cytoplasm, with glycolysis.

Fermentation concludes in cytoplasm.

In eukaryotes, aerobic respiration concludes inside mitochondria.

Figure 7.4 Animated Where the different pathways of carbohydrate breakdown start and end.

mal guts, improperly canned food, sewage treatment ponds, or deep mud. Some of these organisms, including the bacteria that cause botulism, cannot tolerate aerobic conditions, and will die when exposed to oxygen. Fermentation also helps cells of aerobic species produce ATP under anaerobic conditions, but aerobic respiration is a much more efficient way of harvesting energy from carbohydrates. You and other multicelled organisms could not live without its higher yield.

Take-Home Message

How do cells access the chemical energy in carbohydrates?

» Most cells convert the chemical energy of carbohydrates to the chemical energy of ATP by aerobic respiration or fermentation. Aerobic respiration and fermentation pathways start in cytoplasm, with glycolysis.

» Fermentation is anaerobic and ends in the cytoplasm.

» Aerobic respiration requires oxygen. In eukaryotes, it ends in mitochondria.

7.3 Glycolysis—Glucose Breakdown Starts

■ The reactions of glycolysis convert one molecule of glucose to two molecules of pyruvate for a net yield of two ATP.

■ An energy investment of ATP is required to start glycolysis.

■ Links to Hydrolysis 3.3, Glucose 3.4, Endergonic reactions and phosphorylation 5.3, Gradients 5.8, Glucose transporter 5.9, Calvin–Benson cycle 6.7

Glycolysis is a series of reactions that begins carbohydrate breakdown pathways. The reactions, which occur in the cytoplasm, convert one molecule of glucose to two molecules of pyruvate:

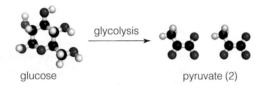

glucose → pyruvate (2)

The word glycolysis (from the Greek words *glyk–*, sweet; and *–lysis*, loosening) refers to the release of chemical energy from sugars. Various sugars can enter glycolysis, but for clarity we focus here on glucose.

Glycolysis begins when a molecule of glucose enters a cell through a glucose transporter, a passive transport protein you encountered in Section 5.9. The cell invests two ATP in the endergonic reactions that begin the pathway (**Figure 7.5**). In the first reaction, a phosphate group is transferred from ATP to the glucose, thus forming glucose-6-phosphate ❶. A model of hexokinase, the enzyme that catalyzes this reaction, is pictured in Section 5.4.

Unlike glucose, glucose-6-phosphate does not pass through glucose transporters in the plasma membrane, so it is trapped inside the cell. Almost all of the glucose that enters a cell is immediately converted to glucose-6-phosphate. This phosphorylation keeps the glucose concentration in the cytoplasm lower than it is in the fluid outside of the cell. By maintaining this concentration gradient across the plasma membrane, the cell favors uptake of even more glucose.

Glycolysis continues as glucose-6-phosphate accepts a phosphate group from another ATP, then splits in two ❷, forming two PGAL (phosphoglyceraldehyde, a phosphorylated sugar that also forms during the Calvin–Benson cycle). A second phosphate group is attached to each PGAL, so two PGA (phosphoglycerate) form ❸. During the reaction, two electrons and a hydrogen ion are transferred from each PGAL to NAD+, so two NADH form.

Next, a phosphate group is transferred from each PGA to ADP, so two ATP form ❹. Two more ATP form when a phosphate group is transferred from another pair of intermediates to two ADP ❺. This

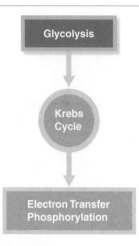

Figure 7.5 Animated Glycolysis.

This first stage of carbohydrate breakdown starts and ends in the cytoplasm of all cells. *Opposite*, for clarity, we track only the six carbon atoms (*black* balls) that enter the reactions as part of glucose. Appendix V has more details for interested students.

Cells invest two ATP to start glycolysis, so the net energy yield from one glucose molecule is two ATP. Two NADH also form, and two pyruvate molecules are the end products.

and any other reaction that transfers a phosphate group directly from a substrate to ADP is called a **substrate-level phosphorylation**.

Glycolysis ends with the formation of two three-carbon pyruvate molecules. These products may now enter the second-stage reactions of either aerobic respiration or fermentation. Remember, two ATP were invested to initiate the reactions of glycolysis. A total of four ATP form, so the net yield is two ATP per molecule of glucose that enters glycolysis ❻. Two NAD+ also pick up hydrogen ions and electrons, thereby becoming reduced to NADH:

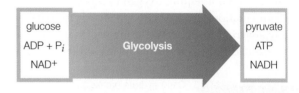

glucose
ADP + P$_i$
NAD+
→ Glycolysis →
pyruvate
ATP
NADH

substrate-level phosphorylation A reaction that transfers a phosphate group from a substrate directly to ADP, thus forming ATP.

Take-Home Message

What is glycolysis?

» Glycolysis is the first stage of carbohydrate breakdown in both aerobic respiration and fermentation.

» The reactions of glycolysis occur in the cytoplasm.

» Glycolysis converts one molecule of glucose to two molecules of pyruvate, with a net energy yield of two ATP. Two NADH also form.

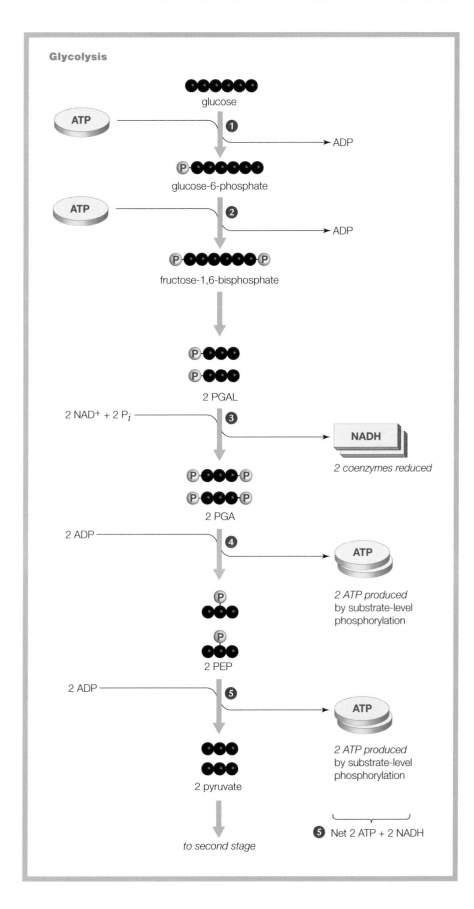

Glycolysis

glucose

ATP

① → ADP

glucose-6-phosphate

ATP

② → ADP

fructose-1,6-bisphosphate

2 PGAL

2 NAD⁺ + 2 P$_i$ — ③

NADH

2 coenzymes reduced

2 PGA

2 ADP — ④

ATP

2 ATP produced by substrate-level phosphorylation

2 PEP

2 ADP — ⑤

ATP

2 ATP produced by substrate-level phosphorylation

2 pyruvate

⑤ Net 2 ATP + 2 NADH

to second stage

ATP-Requiring Steps

① An enzyme (hexokinase) transfers a phosphate group from ATP to glucose, forming glucose-6-phosphate.

② A phosphate group from a second ATP is transferred to the glucose-6-phosphate. The resulting molecule is unstable, and it splits into two three-carbon molecules. The molecules are interconvertible, so we will call them both PGAL (phosphoglyceraldehyde).

Two ATP have now been invested in the reactions.

ATP-Generating Steps

③ Enzymes attach a phosphate to the two PGAL, and transfer two electrons and a hydrogen ion from each PGAL to NAD⁺. Two PGA (phosphoglycerate) and two NADH are the result.

④ Enzymes transfer a phosphate group from each PGA to ADP. Thus, two ATP have formed by substrate-level phosphorylation.

The original energy investment of two ATP has now been recovered.

⑤ Enzymes transfer a phosphate group from each of two intermediates to ADP. Two more ATP have formed by substrate-level phosphorylation.

Two molecules of pyruvate form at this last reaction step.

⑥ Summing up, glycolysis yields two NADH, two ATP (net), and two pyruvate for each glucose molecule.

Depending on the type of cell and environmental conditions, the pyruvate may enter the second stage of aerobic respiration or it may be used in other ways, such as in fermentation.

7.4 Second Stage of Aerobic Respiration

■ The second stage of aerobic respiration completes the breakdown of glucose that began in glycolysis.
■ Links to Acetyl groups 3.3, Mitochondria 4.9, Cyclic pathways 5.5

The second stage of aerobic respiration occurs in mitochondria (**Figure 7.6**). It includes two sets of reactions, acetyl–CoA formation and the Krebs cycle, that break down pyruvate, the product of glycolysis. All of the carbon atoms that were once part of glucose end up in CO_2, which departs the cell. Only two ATP form, but the reactions reduce many coenzymes. What is so important about reduced coenzymes? A molecule becomes reduced when it receives electrons (Section 5.5), and electrons carry energy that can be used to drive endergonic reactions. In this case, the electrons picked up by coenzymes during the first two stages of aerobic respiration carry energy that drives the reactions of the third stage.

The second stage begins when the two pyruvate molecules formed by glycolysis enter a mitochondrion. Pyruvate is transported across the mitochondrion's inner membrane and into the inner compartment, which is called the matrix (**Figure 7.7**). In the first reaction, an enzyme splits each molecule of pyruvate into a molecule of CO_2 and a two-carbon acetyl group ❶. The CO_2 diffuses out of the cell, and the acetyl group combines with a molecule called coenzyme A (abbreviated CoA). The product of this reaction is acetyl–CoA. Electrons and hydrogen ions released by the reaction combine with NAD^+, so NADH also forms.

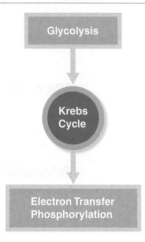

Figure 7.7 Animated Aerobic respiration's second stage: formation of acetyl–CoA and the Krebs cycle. The reactions occur in the mitochondrion's matrix.

Opposite, it takes two cycles of Krebs reactions to break down two pyruvate molecules. After two cycles, all six carbons that entered glycolysis in one glucose molecule have left the cell, in six CO_2. Two ATP, eight NADH, and two $FADH_2$ form during the two cycles. See Appendix V for details of the reactions.

The Krebs Cycle

The **Krebs cycle** breaks down acetyl–CoA to CO_2. Remember from Section 5.5 that a cyclic pathway is not a physical object, such as a wheel. It is called a cycle because the last reaction in the pathway regenerates the substrate of the first. In the Krebs cycle, a substrate of the first reaction—and a product of the last—is four-carbon oxaloacetate.

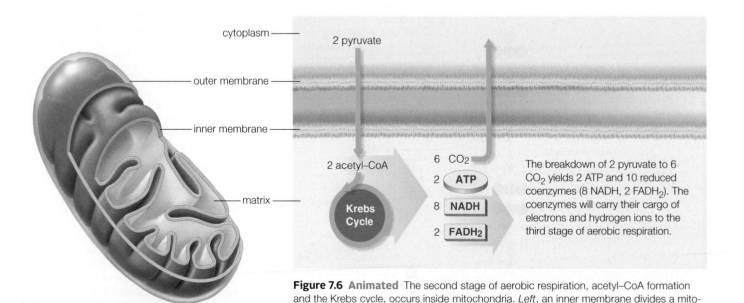

The breakdown of 2 pyruvate to 6 CO_2 yields 2 ATP and 10 reduced coenzymes (8 NADH, 2 $FADH_2$). The coenzymes will carry their cargo of electrons and hydrogen ions to the third stage of aerobic respiration.

Figure 7.6 Animated The second stage of aerobic respiration, acetyl–CoA formation and the Krebs cycle, occurs inside mitochondria. *Left*, an inner membrane divides a mitochondrion's interior into two fluid-filled compartments. *Right*, the second stage of aerobic respiration takes place in the mitochondrion's innermost compartment, or matrix.

Acetyl–CoA Formation and the Krebs Cycle

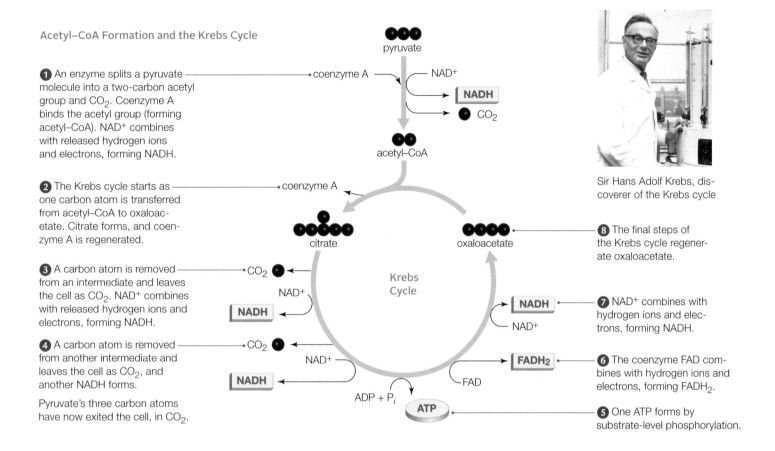

① An enzyme splits a pyruvate molecule into a two-carbon acetyl group and CO_2. Coenzyme A binds the acetyl group (forming acetyl–CoA). NAD^+ combines with released hydrogen ions and electrons, forming NADH.

② The Krebs cycle starts as one carbon atom is transferred from acetyl–CoA to oxaloacetate. Citrate forms, and coenzyme A is regenerated.

③ A carbon atom is removed from an intermediate and leaves the cell as CO_2. NAD^+ combines with released hydrogen ions and electrons, forming NADH.

④ A carbon atom is removed from another intermediate and leaves the cell as CO_2, and another NADH forms.

Pyruvate's three carbon atoms have now exited the cell, in CO_2.

pyruvate

coenzyme A

NAD^+

NADH

CO_2

acetyl–CoA

coenzyme A

citrate

oxaloacetate

Krebs Cycle

NAD^+

NADH

CO_2

NAD^+

NADH

CO_2

NADH

NAD^+

NAD^+

FADH$_2$

FAD

ADP + P$_i$

ATP

Sir Hans Adolf Krebs, discoverer of the Krebs cycle

⑧ The final steps of the Krebs cycle regenerate oxaloacetate.

⑦ NAD^+ combines with hydrogen ions and electrons, forming NADH.

⑥ The coenzyme FAD combines with hydrogen ions and electrons, forming FADH$_2$.

⑤ One ATP forms by substrate-level phosphorylation.

During each cycle of Krebs reactions, two carbon atoms of acetyl–CoA are transferred to four-carbon oxaloacetate, forming citrate, the ionized form of citric acid **②**. The Krebs cycle is also called the citric acid cycle after this first intermediate. In later reactions, two CO_2 form and depart the cell. Two NAD^+ are reduced when they accept hydrogen ions and electrons, so two NADH form **③** and **④**. ATP forms by substrate-level phosphorylation **⑤**, and FAD **⑥** and another NAD^+ **⑦** are reduced. The final steps of the pathway regenerate oxaloacetate **⑧**.

Remember, glycolysis converted one glucose molecule to two pyruvate, and these were converted to two acetyl–CoA when they entered the matrix of a mitochondrion. There, the Krebs cycle reactions convert the two molecules of acetyl–CoA to six CO_2. At this point in aerobic respiration, the carbon backbone of one glucose molecule has been broken down completely: Six carbon atoms have left the cell, in six CO_2. Two ATP formed, which adds to the small net yield of glycolysis. However, six NAD^+ were reduced to six NADH, and two FAD were reduced to two FADH$_2$.

In total, two ATP form and ten coenzymes (eight NAD^+ and two FAD) are reduced during aerobic respiration's second stage. Add the two NAD^+ reduced in

glycolysis, and the breakdown of a glucose molecule has a big potential payoff. Twelve reduced coenzymes will deliver electrons (and the energy they carry) to the third stage reactions of aerobic respiration:

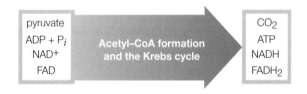

| pyruvate
ADP + P$_i$
NAD^+
FAD | **Acetyl–CoA formation
and the Krebs cycle** | CO_2
ATP
NADH
FADH$_2$ |

Krebs cycle Cyclic pathway that, along with acetyl–CoA formation, breaks down pyruvate to carbon dioxide during aerobic respiration.

Take-Home Message

What happens during the second stage of aerobic respiration?

» The second stage of aerobic respiration, acetyl–CoA formation and the Krebs cycle, occurs in the inner compartment (matrix) of mitochondria.

» The pyruvate that formed in glycolysis is converted to acetyl–CoA and CO_2. Krebs cycle reactions break down the acetyl–CoA to CO_2.

» For two pyruvate molecules broken down in the second-stage reactions, two ATP form, and ten coenzymes (eight NAD^+; two FAD) are reduced.

7.5 Aerobic Respiration's Big Energy Payoff

■ Many ATP are formed during the third and final stage of aerobic respiration.

■ Links to Thermodynamics 5.2, Electron transfer chains 5.5, Membrane proteins 5.7, Selective permeability of cell membranes 5.8, Electron transfer phosphorylation 6.5

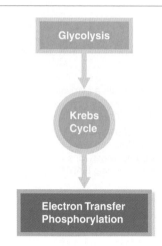

Figure 7.8 The third and final stage of aerobic respiration, electron transfer phosphorylation, occurs at the inner mitochondrial membrane.

1 NADH and FADH₂ deliver electrons to electron transfer chains in the inner mitochondrial membrane.

2 Electron flow through the chains causes hydrogen ions (H⁺) to be pumped from the matrix to the intermembrane space.

3 The activity of the electron transfer chains causes a hydrogen ion gradient to form across the inner mitochondrial membrane.

4 Hydrogen ion flow back to the matrix through ATP synthases drives the formation of ATP from ADP and phosphate (P_i).

5 Oxygen combines with electrons and hydrogen ions at the end of the electron transfer chains, so water forms.

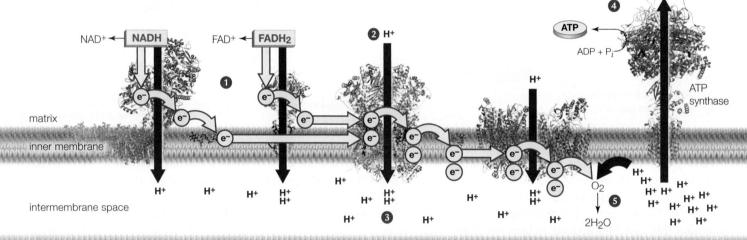

The third stage of aerobic respiration, electron transfer phosphorylation, occurs at the inner mitochondrial membrane (**Figure 7.8**). The reactions begin with the coenzymes NADH and FADH₂, which became reduced in the first two stages of aerobic respiration. These coenzymes donate their cargo of electrons and hydrogen ions to electron transfer chains embedded in the inner mitochondrial membrane **1**. As the electrons pass through the chains, they give up energy little by little (Section 5.5). Some molecules of the transfer chains harness that energy to actively transport hydrogen ions across the inner membrane, from the matrix to the intermembrane space **2**. The ions accumulate in the intermembrane space, so a hydrogen ion gradient forms across the inner mitochondrial membrane **3**.

This gradient attracts hydrogen ions back toward the matrix. However, the ions cannot diffuse through a lipid bilayer on their own (Section 5.8). Hydrogen ions cross the inner mitochondrial membrane only by flowing through ATP synthases embedded in the membrane. The flow of hydrogen ions through ATP synthases causes these proteins to attach phosphate groups to ADP, so ATP forms **4**.

Oxygen accepts electrons at the end of the mitochondrial electron transfer chains **5**. Aerobic respiration, which literally means "taking a breath of air," refers to oxygen as the final electron acceptor in this pathway. When oxygen accepts electrons, it combines with H⁺ to form water, which is one product of the third stage.

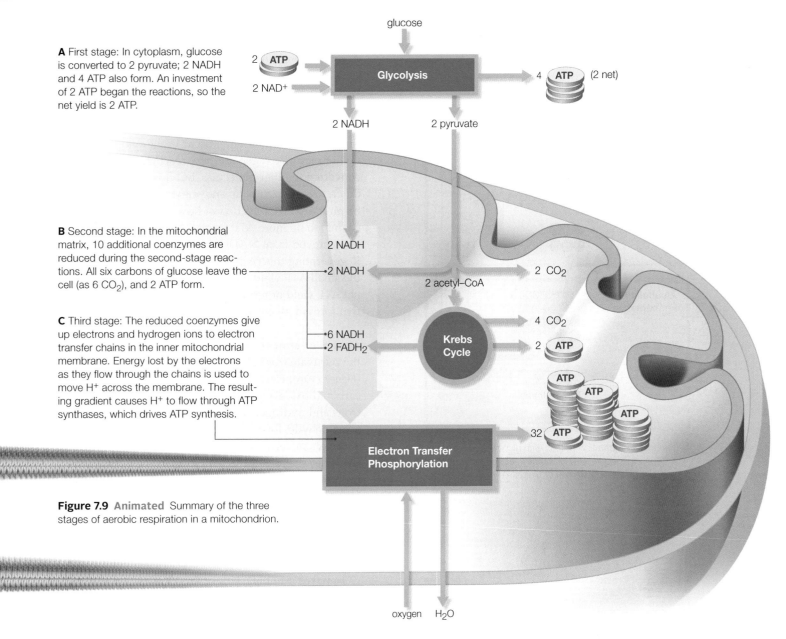

A First stage: In cytoplasm, glucose is converted to 2 pyruvate; 2 NADH and 4 ATP also form. An investment of 2 ATP began the reactions, so the net yield is 2 ATP.

2 ATP

2 NAD⁺

glucose

Glycolysis

4 ATP (2 net)

2 NADH   2 pyruvate

B Second stage: In the mitochondrial matrix, 10 additional coenzymes are reduced during the second-stage reactions. All six carbons of glucose leave the cell (as 6 CO_2), and 2 ATP form.

2 NADH

2 NADH   2 CO_2

2 acetyl–CoA

C Third stage: The reduced coenzymes give up electrons and hydrogen ions to electron transfer chains in the inner mitochondrial membrane. Energy lost by the electrons as they flow through the chains is used to move H⁺ across the membrane. The resulting gradient causes H⁺ to flow through ATP synthases, which drives ATP synthesis.

6 NADH
2 FADH₂

4 CO_2

Krebs Cycle

2 ATP

ATP
ATP
ATP
ATP
32 ATP

Electron Transfer Phosphorylation

Figure 7.9 Animated Summary of the three stages of aerobic respiration in a mitochondrion.

oxygen H_2O

For each glucose molecule that enters aerobic respiration, four ATP form in the first- and second-stage reactions. The twelve coenzymes reduced in these two stages deliver enough H⁺ and electrons to fuel the synthesis of about thirty-two ATP in the third stage. Thus, the breakdown of one glucose molecule typically yields thirty-six ATP (**Figure 7.9**). However, the overall yield varies. Factors such as cell type affect it. For example, the typical yield of aerobic respiration in brain and skeletal muscle cells is thirty-eight ATP, not thirty-six. Remember that some energy dissipates with every transfer (Section 5.2). Even though aerobic respiration is a very efficient way of retrieving energy from carbohydrates, about 60 percent of the energy harvested in this pathway disperses as metabolic heat.

Take-Home Message

What happens during the third stage of aerobic respiration?

» In aerobic respiration's third stage, electron transfer phosphorylation, energy released by electrons flowing through electron transfer chains is ultimately captured in the attachment of phosphate to ADP.

» The third stage reactions begin when coenzymes that were reduced in the first and second stages deliver electrons and hydrogen ions to electron transfer chains in the inner mitochondrial membrane.

» Energy released by electrons as they pass through electron transfer chains is used to pump H⁺ from the mitochondrial matrix to the intermembrane space. The H⁺ gradient that forms across the inner mitochondrial membrane drives the flow of hydrogen ions through ATP synthases, which results in ATP formation.

» About thirty-two ATP form during the third stage reactions, so a typical net yield of all three stages of aerobic respiration is thirty-six ATP per glucose.

7.6 Fermentation

■ Fermentation pathways break down carbohydrates without using oxygen. The final steps in these pathways regenerate NAD+ but do not produce ATP.

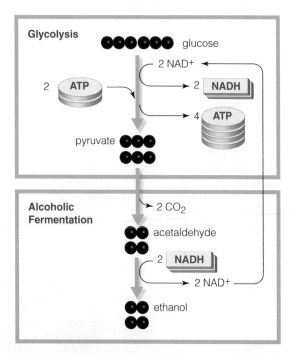

A Alcoholic fermentation begins with glycolysis, and the final steps regenerate NAD+. The net yield of these reactions is two ATP per molecule of glucose (from glycolysis).

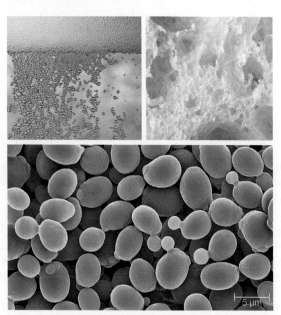

B One product of alcoholic fermentation in *Saccharomyces* cells (ethanol) makes beer alcoholic; another (CO_2) makes it bubbly. Holes in bread are pockets where CO_2 released by fermenting *Saccharomyces* cells accumulated in the dough. The micrograph shows budding *Saccharomyces* cells.

Figure 7.10 Animated Alcoholic fermentation.

Two Fermentation Pathways

Aerobic respiration and fermentation begin with the same set of glycolysis reactions in cytoplasm. Again, two pyruvate, two NADH, and two ATP form during the reactions of glycolysis. However, after that, fermentation and aerobic respiration pathways differ. The final steps of fermentation occur in the cytoplasm and do not require oxygen. In these reactions, pyruvate is converted to other molecules, but it is not fully broken down to carbon dioxide and water as occurs in aerobic respiration. Electrons do not flow through transfer chains, so no more ATP forms. However, electrons are removed from NADH, so NAD+ is regenerated. Regenerating this coenzyme allows glycolysis—and the small ATP yield it offers—to continue. Thus, the net ATP yield of fermentation consists of the two ATP that form in glycolysis.

Alcoholic Fermentation In **alcoholic fermentation**, the pyruvate from glycolysis is converted to ethanol (**Figure 7.10A**). First, 3-carbon pyruvate is split into carbon dioxide and 2-carbon acetaldehyde. Then, electrons and hydrogen are transferred from NADH to the acetaldehyde, forming NAD+ and ethanol.

Alcoholic fermentation in a fungus, *Saccharomyces cerevisiae*, sustains these yeast cells as they grow and reproduce. It also helps us produce beer, wine, and bread (**Figure 7.10B**).

Beer brewers typically use germinated, roasted, and crushed barley as a carbohydrate source for *Saccharomyces* fermentation. Ethanol produced by the fermenting yeast cells makes the beer alcoholic, and CO_2 makes it bubbly. Hops plant flowers add flavor and help preserve the finished product. Winemakers start with crushed grapes for *Saccharomyces* fermentation. The yeast cells convert sugars in the grape juice to ethanol.

Bakers take advantage of alcoholic fermentation by *Saccharomyces* cells to make bread from flour, which contains starches and a protein called gluten. When flour is kneaded with water, the gluten polymerizes in long, interconnected strands that make the resulting dough stretchy and resilient. Yeast cells in the dough produce CO_2 as they ferment the starches. The gas accumulates in bubbles that are trapped by the gluten mesh. As the bubbles expand, they cause the dough to rise. The ethanol produced by the fermentation reactions evaporates during baking.

Lactate Fermentation In **lactate fermentation**, the electrons and hydrogen ions carried by NADH are transferred directly to pyruvate. This reaction converts pyruvate to 3-carbon lactate (the ionized form of lactic acid), and also converts NADH to NAD^+ (**Figure 7.11A**).

Some lactate fermenters spoil food, but we use others to preserve it. For instance, *Lactobacillus* bacteria break down lactose in milk by fermentation. We use this bacteria to produce dairy products such as buttermilk, cheese, and yogurt. Yeast species ferment and preserve pickles, corned beef, sauerkraut, and kimchi.

Animal skeletal muscles, which move bones, consist of cells fused as long fibers. The fibers differ in how they make ATP. Red fibers have many mitochondria and produce ATP by aerobic respiration. These fibers sustain prolonged activity such as marathon runs. They are red because they have an abundance of myoglobin, a protein that stores oxygen for aerobic respiration in these fibers (**Figure 7.11B**).

White muscle fibers contain few mitochondria and no myoglobin, so they do not carry out a lot of aerobic respiration. Instead, they make most of their ATP by lactate fermentation. This pathway makes ATP quickly but not for long, so it is useful for quick, strenuous activities such as sprinting (**Figure 7.11C**) or weight lifting. The low ATP yield does not support prolonged activity. That is one reason why chickens cannot fly very far: Their flight muscles consist mostly of white fibers (thus, the "white" breast meat). Chickens fly only in short bursts. More often, a chicken walks or runs. Its leg muscles consist mostly of red muscle fibers, the "dark meat." Most human muscles consist of a mixture of white and red fibers, but the proportions vary among muscles and among individuals. Great sprinters tend to have more white fibers. Great marathon runners tend to have more red fibers. Section 31.5 offers a closer look at skeletal muscle fibers and how they work.

alcoholic fermentation Anaerobic carbohydrate breakdown pathway that produces ATP and ethanol. Begins with glycolysis; end reactions regenerate NAD^+ so glycolysis can continue.
lactate fermentation Anaerobic carbohydrate breakdown pathway that produces ATP and lactate.

Take-Home Message

What is fermentation?

» ATP can form by carbohydrate breakdown in fermentation pathways, which are anaerobic.

» The end product of lactate fermentation is lactate. The end product of alcoholic fermentation is ethanol. Both pathways have a net yield of two ATP per glucose molecule. The ATP forms during glycolysis.

» Fermentation reactions regenerate the coenzyme NAD^+, without which glycolysis (and ATP production) would stop.

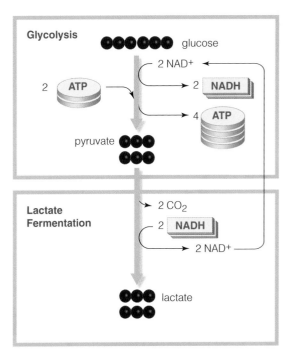

A Lactate fermentation begins with glycolysis, and the final steps regenerate NAD^+. The net yield of these reactions is two ATP per molecule of glucose (from glycolysis).

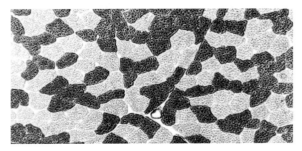

B Lactate fermentation occurs in white muscle fibers, visible in this micrograph of human thigh muscle. The red fibers, which make ATP by aerobic respiration, sustain endurance activities.

C Intense activity such as sprinting quickly depletes oxygen in muscles. Under anaerobic conditions, ATP is produced mainly by lactate fermentation in white muscle fibers. Fermentation does not make enough ATP to sustain this type of activity for long.

Figure 7.11 Animated Lactate fermentation.

7.7 Alternative Energy Sources in Food

- Aerobic respiration can produce ATP from the breakdown of fats and proteins.
- Links to Metabolic reactions 3.3, Carbohydrates 3.4, Lipids 3.5, Proteins 3.6, Redox reactions 5.5

Energy From Dietary Molecules

Glycolysis converts glucose to pyruvate, and the Krebs cycle reactions transfer electrons from pyruvate to coenzymes. In other words, glucose becomes oxidized (it gives up electrons) and coenzymes become reduced (they accept electrons). Oxidizing an organic molecule can break the covalent bonds of its carbon backbone. Aerobic respiration fully oxidizes glucose, completely dismantling it carbon by carbon.

Cells also dismantle other organic molecules by oxidizing them. Complex carbohydrates, fats, and proteins in food can be converted to molecules that enter glycolysis or the Krebs cycle (**Figure 7.12**). As in glucose metabolism, many coenzymes are reduced, and the

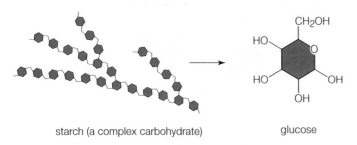

starch (a complex carbohydrate)   glucose

A Complex carbohydrates are broken down to their monosaccharide subunits, which can enter glycolysis **1**.

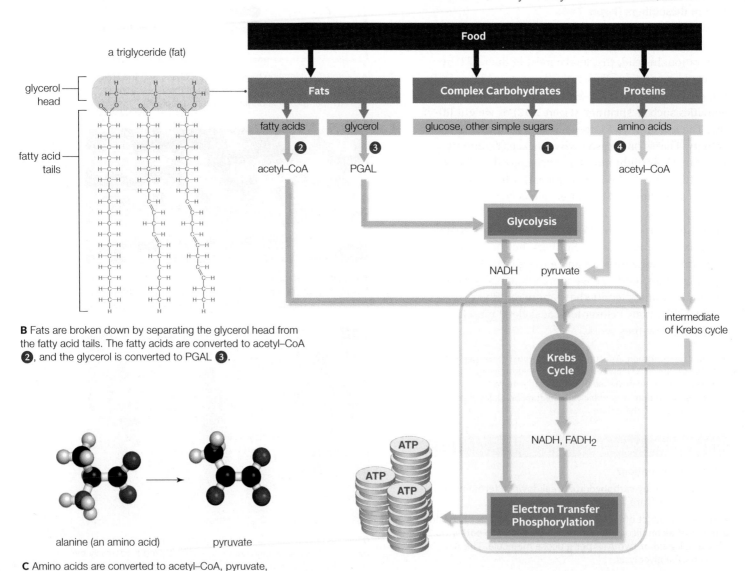

B Fats are broken down by separating the glycerol head from the fatty acid tails. The fatty acids are converted to acetyl–CoA **2**, and the glycerol is converted to PGAL **3**.

alanine (an amino acid)   pyruvate

C Amino acids are converted to acetyl–CoA, pyruvate, or an intermediate of the Krebs cycle **4**.

Figure 7.12 Animated A variety of organic compounds from food can enter the reactions of aerobic respiration.

Mighty Mitochondria (revisited)

At least 83 proteins are directly involved in the electron transfer chains of electron transfer phosphorylation in mitochondria. A defect in any one of them—or in any of the thousands of other proteins used by mitochondria, such as frataxin (*right*)—can wreak havoc in the body. About one in 5,000 people suffer from a known mitochondrial disorder. New research is showing that mitochondrial defects may be involved in many other illnesses such as diabetes, hypertension, Alzheimer's and Parkinson's disease, and even aging.

How would you vote? Developing new drugs is costly, which is why pharmaceutical companies tend to ignore Friedreich's ataxia and other disorders that affect relatively few people. Should governments fund private companies that research potential treatments for rare disorders?

---

energy of the electrons they carry ultimately drives the synthesis of ATP in electron transfer phosphorylation.

Complex Carbohydrates The digestive system breaks down starch and other complex carbohydrates to monosaccharide subunits (**Figure 7.12A**). These sugars are quickly taken up by cells and converted to glucose-6-phosphate, which continues in glycolysis ❶.

Unless ATP is being used quickly, its concentration rises in the cytoplasm. A high concentration of ATP causes glucose-6-phosphate to be diverted away from glycolysis and into a pathway that forms glycogen. Liver and muscle cells especially favor the conversion of glucose to glycogen, and these cells contain the body's largest stores of it. Between meals, the liver maintains the glucose level in blood by converting the stored glycogen to glucose.

Fats A fat molecule has a glycerol head and one, two, or three fatty acid tails (Section 3.5). Cells dismantle fat molecules by first breaking the bonds that connect the glycerol with the fatty acids (**Figure 7.12B**). Nearly all cells in the body oxidize free fatty acids by splitting their long backbones into two-carbon fragments. These fragments are converted to acetyl–CoA, which can enter the Krebs cycle ❷. The glycerol is converted to PGAL by enzymes in liver cells. PGAL, remember, is an intermediate of glycolysis ❸.

On a per carbon basis, fats are a richer source of energy than carbohydrates. Carbohydrate backbones have many oxygen atoms, so they are partially oxidized. A fat's long fatty acid tails are hydrocarbon chains that typically have no oxygen atoms bonded to them, so they have a longer way to go to become oxidized—more reactions are required to fully break them down. Coenzymes accept electrons in these oxidation reactions. The more reduced coenzymes that form, the more electrons can be delivered to the ATP-forming machinery of electron transfer phosphorylation.

What happens if you eat too many carbohydrates? When the blood level of glucose gets too high, acetyl–CoA is diverted away from the Krebs cycle and into a pathway that makes fatty acids. That is why excess dietary carbohydrate ends up as fat.

Proteins Some enzymes in your digestive system split dietary proteins into their amino acid subunits, which are absorbed into the bloodstream. Cells use the amino acids to build proteins or other molecules. Even so, when you eat more protein than your body needs, the amino acids become broken down further. The amino (NH_3^+) group is removed, and it becomes ammonia (NH_3), a waste product that the body eliminates in urine. The carbon backbone is split, and acetyl–CoA, pyruvate, or an intermediate of the Krebs cycle forms, depending on the amino acid (**Figure 7.12C**). Cells can divert these organic molecules into the Krebs cycle ❹.

Take-Home Message

Can the body use organic molecules other than glucose for energy?

» Complex carbohydrates, fats, and proteins can be oxidized in aerobic respiration to yield ATP.

» First the digestive system and then individual cells convert molecules in food into intermediates of glycolysis or the Krebs cycle.

Summary

Section 7.1 Photosynthesis by early photoautotrophs changed the composition of Earth's early atmosphere, with profound effects on life's evolution. Organisms that could not tolerate the increased oxygen content persisted only in **anaerobic** habitats. Oxygen-detoxifying pathways evolved, allowing organisms to thrive in **aerobic** conditions. Defects in these pathways, such as occurs in Friedreich's ataxia, can cause serious health problems.

Section 7.2 Most organisms convert chemical energy of carbohydrates to the chemical energy of ATP. Carbohydrate breakdown pathways start in the cytoplasm with the same set of reactions,

glycolysis, which converts glucose and other sugars to **pyruvate**. Anaerobic **fermentation** pathways end in cytoplasm and yield two ATP per molecule of glucose. **Aerobic respiration** uses oxygen and yields much more ATP than fermentation. In modern eukaryotes, aerobic respiration is completed inside mitochondria.

Section 7.3 Glycolysis, the first stage of aerobic respiration and fermentation, occurs in the cytoplasm. In the reactions, enzymes use two ATP to convert one molecule of glucose or another six-carbon sugar to two molecules of pyruvate. Two NAD^+ are reduced to NADH. Four ATP also form by **substrate-level phosphorylation**, the direct transfer of a phosphate group from a reaction intermediate to ADP.

The net yield of glycolysis is two pyruvate, two ATP, and two NADH per glucose molecule. The pyruvate may continue in fermentation in the cytoplasm, or it may enter mitochondria and continue in the next steps of aerobic respiration.

Section 7.4 A mitochondrion's inner membrane divides its interior into two fluid-filled spaces: the inner compartment, or matrix, and the intermembrane space. The second stage of aerobic respiration, acetyl–CoA formation and the **Krebs cycle**, takes place in the matrix. The first steps convert two pyruvate from glycolysis to two acetyl–CoA and two CO_2. The acetyl–CoA enters the Krebs cycle.

It takes two cycles of Krebs reactions to dismantle the two acetyl–CoA. At this stage, all of the carbon atoms in the glucose molecule that entered glycolysis have left the cell in CO_2. During these reactions, electrons and hydrogen ions are transferred to NAD^+ and FAD, which are thereby reduced to NADH and $FADH_2$. ATP forms by substrate-level phosphorylation.

In total, the breakdown of two pyruvate molecules in the second stage of aerobic respiration yields ten reduced coenzymes and two ATP.

Section 7.5 Aerobic respiration ends in mitochondria. In the third stage of reactions, electron transfer phosphorylation, coenzymes that were reduced in the first two stages deliver their cargo of electrons and hydrogen ions to electron transfer chains in the inner mitochondrial membrane. Electrons moving through the chains release energy bit by bit; molecules of the chain use that energy to move H^+ from the matrix to the intermembrane space.

Hydrogen ions that accumulate in the intermembrane space form a gradient across the inner membrane. The ions follow the gradient back to the matrix through ATP synthases. H^+ flow through these transport proteins drives ATP synthesis.

Oxygen combines with electrons and H^+ at the end of the transfer chains, forming water.

Overall, aerobic respiration typically yields thirty-six ATP for each glucose molecule.

Section 7.6 Anaerobic fermentation pathways begin with glycolysis and finish in the cytoplasm. A molecule other than oxygen accepts electrons at the end of these reactions. The end product of **alcoholic fermentation** is ethyl alcohol, or ethanol. The end product of **lactate fermentation** is lactate. The final steps of fermentation serve to regenerate NAD^+, which is required for glycolysis to continue, but they produce no ATP. Thus, the breakdown of one glucose molecule in either alcoholic or lactate fermentation yields only the two ATP that form in glycolysis reactions.

Skeletal muscle has two types of fibers: red and white. ATP is produced primarily by aerobic respiration in red muscle fibers, so these fibers sustain activities that require endurance. Lactate fermentation in white fibers supports activities that occur in short, intense bursts.

Section 7.7 In humans and other mammals, simple sugars from carbohydrate breakdown, glycerol and fatty acids from fat breakdown, and carbon backbones of amino acids from protein breakdown may enter aerobic respiration at various reaction steps.

Self-Quiz

Answers in Appendix III

1. Is the following statement true or false? Unlike animals, which make many ATP by aerobic respiration, plants make all of their ATP by photosynthesis.

2. Glycolysis starts and ends in the _____ .
   a. nucleus   c. plasma membrane
   b. mitochondrion   d. cytoplasm

3. Which of the following metabolic pathways require(s) molecular oxygen (O_2)?
   a. aerobic respiration
   b. lactate fermentation
   c. alcoholic fermentation
   d. all of the above

4. Which molecule does not form during glycolysis?
   a. NADH   b. pyruvate   c. $FADH_2$ d. ATP

5. In eukaryotes, aerobic respiration is completed in the _____ .
   a. nucleus   c. plasma membrane
   b. mitochondrion   d. cytoplasm

6. Which of the following reaction pathways is not part of the second stage of aerobic respiration?
   a. electron transfer   c. Krebs cycle
      phosphorylation   d. glycolysis
   b. acetyl–CoA formation   e. a and d

7. After Krebs reactions run through _____ cycle(s), one glucose molecule has been completely broken down to CO_2.
   a. one   b. two   c. three   d. six

8. In the third stage of aerobic respiration, _____ is the final acceptor of electrons.
   a. water   b. hydrogen   c. oxygen   d. NADH

9. _____ is the final acceptor of electrons in alcoholic fermentation.
   a. Oxygen   c. Acetaldehyde
   b. Pyruvate   d. Sulfate

Mitochondrial Abnormalities in Tetralogy of Fallot Tetralogy of Fallot (TF) is a genetic disorder characterized by four major malformations of the heart. The circulation of blood is abnormal, so TF patients have too little oxygen in their blood. Inadequate oxygen levels result in damaged mitochondrial membranes, which in turn cause cells to self-destruct.

In 2004, Sarah Kuruvilla and her colleages looked at abnormalities in the mitochondria of heart muscle in TF patients. Some of their results are shown in **Figure 7.13**.

1. In this study, which abnormality was most strongly associated with TF?

2. What percentage of the TF patients had mitochondria that were abnormal in size?

3. Can you make any correlations between blood oxygen content and mitochondrial abnormalities in these patients?

Figure 7.13 Mitochondrial changes in tetralogy of Fallot (TF).

(A) Normal heart muscle. Many mitochondria between the fibers provide muscle cells with ATP for contraction. **(B)** Heart muscle from a person with TF shows swollen, broken mitochondria.

(C) Mitochondrial abnormalities in TF patients. SPO_2 is oxygen saturation of the blood. A normal value of SPO_2 is 96%. Abnormalities are marked "+".

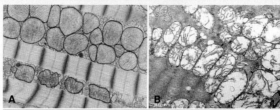

Patient (age)	SPO_2 (%)	Mitochondrial Abnormalities in TF			
		Number	Shape	Size	Broken
1 (5)	55	+	+	−	−
2 (3)	69	+	+	−	−
3 (22)	72	+	+	−	−
4 (2)	74	+	+	−	−
5 (3)	76	+	+	−	+
6 (2.5)	78	+	+	−	+
7 (1)	79	+	+	−	−
8 (12)	80	+	−	+	−
9 (4)	80	+	+	−	−
10 (8)	83	+	−	+	−
11 (20)	85	+	+	−	−
12 (2.5)	89	+	−	+	−

10. Fermentation pathways make no more ATP beyond the small yield from glycolysis. The remaining reactions serve to regenerate _____ .
   a. FAD
   b. NAD^+
   c. glucose
   d. oxygen

11. Most of the energy that is released by the full breakdown of glucose to CO_2 and water ends up in _____ .
   a. NADH
   b. ATP
   c. heat
   d. electrons

12. Your body cells can use _____ as an alternative energy source when glucose is in short supply.
   a. fatty acids
   b. glycerol
   c. amino acids
   d. all of the above

13. Which of the following is not produced by an animal muscle cell operating under anaerobic conditions?
   a. heat
   b. pyruvate
   c. NAD^+
   d. ATP
   e. lactate
   f. all are produced

14. Match the event with its most suitable description.
   ___ glycolysis
   ___ fermentation
   ___ Krebs cycle
   ___ electron transfer phosphorylation
   a. ATP, NADH, $FADH_2$, and CO_2 form
   b. glucose to two pyruvate
   c. NAD^+ regenerated, little ATP
   d. H^+ flows via ATP synthases

15. Match the term with the best description.
   ___ matrix
   ___ pyruvate
   ___ NAD^+
   ___ mitochondrion
   ___ intermembrane space
   ___ NADH
   ___ anaerobic
   a. needed for glycolysis
   b. inner space
   c. makes many ATP
   d. end of glycolysis
   e. reduced coenzyme
   f. hydrogen ions accumulate here
   g. no oxygen required

Critical Thinking

1. The higher the altitude, the lower the oxygen level in air. Climbers of very tall mountains risk altitude sickness, which is characterized by shortness of breath, weakness, dizziness, and confusion.

The early symptoms of cyanide poisoning are the same as those for altitude sickness. Cyanide binds tightly to cytochrome *c* oxidase, a protein complex that is the last component of mitochondrial electron transfer chains. Cytochrome *c* oxidase with bound cyanide can no longer transfer electrons. Explain why cyanide poisoning starts with the same symptoms as altitude sickness.

2. As you learned, membranes impermeable to hydrogen ions are required for electron transfer phosphorylation. Membranes in mitochondria serve this purpose in eukaryotes. Bacteria do not have this organelle, but they do make ATP by electron transfer phosphorylation. How do you think they do it, given that they have no mitochondria?

3. The bar-tailed godwit is a type of shorebird that makes an annual migration from Alaska to New Zealand and back. The birds make each 11,500-kilometer (7,145-mile) trip by flying over the Pacific Ocean in about nine days, depending on weather, wind speed, and direction of travel. One bird was observed to make the entire journey uninterrupted, a feat that is comparable to a human running a nonstop seven-day marathon at 70 kilometers per hour (43.5 miles per hour). Would you expect the flight (breast) muscles of bar-tailed godwits to be light or dark colored? Explain your answer.

4. The bacterium *Escherichia coli* can produce ATP by aerobic respiration and fermentation. Two cultures of *E. coli* are started and held under identical conditions, except that one is kept under anaerobic conditions. At the end of a week, which culture has more bacteria? Why?

8 DNA Structure and Function

LEARNING ROADMAP

Where you have been Radioisotope tracers (Section 2.2) were used in research that led to the discovery that DNA (3.8), not protein (3.6), is the hereditary material of all organisms. This chapter revisits free radicals (2.3), the cell nucleus (4.6) and metabolism (5.4–5.6). Your knowledge of carbohydrate ring numbering (3.3) will help you understand DNA replication.

Where you are now

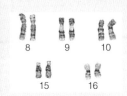

Chromosomes
The DNA of a eukaryotic cell is divided among a characteristic number of chromosomes that differ in length and shape. Sex chromosomes determine an individual's gender.

Discovery of DNA's Function
The work of many scientists over more than a century led to the discovery that DNA is the molecule that stores hereditary information.

Structure of DNA
A DNA molecule consists of two long chains of nucleotides coiled into a double helix. The order of nucleotide bases in the chains differs among individuals and among species.

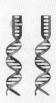

DNA Replication
Before a cell divides, it copies its DNA. Newly forming DNA is monitored for errors, most of which are corrected quickly. Uncorrected errors are mutations.

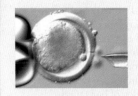

Cloning Animals
Clones of adult animals are commonly produced for research and agriculture. The techniques are far from perfect, and the practice continues to raise serious ethical questions.

Where you are going You will revisit concepts of DNA structure and function presented in this chapter many times, particularly when you learn about how genetic information is converted into structural and functional parts of a cell (Chapter 9), and how cells control that process (Chapter 10). Chromosome structure will turn up again in the context of cell division in Chapters 11 and 12, and in human inheritance and disease in Chapter 14. Viruses such as bacteriophage will be explained in more detail in Chapter 20, and stem cells in Chapter 31.

8.1 A Hero Dog's Golden Clones

On September 11, 2001, an off-duty Canadian police officer, Constable James Symington, drove his search dog Trakr from Nova Scotia to Manhattan. Within hours of arriving, the dog led rescuers to the spot where the fifth and final survivor of the World Trade Center attacks was buried. She had been clinging to life, pinned under rubble from the building where she had worked. Symington and Trakr helped with the search and rescue efforts for three days nonstop, until Trakr collapsed from smoke and chemical inhalation, burns, and exhaustion (**Figure 8.1**).

Trakr survived the ordeal, but later lost the use of his limbs from a degenerative neurological disease probably linked to toxic smoke exposure at Ground Zero. The hero dog died in April 2009, but his DNA lives on in his genetic copies—his **clones**. Symington's essay about Trakr's superior nature and abilities as a search and rescue dog won the Golden Clone Giveaway, a contest to find the world's most clone-worthy dog. Trakr's DNA was shipped to Korea, where it was inserted into donor dog eggs, which were then implanted into surrogate mother dogs. Five puppies, all clones of Trakr, were delivered to Symington in July 2009.

Many adult animals have been cloned besides Trakr, but cloning mammals is still unpredictable and far from routine. Typically, less than 2 percent of the implanted embryos result in a live birth. Of the clones that survive, many have serious health problems.

Why the difficulty? Even though all cells of an individual inherit the same DNA, an adult cell uses only a fraction of it compared with an embryonic cell. To make a clone from an adult cell, researchers must reprogram its DNA to function like the DNA of an egg. Even though we are getting better at doing that, we still have a lot to learn.

So why do we keep trying? The potential benefits are enormous. Already, cells of cloned human embryos are helping researchers unravel the molecular mechanisms of human genetic diseases. Such cells may one day be induced to form replacement tissues or organs for people with incurable diseases. Endangered animals might be saved from extinction; extinct animals may be brought back. Livestock and pets are already being cloned commercially.

Perfecting the methods for cloning animals brings us closer to the possibility of cloning humans, both technically and ethically. For example, if cloning a lost cat for a grieving pet owner is acceptable, why would it not be acceptable to clone a lost child for a grieving parent? Different people have very different answers to such questions, so controversy over cloning continues to rage even as the techniques improve. Understanding the basis of heredity—what DNA is and how it works—will help inform your own opinions about the issues surrounding cloning.

clone Genetically identical copy of an organism.

Figure 8.1
James Symington and his dog Trakr at Ground Zero, September 2001.

8.2 Eukaryotic Chromosomes

2 μm

■ The DNA in a eukaryotic cell nucleus is organized as one or more chromosomes.
■ Links to DNA 3.8, Cell nucleus 4.6

Stretched out end to end, the DNA molecules in a human cell would be about 2 meters (6.5 feet) long. That is a lot of DNA to fit into a nucleus that is less than 10 micrometers in diameter! Proteins structurally organize the DNA and help it pack tightly into a small nucleus.

Each DNA molecule is organized together with proteins as a structure called a **chromosome** (Figure 8.2). The DNA of a typical eukaryotic cell is divided among several chromosomes that differ in length and shape ❶. During most of the cell's life, each chromosome consists of one DNA strand. When the cell prepares to divide, it duplicates all of its chromosomes, so that both of its offspring will get a full set (we return to details of DNA replication in Section 8.5). At that point, each (duplicated) chromosome has two DNA strands—**sister chromatids**—attached to one another at a constricted region called the **centromere**:

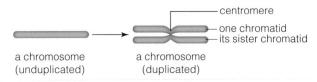

centromere
one chromatid
its sister chromatid

a chromosome
(unduplicated)

a chromosome
(duplicated)

A duplicated chromosome in its most condensed form consists of two long, tangled filaments (the sister chromatids) bunched into a characteristic X shape ❷. A closer look reveals that each filament is actually a hollow tube formed by coils, like an old-style telephone cord ❸. The coils consist of a twisted fiber ❹ of DNA wrapped twice at regular intervals around "spools" of proteins called **histones** ❺. In micrographs, these DNA–histone spools look like beads on a string. Each "bead" is a **nucleosome**, the smallest unit of chromosomal organization in eukaryotes. As you will see in Section 8.4, the DNA molecule consists of two strands twisted into a double helix ❻.

Chromosome Number

The total number of chromosomes in a eukaryotic cell—the **chromosome number**—is a characteristic of the species. For example, the chromosome number of oak trees is 12, so the nucleus of a cell from an oak tree contains 12 chromosomes. The chromosome number of king crab cells is 208, so they have 208 chromosomes. That of human body cells is 46, so human body cells have 46 chromosomes.

Actually, human body cells have two of each type of chromosome, which means that their chromosome number is **diploid** (2n). A technique called karyotyping reveals an individual's diploid complement of chromosomes. With this technique, cells taken from the individual are treated to make the chromosomes condense, and then stained so the individual chromosomes can be seen under a microscope. A micrograph of a single cell is digitally rearranged so the images of the chromosomes are lined up by centromere location,

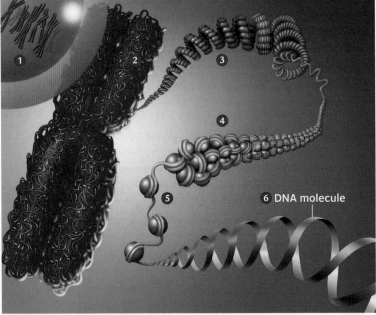

❻ DNA molecule

❶ The DNA in the nucleus of a eukaryotic cell is typically divided up into a number of chromosomes.

❷ At its most condensed, a duplicated chromosome packs tightly into an X shape.

❸ A chromosome unravels as a single fiber, a hollow cylinder formed by coiled coils.

❹ The coiled coils consist of a long molecule of DNA and proteins associated with it.

❺ At regular intervals, the DNA molecule is wrapped twice around a core of histone proteins. In this "beads-on-a-string" structure, the "string" is the DNA, and each "bead" is called a nucleosome.

❻ The DNA molecule itself has two strands twisted into a double helix.

Figure 8.2 Animated Zooming in on chromosome structure. Tight packing allows a lot of DNA to fit into a very small nucleus.

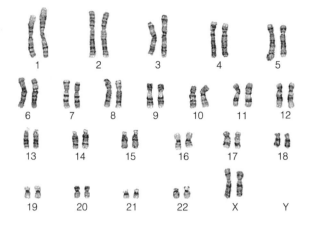

A A karyotype. This one shows 22 pairs of autosomes and a pair of X chromosomes.

Figure 8.3 Human chromosomes. **Figure It Out:** Is the karyotype in (A) from the cell of a female or a male?

Answer: A female

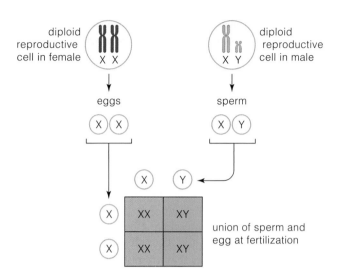

B Pattern of sex determination in humans. The grid shows how the different combinations of sex chromosomes result in offspring that are female (*pink*) or male (*blue*).

and arranged according to size, shape, and length. The finished array constitutes the individual's **karyotype** (**Figure 8.3A**). A karyotype shows how many chromosomes are in the individual's cells, and can also reveal major structural abnormalities.

Autosomes and Sex Chromosomes

All except one pair of a diploid cell's chromosomes are **autosomes**, which are the same in both females and males. The two autosomes of each pair have the same length, shape, and centromere location. They also hold information about the same traits. Think of them as two sets of books on how to build a house. Your father gave you one set. Your mother had her own ideas

autosome Any chromosome other than a sex chromosome.
centromere Constricted region in a eukaryotic chromosome where sister chromatids are attached.
chromosome A structure that consists of DNA and associated proteins; carries part or all of a cell's genetic information.
chromosome number The sum of all chromosomes in a cell of a given type.
diploid Having two of each type of chromosome characteristic of the species (2*n*).
histone Type of protein that structurally organizes eukaryotic chromosomes.
karyotype Image of an individual's complement of chromosomes arranged by size, length, shape, and centromere location.
nucleosome A length of DNA wound twice around a spool of histone proteins.
sex chromosome Member of a pair of chromosomes that differs between males and females.
sister chromatid One of two attached DNA molecules of a duplicated eukaryotic chromosome.

about wiring, plumbing, and so on. She gave you an alternate set that says slightly different things about many of those tasks.

Members of a pair of **sex chromosomes** differ between females and males. The differences determine an individual's sex. The sex chromosomes of humans are called X and Y. Inheriting two X chromosomes makes an individual female, and inheriting one X and one Y makes an individual male (**Figure 8.3B**). Thus, the body cells of human females contain two X chromosomes (XX); those of human males contain one X and one Y chromosome (XY).

XX females and XY males are the rule among fruit flies, mammals, and many other animals, but there are other patterns. In butterflies, moths, birds, and certain fishes, males have two identical sex chromosomes, and females do not. Environmental factors (not sex chromosomes) determine sex in some species of invertebrates, turtles, and frogs. As an example, the temperature of the sand in which sea turtle eggs are buried determines the sex of the hatchlings.

Take-Home Message

What are chromosomes?

» A chromosome consists of a molecule of DNA that is structurally organized by proteins. The organization allows the DNA to pack tightly.

» A eukaryotic cell's DNA is divided among some characteristic number of chromosomes, which differ in length and shape.

» Members of a pair of sex chromosomes differ between males and females. Chromosomes that are the same in males and females are called autosomes.

8.3 The Discovery of DNA's Function

■ Investigations that led to our understanding that DNA is the molecule of inheritance reveal how science advances.
■ Links to Radioisotopes 2.2, Proteins 3.6, DNA 3.8

In 1869, chemist Johannes Miescher isolated an acidic substance from nuclei (**Figure 8.4**). That substance, which was composed mostly of nitrogen and phosphorus, was deoxyribonucleic acid, or DNA. Sixty years later, Frederick Griffith was trying to make a vaccine for pneumonia. He isolated two strains (types) of *Streptococcus pneumoniae*, a bacteria that causes pneumonia. He named one strain R, because it grows in Rough colonies. He named the other strain S, because it grows in Smooth colonies. Griffith used both strains in a series of experiments that did not lead to the development of a vaccine, but did reveal a clue about inheritance (**Figure 8.5**).

First, he injected mice with live R cells ❶. The mice remained healthy despite having live R cells in their blood, so the R strain was harmless.

Second, he injected other mice with live S cells ❷. The mice died. The S strain caused pneumonia.

Third, he exposed S cells to high temperature. Mice injected with the heat-killed S cells did not die ❸.

Fourth, he mixed live R cells with heat-killed S cells. Mice injected with the mixture died ❹. Blood samples drawn from them teemed with live S cells.

What happened in the fourth experiment? If heat-killed S cells in the mix were not really dead, then mice injected with them in the third experiment would have died. If the harmless R cells had changed into killer cells on their own, then mice injected with R cells in experiment 1 would have died.

The simplest explanation was that heat had killed the S cells, but had not destroyed their hereditary material, including whatever specified "kill mice." That material had been transferred from the dead S cells to the live R cells, which put it to use. The trans-

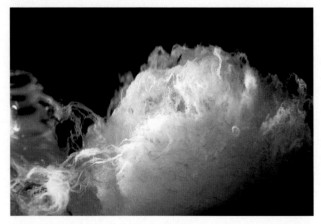

Figure 8.4 DNA, the substance. Eighty years after its discovery, DNA was determined to be the material of heredity. The photo shows DNA extracted from human cells.

formation was permanent and heritable. Even after hundreds of generations, the descendants of transformed R cells were infectious.

What substance had caused the transformation? In 1940, Oswald Avery and Maclyn McCarty set out to identify that substance, which they termed the "transforming principle," by a process of elimination. The researchers made an extract of S cells that contained only lipid, protein, and nucleic acids. The S cell extract could still transform R cells after it had been treated with lipid- and protein-destroying enzymes. Thus, the transforming principle could not be lipid or protein, so Avery and McCarty realized that the substance they were seeking must be nucleic acid—RNA or DNA. The S cell extract could still transform R cells after treatment with RNA-degrading enzymes, but not after treatment with DNA-degrading enzymes. DNA had to be the transforming principle.

The result surprised Avery and McCarty, who, along with most other scientists, had assumed that proteins were the substance of heredity. After all,

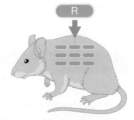

❶ Mice injected with live cells of harmless strain R do not die. Live R cells in their blood.

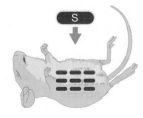

❷ Mice injected with live cells of killer strain S die. Live S cells in their blood.

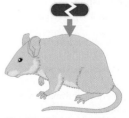

❸ Mice injected with heat-killed S cells do not die. No live S cells in their blood.

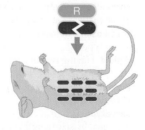

❹ Mice injected with live R cells plus heat-killed S cells die. Live S cells in their blood.

Figure 8.5 Animated Summary of results from Fred Griffith's experiments. The hereditary material of harmful *Streptococcus pneumoniae* cells transformed harmless cells into killers.

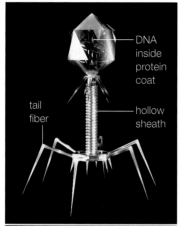

A *Top*, model of a bacteriophage. *Bottom*, micrograph of three viruses injecting DNA into an *E. coli* cell.

DNA inside protein coat

tail fiber

hollow sheath

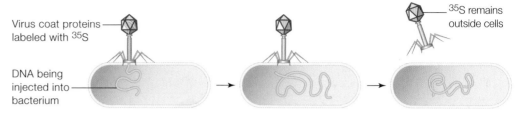

Virus coat proteins labeled with ^{35}S

DNA being injected into bacterium

^{35}S remains outside cells

B In one experiment, bacteria were infected with virus particles that had been labeled with a radio-isotope of sulfur (^{35}S). The sulfur had labeled only viral proteins. The viruses were dislodged from the bacteria by whirling the mixture in a kitchen blender. Most of the radioactive sulfur was detected in the viruses, not in the bacterial cells. The viruses had not injected protein into the bacteria.

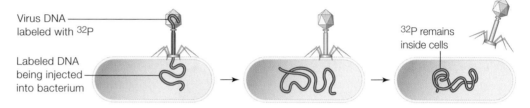

Virus DNA labeled with ^{32}P

Labeled DNA being injected into bacterium

^{32}P remains inside cells

C In another experiment, bacteria were infected with virus particles that had been labeled with a radioisotope of phosphorus (^{32}P). The phosphorus had labeled only viral DNA. When the viruses were dislodged from the bacteria, the radioactive phosphorus was detected mainly inside the bacterial cells. The viruses had injected DNA into the cells—evidence that DNA is the genetic material of this virus.

Figure 8.6 Animated The Hershey–Chase experiments. Alfred Hershey and Martha Chase tested whether the genetic material injected by bacteriophage into bacteria is DNA, protein, or both. The experiments were based on the knowledge that proteins contain more sulfur (S) than phosphorus (P), and DNA contains more phosphorus than sulfur.

traits are diverse, and proteins were thought to be the most diverse biological molecules. Other molecules seemed too uniform. The two scientists were so skeptical that they published their results only after they had convinced themselves, by years of painstaking experiments, that DNA was hereditary material. They were also careful to point out that they had not proven DNA was the *only* hereditary material.

Confirmation of DNA's Function

By 1950, researchers had discovered **bacteriophage**, a type of virus that infects bacteria (**Figure 8.6A**). Like all viruses, these infectious particles carry hereditary information about how to make new viruses. After a virus infects a cell, the cell starts making new virus particles. Bacteriophages inject genetic material into bacteria, but was that material DNA, protein, or both?

Alfred Hershey and Martha Chase found the answer to that question by exploiting the long-known properties of protein (high sulfur content) and DNA (high phosphorus content). They cultured bacteria in growth medium containing a sulfur isotope, ^{35}S. In this medium, the protein (but not the DNA) of bacteriophage that infected the bacteria became labeled

with the ^{35}S tracer. Hershey and Chase allowed the labeled viruses to infect a fresh culture of unlabeled bacteria. They knew from electron micrographs that bacteriophages attach to bacteria by their slender tails. They reasoned it would be easy to break this precarious attachment, so they whirled the virus–bacteria mixture in a kitchen blender. The researchers then separated the bacteria from the virus-containing fluid, and measured the ^{35}S content of each separately. The fluid contained most of the ^{35}S. Thus, the viruses had not injected protein into the bacteria (**Figure 8.6B**).

Hershey and Chase repeated the experiment using an isotope of phosphorus, ^{32}P, which labeled the DNA (but not the proteins) of the bacteriophage. This time, infected bacteria contained most of the isotope. The viruses had injected DNA into the bacteria (**Figure 8.6C**). This experiment confirmed that DNA, not protein, carries hereditary information.

bacteriophage Virus that infects bacteria.

Take-Home Message

What is the molecular basis of inheritance?

» DNA is the material of heredity common to all life on Earth.

8.4 The Discovery of DNA's Structure

■ Watson and Crick's discovery of DNA's structure was based on 150 years of research by other scientists.

■ Links to Numbering carbohydrate rings 3.3, Protein structure 3.6, Nucleic acids 3.8

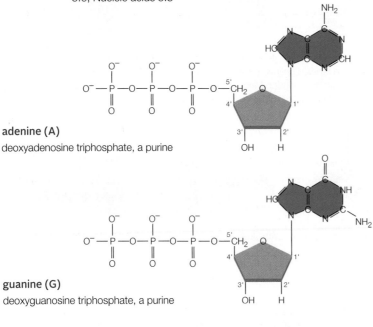

adenine (A)
deoxyadenosine triphosphate, a purine

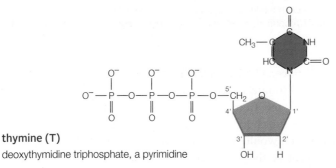

guanine (G)
deoxyguanosine triphosphate, a purine

thymine (T)
deoxythymidine triphosphate, a pyrimidine

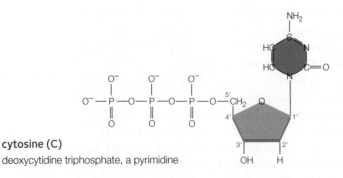

cytosine (C)
deoxycytidine triphosphate, a pyrimidine

Figure 8.7 Animated The four nucleotides in DNA. Adenine and guanine are called purines; thymine and cytosine, pyrimidines. All have three phosphate groups, a deoxyribose sugar (*orange*), and a nitrogen-containing base (*blue*) after which it is named. Biochemist Phoebus Levene identified the structure of the bases and how they are connected in nucleotides in the early 1900s. Levene worked with DNA for almost 40 years.

Numbering the carbons in the sugar rings (Section 3.3) allows us to keep track of the orientation of nucleotide chains (see **Figure 8.8**). The orientation is important in processes such as DNA replication.

A DNA nucleotide has a five-carbon sugar, three phosphate groups, and a nitrogen-containing base (**Figure 8.7**). The four kinds of nucleotides found in DNA—adenine (A), guanine (G), thymine (T), and cytosine (C)—differ only in their base. Just how these nucleotides are arranged in DNA was a puzzle that took over 50 years to solve. As molecules go, DNA is gigantic, and chromosomal DNA has a complex structural organization. Both factors made the molecule difficult to work with, given the laboratory methods of the time.

In 1950, Erwin Chargaff, one of many researchers trying to determine the structure of DNA, made two discoveries about the molecule. First, the amounts of thymine and adenine are identical, as are the amounts of cytosine and guanine (A = T and G = C). We call this discovery Chargaff's first rule. Chargaff's second discovery, or rule, is that the DNA of different species differs in its proportions of adenine and guanine.

Meanwhile, American biologist James Watson and British biophysicist Francis Crick, both at Cambridge University, had been sharing ideas about the structure of DNA. Linus Pauling, Robert Corey, and Herman Branson had just discovered the helical secondary structure that occurs in many proteins, and Watson and Crick suspected that the DNA molecule was also a helix. The two spent many hours arguing about the size, shape, and bonding requirements of the four kinds of nucleotides that make up DNA. They pestered chemists to help them identify bonds they might have overlooked, fiddled with cardboard cutouts, and made models from scraps of metal connected by suitably angled "bonds" of wire.

At King's College in London, biochemist Rosalind Franklin had also been working on the structure of DNA. Like Crick, Franklin specialized in x-ray crystallography, a technique in which x-rays are directed through a purified and crystallized substance. Atoms in the substance's molecules scatter the x-rays in a pattern that can be captured as an image. Researchers can use the pattern to calculate the size, shape, and spacing between any repeating elements of the molecules—all of which are details of molecular structure.

Franklin discovered that "wet" and "dry" DNA samples have different shapes. She made the first clear x-ray diffraction image of "wet" DNA, the form that occurs in cells. From the information in that image, she calculated that DNA is very long compared to its 2-nanometer diameter. She also identified a repeating pattern every 0.34 nanometer along its length, and another every 3.4 nanometers.

Franklin's image and data came to the attention of Watson and Crick, who now had all the information they needed to build a model of the DNA helix—one with two nucleotide chains running in opposite directions, and paired bases inside (**Figure 8.8**). Bonds between the sugar of one nucleotide and the phosphate of the next form the backbone of each chain. Hydrogen bonds between the internally positioned bases hold the two strands together. Only two kinds of base pairings form: A to T, and G to C, which explains the first of Chargaff's rules. Most scientists had assumed (incorrectly) that the bases had to be on the outside of the helix, because they would be more accessible to DNA-copying enzymes that way. You will see in the next section how DNA replication enzymes access the bases on the inside of a double helix.

DNA's Base-Pair Sequence

Just two kinds of base pairings give rise to the incredible diversity of traits we see among living things. How? Even though DNA is composed of only four bases, the order in which one base pair follows the next—the **DNA sequence**—varies tremendously among species (which explains Chargaff's second rule). For example, a piece of DNA from any organism might be:

Note how the two strands match. They are complementary, which means each base on one strand pairs with a suitable partner base on the other. This bonding pattern (A to T, G to C) is the same in all molecules of DNA. However, the DNA sequence differs among species, and even among individuals of the same species. The information encoded by that sequence is the basis of traits that define species and distinguish individuals. Thus DNA, the molecule of inheritance in every cell, is the basis of life's unity. Variations in its base sequence are the foundation of life's diversity.

DNA sequence Order of nucleotide bases in a strand of DNA.

Take-Home Message

What is the structure of DNA?

» A DNA molecule consists of two nucleotide chains (strands) running in opposite directions and coiled into a double helix. Internally positioned nucleotide bases hydrogen-bond between the two strands. A pairs with T, and G with C.

» The sequence of bases along a DNA strand is genetic information. DNA sequences vary among species and among individuals. This variation is the basis of life's diversity.

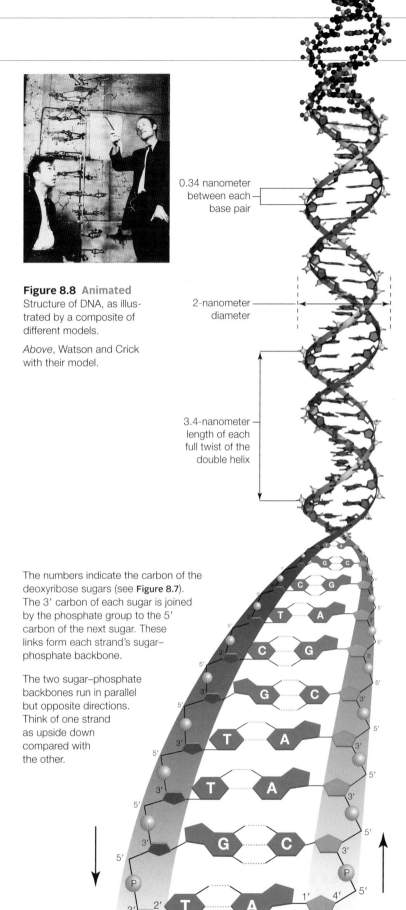

Figure 8.8 Animated Structure of DNA, as illustrated by a composite of different models.

Above, Watson and Crick with their model.

0.34 nanometer between each base pair

2-nanometer diameter

3.4-nanometer length of each full twist of the double helix

The numbers indicate the carbon of the deoxyribose sugars (see **Figure 8.7**). The 3' carbon of each sugar is joined by the phosphate group to the 5' carbon of the next sugar. These links form each strand's sugar–phosphate backbone.

The two sugar–phosphate backbones run in parallel but opposite directions. Think of one strand as upside down compared with the other.

8.5 DNA Replication

- A cell copies its DNA before it reproduces. Each of the two strands of DNA in the double helix is replicated.
- DNA replication is an energy-intensive process that requires many enzymes, including DNA polymerase, and other molecules.
- Links to Functional groups 3.3, Enzymes 5.4, Metabolic pathways 5.5, Phosphate groups and energy transfers 5.6

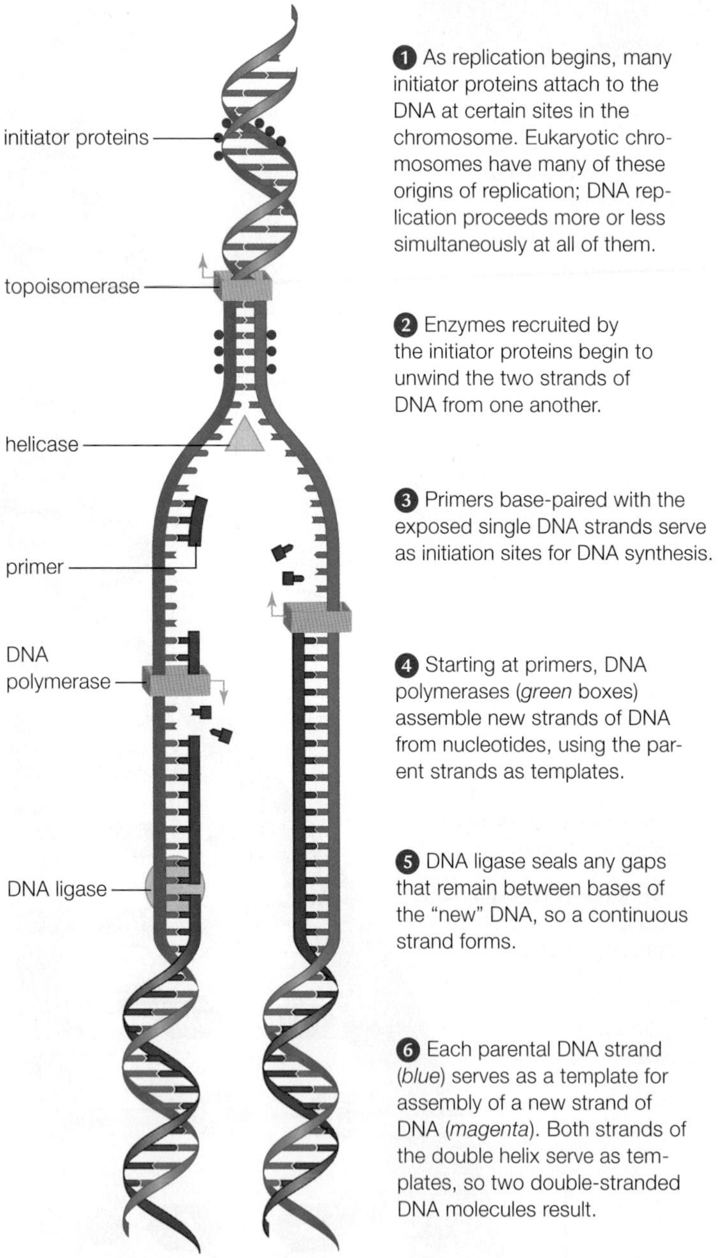

During most of its life, a cell contains only one set of chromosomes. However, when the cell reproduces, it must contain two sets of chromosomes: one for each of its future offspring. **DNA replication** is the energy-intensive process by which a cell copies its DNA. The process requires the coordinated action of a host of enzymes and other molecules that first open the double helix to expose the internally positioned bases, and then string together free nucleotides into new strands of DNA based on the sequence of those bases.

A cell's genetic information consists of the order of nucleotide bases—the DNA sequence—of its chromosomes. Descendant cells must get an exact copy of that information, or inheritance will go awry. Thus, each chromosome is copied in entirety, and the two chromosomes that result from replication are duplicates of the parent molecule.

A Host of Participants

In eukaryotes, DNA replication necessarily takes place in the nucleus. An enzyme called **DNA polymerase** is central to the process. There are several types of DNA polymerases, but all can assemble a new strand of DNA from free nucleotides. Each type requires one strand of DNA that serves as a template, or guide, for the synthesis of another. Each also requires a primer in order to initiate DNA synthesis. A **primer** is a short, single strand of DNA or RNA that is complementary to a targeted DNA sequence.

Before DNA replication, a chromosome consists of one molecule of DNA—one double helix (**Figure 8.9**). As replication begins, a large number of proteins called initiators bind to certain sequences of nucleotide bases in the DNA ❶. Initiator proteins recruit molecules that pry apart the two strands of the DNA. An enzyme called topoisomerase untwists the double helix, and another enzyme, helicase, breaks the hydrogen bonds holding the double helix together. The two DNA strands begin to unwind from one another as these enzymes move along the double helix ❷.

Another enzyme constructs short RNA primers that base-pair with the opened, single-stranded areas of the DNA ❸. The primers serve as attachment points for DNA polymerases, which begin assembling new strands of DNA on each of the parent strands. As a DNA polymerase moves along a strand of DNA, it uses the sequence of bases as a template, or guide, to assemble a new strand of DNA from free nucleotides ❹. The polymerase follows base-pairing rules: It adds a T to the end of the new DNA strand when it reaches an A in the template DNA strand; it adds a G when

❶ As replication begins, many initiator proteins attach to the DNA at certain sites in the chromosome. Eukaryotic chromosomes have many of these origins of replication; DNA replication proceeds more or less simultaneously at all of them.

❷ Enzymes recruited by the initiator proteins begin to unwind the two strands of DNA from one another.

❸ Primers base-paired with the exposed single DNA strands serve as initiation sites for DNA synthesis.

❹ Starting at primers, DNA polymerases (*green* boxes) assemble new strands of DNA from nucleotides, using the parent strands as templates.

❺ DNA ligase seals any gaps that remain between bases of the "new" DNA, so a continuous strand forms.

❻ Each parental DNA strand (*blue*) serves as a template for assembly of a new strand of DNA (*magenta*). Both strands of the double helix serve as templates, so two double-stranded DNA molecules result.

Figure 8.9 Animated DNA replication, in which a double-stranded molecule of DNA is copied in entirety. Two double-stranded DNA molecules form; one strand of each is parental (old), and the other is new, so DNA replication is said to be semiconservative. *Green* arrows show the direction of synthesis for each strand. The Y-shaped structure of a DNA molecule undergoing replication is called a replication fork.

it reaches a C; and so on. Thus, the base sequence of each new strand of DNA is complementary to its template (parental) strand. The enzyme **DNA ligase** seals any gaps, so the new DNA strands are continuous.

Numbering the carbons in nucleotides allows us to keep track of the DNA strands in a double helix, because each strand has an unbonded 5′ carbon at one end and an unbonded 3′ carbon at the other:

5′ ▬▬▬▬▬▬▬▬ 3′
3′ ▥▥▥▥▥▥▥▥ 5′

DNA polymerases can attach a free nucleotide only to the 3′ end of a DNA strand. Thus, only one of the two new strands of DNA can be synthesized continuously during DNA replication. Synthesis of the other strand occurs in segments, in the direction opposite that of unwinding (**Figure 8.10**). DNA ligase joins those segments into a continuous strand of DNA ❺.

Each nucleotide provides energy for its own attachment to the end of a growing strand of DNA. Remember from Section 5.6 that the bonds between phosphate groups hold a lot of free energy. Two of a nucleotide's three phosphate groups are removed when it is added to a DNA strand. Breaking those bonds releases enough energy to drive the attachment.

As each new DNA strand lengthens, it winds with its template strand into a double helix. So, after replication, two double-stranded molecules of DNA have formed ❻. One strand of each molecule is parental (old), and the other is new; hence the name of the process, **semiconservative replication.** Each new strand of DNA is complementary in sequence to one of the two parent strands, so both double-stranded molecules produced by DNA replication are duplicates of the parent molecule.

DNA ligase Enzyme that seals gaps in double-stranded DNA.
DNA polymerase DNA replication enzyme. Uses a DNA template to assemble a complementary strand of DNA.
DNA replication Process by which a cell duplicates its DNA before it divides.
primer Short, single strand of DNA that base-pairs with a targeted DNA sequence.
semiconservative replication Describes the process of DNA replication, which produces two copies of a DNA molecule: one strand of each copy is new, and the other is a strand of the original DNA.

Take-Home Message

How is DNA copied?

» DNA replication is an energy-intensive process by which a cell copies its chromosomes.

» Each strand of the double helix serves as a template for synthesis of a new, complementary strand of DNA.

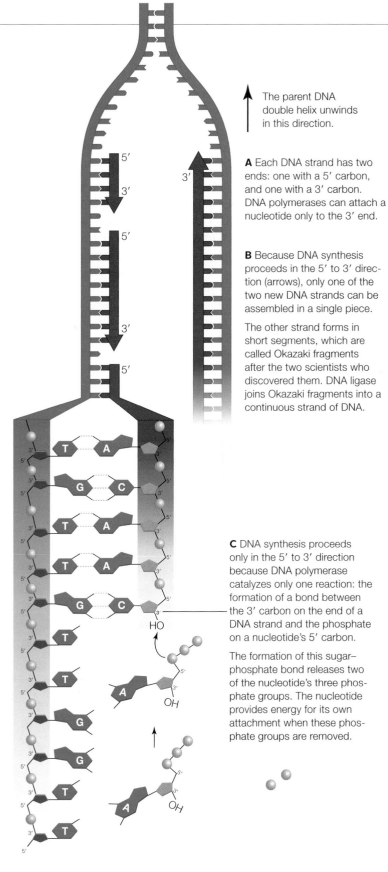

The parent DNA double helix unwinds in this direction.

A Each DNA strand has two ends: one with a 5′ carbon, and one with a 3′ carbon. DNA polymerases can attach a nucleotide only to the 3′ end.

B Because DNA synthesis proceeds in the 5′ to 3′ direction (arrows), only one of the two new DNA strands can be assembled in a single piece.

The other strand forms in short segments, which are called Okazaki fragments after the two scientists who discovered them. DNA ligase joins Okazaki fragments into a continuous strand of DNA.

C DNA synthesis proceeds only in the 5′ to 3′ direction because DNA polymerase catalyzes only one reaction: the formation of a bond between the 3′ carbon on the end of a DNA strand and the phosphate on a nucleotide's 5′ carbon.

The formation of this sugar–phosphate bond releases two of the nucleotide's three phosphate groups. The nucleotide provides energy for its own attachment when these phosphate groups are removed.

Figure 8.10 Discontinuous synthesis of DNA. This close-up of a replication fork shows that only one new DNA strand is assembled continuously.

Figure It Out: What do the yellow balls represent? Answer: Phosphate groups

8.6 Mutations: Cause and Effect

■ DNA repair mechanisms correct most replication errors.

■ In science, as in other professions, public recognition for a discovery does not always include all contributors.

■ Links to Electron energy levels and free radicals 2.3, Radiant energy 6.2

DNA Repair Mechanisms

A DNA molecule is not always replicated with perfect fidelity. Sometimes the wrong base is added to a growing DNA strand; at other times, bases get lost, or extra ones are added. Either way, the new DNA strand will no longer match up perfectly with its parent. Some of these replication errors occur after the DNA becomes damaged by exposure to radiation or toxic chemicals, because DNA polymerases do not copy damaged DNA very well. However, most replication errors occur simply because DNA polymerases catalyze a tremendous number of reactions very quickly—about 50 nucleotides per second in eukaryotes, and up to 1,000 bases per second in bacteria. Mistakes are inevitable, and some types of DNA polymerases make many of them.

Luckily, most DNA polymerases proofread their own work. They correct mismatches by reversing the synthesis reaction to remove a mispaired nucleotide, then resuming synthesis in the forward direction. In addition, a set of enzymes and other proteins function as a DNA repair mechanism by excising and replacing

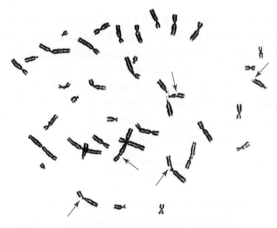

Figure 8.11 DNA damage. *Red* arrows show major breaks in chromosomes of a human white blood cell after exposure to ionizing radiation. Pieces of broken chromatids are often lost during DNA replication.

damaged or mismatched bases in DNA before replication begins.

When proofreading and repair mechanisms fail, an error becomes a **mutation**, a permanent change in the DNA sequence. Repair enzymes cannot recognize a mutation after it has been replicated, because it will base-pair properly with the new strand of DNA. Thus, the mutation becomes perpetuated with every subsequent cell division. A mutation in a body cell may cause that cell to become cancerous; mutations in cells that form eggs or sperm can be passed to offspring. However, not all mutations are dangerous. As you will see in later chapters, mutations that give rise to variations in traits are the raw material of evolution.

Environmental Causes of Mutations

Many environmental agents cause mutations. For example, electromagnetic energy with a wavelength shorter than 320 nanometers—gamma rays, x-rays, and most ultraviolet (UV) light—can knock electrons out of atoms. Such ionizing radiation breaks chromosomes into pieces that get lost during DNA replication (**Figure 8.11**). It also damages DNA indirectly when it penetrates living tissue, because it leaves a trail of destructive free radicals in its wake (Section 2.3).

UV light in the range of 320–400 nanometers can boost electrons to a higher energy level, but not enough to knock them out of atoms. UV light in this range is still dangerous, because it has enough energy to open up the double bond in the ring of a pyrimidine base (thymine or cytidine). The open ring can form a covalent bond with the ring of an adjacent pyrimidine base. The resulting pyrimidine dimer makes a kink in the DNA strand (**Figure 8.12**). DNA polymerase tends to

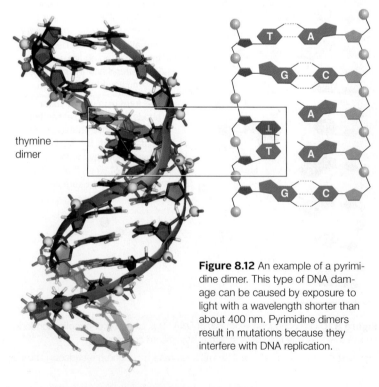

thymine dimer

T — A

G = C

A

T
T

A

G = C

Figure 8.12 An example of a pyrimidine dimer. This type of DNA damage can be caused by exposure to light with a wavelength shorter than about 400 nm. Pyrimidine dimers result in mutations because they interfere with DNA replication.

copy the kinked part incorrectly during replication, so a mutation becomes introduced into the DNA.

Mutations that arise as a result of pyrimidine dimers cause skin cancer. Exposing unprotected skin to sunlight increases the risk of cancer because its UV wavelengths cause pyrimidine dimers to form. For every second a skin cell spends in the sun, 50–100 pyrimidine dimers form in its DNA.

Some natural or synthetic chemicals also cause mutations. For instance, some of the fifty-five or more carcinogenic (cancer-causing) chemicals in tobacco smoke transfer small hydrocarbon groups to the nucleotide bases in DNA. The altered bases mispair during replication, or prevent replication entirely. Both events increase the chance of mutation. Many environmental pollutants, including those found in tobacco smoke, are converted by the body to other compounds that are easier to excrete. Some of the intermediates in these pathways bind irreversibly to DNA, thus causing replication errors that lead to mutation.

The Short Story of Rosalind Franklin

By the time Rosalind Franklin arrived at King's College, she was an expert x-ray crystallographer. She had solved the structure of coal, which is complex and unorganized (as is DNA). Like Pauling, she had built three-dimensional molecular models. Her assignment was to investigate DNA's structure. No one had told her Maurice Wilkins was already doing the same thing just down the hall. No one had told Wilkins about Franklin's assignment; he assumed she was a technician hired to do his x-ray crystallography work. And so a clash began. Wilkins thought Franklin displayed an appalling lack of deference that technicians of the era usually accorded researchers. To Franklin, Wilkins seemed oddly overinterested in her work and inexplicably prickly.

Wilkins and Franklin had been given identical samples of DNA carefully prepared by Rudolf Signer. Franklin's meticulous work with hers yielded the first clear x-ray diffraction image of DNA as it occurs in cells (**Figure 8.13**). She gave a presentation on this work in 1952. DNA, she said, had two chains twisted into a double helix, with a backbone of phosphate groups on the outside, and bases arranged in an unknown way on the inside. She had calculated DNA's diameter, the distance between its chains and between its bases, the angle of the helix, and the number of bases in each coil. Crick, with his crystallography background, would have recognized the significance of the work—if he had been there. Watson was in the audience but he

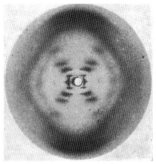

Figure 8.13 Rosalind Franklin and her x-ray diffraction image of DNA. This image was the final link in a long chain of clues that led to the discovery of DNA's structure. Making the image likely contributed to the cancer that took Franklin's life.

was not a crystallographer, and he did not understand the implications of Franklin's x-ray diffraction image or her calculations.

Franklin started to write a research paper on her findings. Meanwhile, and perhaps without her knowledge, Watson reviewed Franklin's x-ray diffraction image with Wilkins, and Watson and Crick read a report detailing Franklin's unpublished data. Crick, who had more experience with molecular modeling than Franklin, immediately understood what the image and the data meant. Watson and Crick used that information to build their model of DNA.

On April 25, 1953, Franklin's paper appeared third in a series of articles about the structure of DNA in the journal *Nature*. It supported with solid experimental evidence Watson and Crick's theoretical model, which appeared in the first article of the series.

Rosalind Franklin died in 1958, at the age of 37, of ovarian cancer probably caused by extensive exposure to x-rays during her work. At the time, the link between x-rays, mutations, and cancer was not fully understood. Because the Nobel Prize is not given posthumously, Franklin did not share in the 1962 honor that went to Watson, Crick, and Wilkins for the discovery of the structure of DNA.

mutation Permanent change in the nucleotide sequence of DNA.

Take-Home Message

What are mutations?

» Permanent changes in a DNA sequence are mutations.

» DNA damage by environmental agents such as UV light and chemicals can result in mutations, because damaged DNA is not replicated very well.

» Proofreading and repair mechanisms usually maintain the integrity of a cell's genetic information by fixing damaged DNA or correcting mispaired bases.

8.7 Animal Cloning

■ Various reproductive interventions produce genetically identical individuals.

The word "cloning," which means making an identical copy of something, can refer to deliberate interventions in reproduction that produce an exact genetic copy of an organism. Genetically identical organisms occur all the time in nature, arising most often by the process of asexual reproduction (which we discuss in Chapter 11). Embryo splitting, another natural process, results in identical twins. The first few divisions of a fertilized egg form a ball of cells that sometimes splits spontaneously. If both halves continue to develop independently, identical twins result. Artificial embryo splitting has been routine in research and animal husbandry for decades. With this technique, a ball of cells is grown from a fertilized egg in a laboratory. The ball is teased apart into two halves, each of which goes on to develop as a separate embryo. The embryos are implanted in surrogate mothers, who give birth to identical twins. Artificial twinning and any other technology that yields genetically identical individuals is called **reproductive cloning**.

Twins get their DNA from two parents that typically differ in their DNA sequence. Thus, although twins produced by embryo splitting are identical to one another, they are not identical to either parent. When animal breeders want an exact copy of a specific individual, they may turn to a cloning method that starts with a single cell taken from an adult organism. Such procedures present more of a technical challenge than embryo splitting. Unlike a fertilized egg, a body cell from an adult will not automatically start dividing. It must first be tricked into rewinding its developmental clock.

All cells descended from a fertilized egg inherit the same DNA. Thus, the DNA in each living cell of an individual is like a master blueprint that contains enough information to build an entirely new individual. As different cells in a developing embryo start using different subsets of their DNA, they differentiate (become different in form and function). Differentiation is usually a one-way path in animal cells. Once a cell specializes, all of its descendant cells will be specialized the same way. By the time a liver cell, muscle cell, or other specialized cell forms, most of its DNA has been turned off, and is no longer used.

To clone an adult, scientists must first transform one of its differentiated cells into an undifferentiated cell by turning its unused DNA back on. In **somatic cell nuclear transfer** (**SCNT**), a researcher removes the nucleus from an unfertilized egg, then inserts into the egg a nucleus from an adult animal cell (**Figure 8.14**). A somatic cell is a body cell, as opposed to a reproductive cell (*soma* is a Greek word for body). If all goes well, the egg's cytoplasm reprograms the transplanted DNA to direct the development of an embryo, which is then implanted into a surrogate mother. The animal

A A cow egg is held in place by suction through a hollow glass tube called a micropipette. DNA is identified by a *purple* stain.

B Another micropipette punctures the egg and sucks out the DNA. All that remains inside the egg's plasma membrane is cytoplasm.

C A new micropipette prepares to enter the egg at the puncture site. The pipette contains a cell grown from the skin of a donor animal.

D The micropipette enters the egg and delivers the skin cell to a region between the cytoplasm and the plasma membrane.

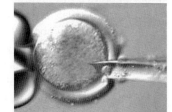

E After the pipette is withdrawn, the donor's skin cell is visible next to the cytoplasm of the egg. The transfer is now complete.

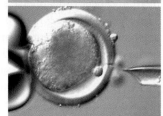

F An electric current causes the foreign cell to fuse with and empty its nucleus into the cytoplasm of the egg. The egg begins to divide, and an embryo forms that may be transplanted into a surrogate mother.

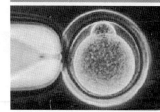

Figure 8.14 **Animated** Somatic cell nuclear transfer, using cattle cells. This series of micrographs was taken by scientists at Cyagra, a company that specializes in cloning livestock.

A Hero Dog's Golden Clones (revisited)

Trakr's clones were produced using SCNT. The ability to clone dogs is a recent development, but the technique is not. Scottish geneticist Ian Wilmut made headlines in 1997 when his team cloned an adult sheep using SCNT. The cloned lamb, Dolly, was genetically identical to the sheep that had donated an udder cell.

At first, Dolly looked and acted like a normal sheep, but she died early. Dolly may have had health problems because she was a clone.

SCNT has also been used to clone mice, rats, rabbits, pigs, cattle, goats, sheep, horses, mules, deer, cats, a camel, a ferret, a monkey, and a wolf. Many of the clones are unusually overweight or have enlarged organs. Cloned mice develop lung and liver problems, and almost all die prematurely. Cloned pigs tend to limp and have heart problems. One never did develop a tail or, even worse, an anus.

How would you vote? Some view sickly or deformed clones as unfortunate but acceptable casualties of animal cloning research that yields medical advances for humans. Should animal cloning be banned?

---

that is born to the surrogate is genetically identical with the donor of the nucleus.

SCNT is now a common practice among people who breed prized livestock. Among other benefits, many more offspring can be produced in a given time frame by cloning than by traditional breeding methods. Cloned animals have the same championship features as their DNA donors (**Figure 8.15**). Offspring can also be produced from a donor animal that is castrated or even dead.

As the techniques become routine, cloning a human is no longer only within the realm of science fiction. SCNT is already being used to produce human embryos for research, a practice called **therapeutic cloning**. Researchers harvest undifferentiated (stem) cells from the cloned human embryos. Cells from these embryos are being used to study, among many other things, how fatal diseases progress. For example, embryos created using cells from people with genetic heart defects will allow researchers to study how the defect causes developing heart cells to malfunction. Such research may ultimately lead to treatments for people who suffer from fatal diseases. (We return to the topic of stem cells and their potential medical benefits in Chapter 31.) Reproductive cloning of humans is not the intent of such research, but if it were, SCNT would indeed be the first step toward that end.

Human eggs are difficult to come by. They also come with a hefty set of ethical dilemmas. Thus, researchers have started trying to make hybrid embryos using adult human cells and eggs from other animals, a technique called interspecies nuclear transfer, or iSCNT.

reproductive cloning Technology that produces genetically identical individuals.
somatic cell nuclear transfer (**SCNT**) Method of reproductive cloning in which genetic material is transferred from an adult somatic cell into an unfertilized, enucleated egg.
therapeutic cloning The use of SCNT to produce human embryos for research purposes.

Figure 8.15 Champion Holstein dairy cow Nelson's Estimate Liz (*right*) and her clone, Nelson's Estimate Liz II (*left*), who was produced by somatic cell nuclear transfer in 2003. Liz II had already begun to win championships by the time she was one year old.

Take-Home Message

What is cloning?

» Reproductive cloning technologies produce clones: genetically identical individuals.

» The DNA inside a living cell contains all the information necessary to build a new individual.

» Somatic cell nuclear transfer (SCNT) is a reproductive cloning technology in which nuclear DNA of an adult donor is transferred to an egg with no nucleus. The hybrid cell develops into an embryo that is genetically identical to the adult donor.

» Therapeutic cloning uses SCNT to produce human embryos for research.

Summary

Section 8.1 Making **clones**, or exact genetic copies, of adult animals is now a common practice. The techniques, while improving, are still far from perfect; many attempts are required to produce a clone, and clones that survive often have health problems. The practice continues to raise ethical questions.

Section 8.2 The DNA of eukaryotes is divided among a characteristic number of **chromosomes** that differ in length and shape. **Histone** proteins organize eukaryotic DNA into **nucleosomes**. When duplicated, a eukaryotic chromosome consists of two **sister chromatids** attached at a **centromere**.

Diploid cells have two of each type of chromosome. **Chromosome number** is the sum of all chromosomes in cells of a given type. A human body cell has twenty-three pairs of chromosomes. Members of a pair of **sex chromosomes** differ among males and females. All others are **autosomes**. Autosomes of a pair have the same length, shape, and centromere location, and they carry the same genes. A **karyotype** can reveal abnormalities in an individual's complement of chromosomes.

Section 8.3 Eighty years of experiments with bacteria and **bacteriophage** offered solid evidence that deoxyribonucleic acid (DNA), not protein, is the hereditary material of life.

Section 8.4 A DNA molecule is a polymer of nucleotides, and consists of two strands coiled into a double helix. A DNA nucleotide has a five-carbon sugar (deoxyribose), three phosphate groups, and one of four nitrogen-containing bases after which the nucleotide is named: adenine, thymine, guanine, or cytosine. The bases pair in a consistent way: adenine with thymine (A–T), and guanine with cytosine (G–C). The order of bases along the strands—the **DNA sequence**—varies among species and among individuals.

Section 8.5 The DNA sequence of an organism's chromosome(s) constitutes genetic information. A cell passes that information to offspring by copying its DNA before it divides. In the process of **DNA replication**, a double-stranded molecule of DNA is copied, and two double-stranded DNA molecules that are identical to the parent are the result. One strand of each molecule is new, and the other is parental; thus the name **semiconservative replication**.

During the replication process, the double helix unwinds and RNA **primers** form on the exposed single strands of DNA. Starting at the primers, **DNA polymerase** uses each strand as a template to assemble new, complementary strands of DNA from free nucleotides. Synthesis of one strand occurs discontinuously. **DNA ligase** seals any gaps to form a continuous strand.

Section 8.6 Proofreading by DNA polymerases corrects most base-pairing errors as they occur. Uncorrected replication errors become **mutations**—permanent changes in the nucleotide sequence of a cell's DNA. Cancer arises by mutations.

Environmental agents such as UV light cause DNA damage that can lead to replication errors, so damaged DNA is normally repaired before replication begins.

Section 8.7 Reproductive cloning technologies produce genetically identical individuals (clones). In **somatic cell nuclear transfer** (**SCNT**), one cell from an adult is fused with an egg that has had its nucleus removed. The hybrid cell is treated with electric shocks or another stimulus that provokes the cell to divide and begin developing into a new individual. SCNT with human cells, which is called **therapeutic cloning**, produces embryos that are used for stem cell research.

Self-Quiz

Answers in Appendix III

1. Chromosome number _____ .
   a. refers to a particular chromosome pair in a cell
   b. is an identifiable feature of a species
   c. is like a set of books
   d. all of the above

2. Sister chromatids connect at the _____ .

3. A karyotype reveals the _____ of a single cell.
   a. base sequences   c. hereditary information
   b. chromosomes   d. clones

4. The basic unit that structurally organizes a eukaryotic chromosome is the _____ .
   a. higher-order coiling   c. base sequence
   b. double helix   d. nucleosome

5. Which is *not* a nucleotide base in DNA?
   a. adenine   c. uracil   e. cytosine
   b. guanine   d. thymine   f. All are in DNA.

6. What are the base-pairing rules for DNA?
   a. A–G, T–C   c. A–U, C–G
   b. A–C, T–G   d. A–T, G–C

7. Energy that drives DNA synthesis comes from _____ .
   a. ATP   c. DNA nucleotides
   b. DNA polymerase   d. a and c

8. When DNA replication begins, _____ .
   a. the two DNA strands unwind from each other
   b. the two DNA strands condense for base transfers
   c. two DNA molecules bond
   d. old strands move to find new strands

9. DNA replication requires _____ .
   a. template DNA   d. primers
   b. free nucleotides   e. a through c
   c. DNA polymerase   f. all are required

10. Show the complementary strand of DNA that forms on this template DNA fragment during replication:

    5'—GGTTTCTTCAAGAGA—3'

Hershey–Chase Experiments The graph in **Figure 8.16** is reproduced from Hershey and Chase's original 1952 publication that showed DNA is the hereditary material of bacteriophage. The data are from the two experiments described in Section 8.3, in which bacteriophage DNA and protein were labeled with radioactive tracers and allowed to infect bacteria. The virus–bacteria mixtures were whirled in a blender to dislodge the viruses, and the tracers were tracked inside and outside of the bacteria.

1. Before blending, what percentage of ^{35}S was outside the bacteria? What percentage of ^{32}P was outside the bacteria?

2. After 4 minutes in the blender, what percentage of ^{35}S was outside the bacteria? What percentage of ^{32}P was outside the bacteria?

3. How did the researchers know that the radioisotopes in the fluid came from outside of the bacterial cells (extracellular) and not from bacteria that had been broken apart by whirling in the blender?

4. The extracellular concentration of which isotope, ^{35}S or ^{32}P, increased the most with blending? DNA contains much more phosphorus than do proteins; proteins contain much more sulfur than does DNA. Do these results imply that the viruses inject DNA or protein into bacteria? Why or why not?

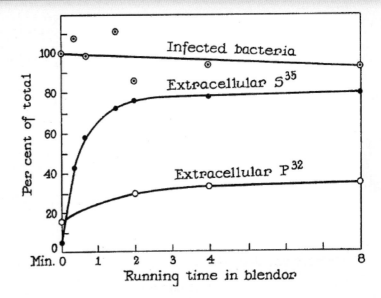

Figure 8.16 Detail of Alfred Hershey and Martha Chase's publication describing their experiments with bacteriophage. "Infected bacteria" refers to the percentage of bacteria that survived the blender. *Source:* "Independent Functions of Viral Protein and Nucleic Acid in Growth of Bacteriophage." *Journal of General Physiology*, 36(1), Sept. 20, 1952.

11. _____ is an example of reproductive cloning.
   a. Somatic cell nuclear transfer (SCNT)
   b. Multiple offspring from the same pregnancy
   c. Artificial embryo splitting
   d. a and c
   e. all of the above

12. The DNA of each species has unique _____ that set it apart from the DNA of all other species.
   a. nucleotides   c. sequences
   b. chromosomes   d. bases

13. Exposure to _____ can cause mutations.
   a. UV light   c. x-rays
   b. cigarette smoke   d. all of the above

14. SCNT with human cells is called _____ .
   a. hCNT   c. iSCNT
   b. therapeutic cloning d. all are correct

15. _____ can be used to produce genetically identical organisms (clones).
   a. SCNT   c. Therapeutic cloning
   b. Embryo splitting   d. all of the above

16. Match the terms appropriately.
   ___ bacteriophage   a. nitrogen-containing base, sugar, phosphate group(s)
   ___ clone
   ___ nucleotide   b. copy of an organism
   ___ diploid   c. does not determine sex
   ___ DNA ligase   d. only DNA and protein
   ___ DNA polymerase   e. fills in gaps, seals breaks in a DNA strand
   ___ autosome
   ___ mutation   f. two chromosomes of each type
                          g. adds nucleotides to a growing DNA strand
                          h. can cause cancer

Critical Thinking

1. Matthew Meselson and Franklin Stahl's experiments supported the semiconservative model of replication. These researchers obtained "heavy" DNA by growing *Escherichia coli* with ^{15}N, a radioactive isotope of nitrogen. They also prepared "light" DNA by growing *E. coli* in the presence of ^{14}N, the more common isotope. An available technique helped them identify which of the replicated molecules were heavy, light, or hybrid (one heavy strand and one light).

   Use different colored pencils to draw the heavy and light strands of DNA. Starting with a DNA molecule having two heavy strands, show the formation of the two molecules that result from replication in a ^{14}N-containing medium. Show the four DNA molecules that would form if those two molecules were replicated a second time in the ^{14}N medium. Would the DNA molecules that result from two replications in this medium be heavy, light, or mixed?

2. Woolly mammoths have been extinct for about 10,000 years, but we occasionally find one that has been preserved in Siberian permafrost. Resurrecting these huge elephant-like mammals may be possible by cloning DNA isolated from such frozen remains. Researchers are now studying the DNA of a remarkably intact baby mammoth recently discovered frozen in a Siberian swamp. What are some of the pros and cons, both technical and ethical, of cloning an extinct animal?

3. *Xeroderma pigmentosum* is an inherited disorder characterized by rapid formation of skin sores (*right*) that can develop into cancers. All forms of radiation trigger these symptoms, including fluorescent light, which contains UV light in the range of 320–400 nm. What normal function has been compromised in affected individuals?

LEARNING ROADMAP

Where you have been Your knowledge of chromosomes (Section 8.2) and base pairing (8.4) will help you understand how cells use nucleic acids (3.8) to build proteins (3.6). You will revisit hydrophobicity (2.5), hemoglobin (3.2), pathogenic bacteria (4.1), organelles (4.4, 4.7), radicals and cofactors (5.6), enzyme function (3.3, 5.4), DNA replication (8.5), and mutations (8.6).

Where you are now

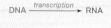

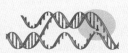

Gene Expression
The sequence of amino acids in a polypeptide is determined by a gene. The conversion of information in DNA to protein occurs in two steps: transcription and translation.

DNA to RNA: Transcription
During transcription, one strand of DNA serves as a template for assembling a single, complementary strand of RNA (a transcript). Each transcript is an RNA copy of a gene.

RNA
Messenger RNA carries DNA's protein-building instructions. Sixty-four mRNA codons represent the genetic code. Two other types of RNA translate that code.

RNA to Protein: Translation
Translation converts a sequence of codons in mRNA to a sequence of amino acids in a polypeptide. Transfer RNAs and ribosomes carry out the process.

Altered Proteins
Some mutations change a gene's DNA sequence so that an altered protein forms, an outcome that can have drastic consequences.

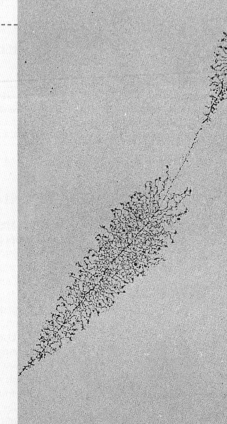

Where you are going What you learn in this chapter about genes will be the foundation for concepts of gene expression (Chapter 10), inheritance (Chapters 13 and 14), and genetic engineering (Chapter 15). Chapters 16 and 17 will show you how mutations are the raw material of natural selection and other processes of evolution. You will also revisit hemoglobin, sickle-cell anemia, and the circulatory system in Chapter 36, and immunity in Chapter 37.

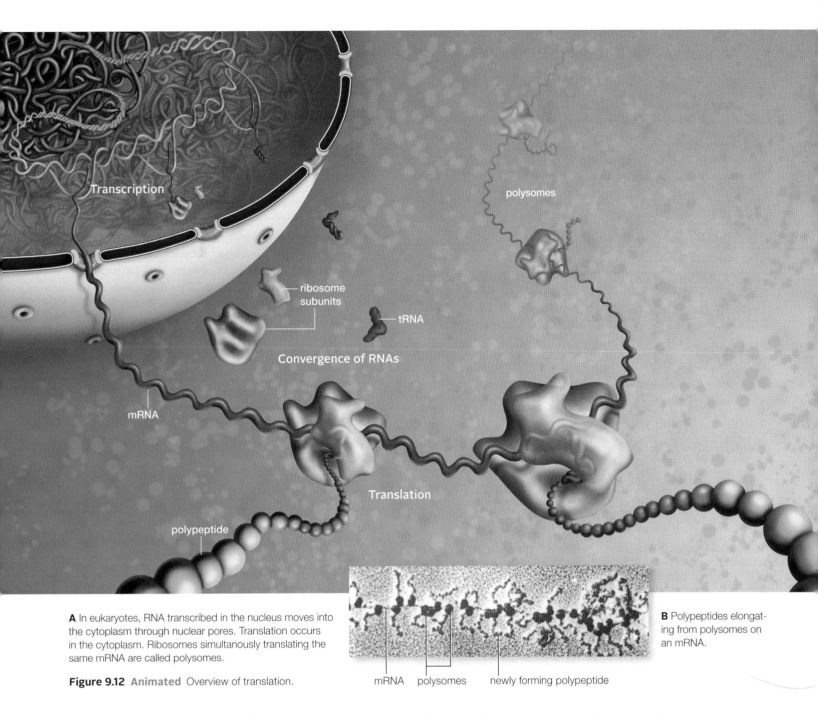

A In eukaryotes, RNA transcribed in the nucleus moves into the cytoplasm through nuclear pores. Translation occurs in the cytoplasm. Ribosomes simultanously translating the same mRNA are called polysomes.

Figure 9.12 Animated Overview of translation.

B Polypeptides elongating from polysomes on an mRNA.

mRNA   polysomes   newly forming polypeptide

In cells that are making a lot of protein, many ribosomes may simultaneously translate the same mRNA, in which case they are called polysomes (**Figure 9.12B**). In bacteria and archaea, transcription and translation both occur in the cytoplasm, and these processes are closely linked in time and space. Translation begins before transcription ends, so a transcription "Christmas tree" is often decorated with polysome "balls."

Given that many polypeptides can be translated from one mRNA, why would any cell also make many copies of an mRNA? Compared with DNA, RNA is not very stable. An mRNA may last only a few minutes before it is disassembled by enzymes in cytoplasm. The fast turnover allows cells to adjust their protein synthesis quickly in response to changing needs.

Translation is energy intensive. That energy is provided mainly in the form of phosphate-group transfers from the RNA nucleotide GTP (shown in **Figure 9.2B**) to molecules involved in the process.

Take-Home Message

How is mRNA translated into protein?

» Translation is an energy-requiring process that converts protein-building information carried by an mRNA into a polypeptide.

» During initiation, an mRNA, an initiator tRNA, and two ribosome subunits join.

» During elongation, amino acids are delivered to the complex by tRNAs in the order dictated by successive mRNA codons. As they arrive, the ribosome joins each to the end of the polypeptide chain.

» Termination occurs when the ribosome reaches a stop codon in the mRNA. The mRNA and the polypeptide are released, and the ribosome disassembles.

9.6 Mutated Genes and Their Protein Products

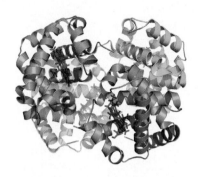

A Hemoglobin, an oxygen-binding protein in red blood cells. This protein consists of four polypeptides: two alpha globins (*blue*) and two beta globins (*green*). Each globin has a pocket that cradles a heme (*red*). Oxygen molecules bind to the iron atom at the center of each heme.

| 16 | 17 | 18 | 19 | 20 | 21 | 22 | 23 | 24 | 25 | 26 | 27 | 28 | 29 | 30 |

G G A C T C C T C T T C A G A
C C U G A G G A G A A G U C U

```
···· pro ──── glu ──── glu ──── lys ──── ser ····
```

B Part of the DNA (*blue*), mRNA (*brown*), and amino acid sequence (*green*) of human beta globin. Numbers indicate the position of the base pair in the coding sequence of the mRNA.

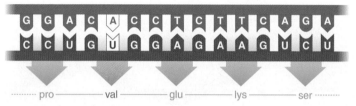

G G A C A C C T C T T C A G A
C C U G U G G A G A A G U C U

```
···· pro ──── val ──── glu ──── lys ──── ser ····
```

C A base-pair substitution replaces a thymine with an adenine. When the altered mRNA is translated, valine replaces glutamic acid as the sixth amino acid of the polypeptide. Hemoglobin with this form of beta globin is called HbS, or sickle hemoglobin.

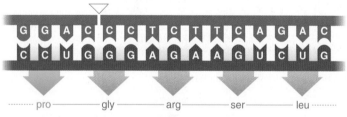

G G A C C T C T T C A G A C
C C U G G G A G A A G U C U G

```
···· pro ──── gly ──── arg ──── ser ──── leu ····
```

D A deletion of one base pair causes the reading frame for the rest of the mRNA to shift, so a different protein product forms. This frameshift results in a defective beta globin chain. The outcome is beta thalassemia, a genetic disorder in which a person has an abnormally low amount of hemoglobin.

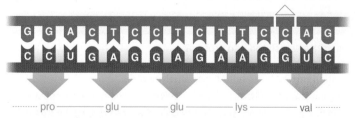

G G A C T C C T C T T C C A G
C C U G A G G A G A A G G U C

```
···· pro ──── glu ──── glu ──── lys ──── val ····
```

E An insertion of one base pair causes the reading frame for the rest of the mRNA to shift, so a different protein product forms. This frameshift results in a defective beta globin chain. The outcome is beta thalassemia.

Figure 9.13 Animated Examples of mutations in the human beta globin gene.

■ If the nucleotide sequence of a gene changes, it may result in an altered gene product, with harmful effects.

■ Links to Hydrophobicity 2.5, Hemoglobin 3.2, Protein structure 3.6, Free radicals and heme cofactors 5.6, Chromosomes 8.2, DNA replication 8.5, Mutations 8.6

Mutations, remember, are permanent changes in a DNA sequence (Section 8.6). A mutation in which one nucleotide and its partner are replaced by a different base pair is called a **base-pair substitution**. Other mutations involve the loss of one or more base pairs (a **deletion**) or the addition of extra base pairs (an **insertion**).

Mutations are relatively uncommon events in a normal cell. For example, the chromosomes in a diploid human cell collectively consist of about 7×10^9 base pairs, any of which may become mutated each time that cell divides. However, only about 175 bases actually do change after a division. On top of that, only about 3 percent of the cell's DNA encodes protein products, so there is a low probability that any of those mutations will be in a protein coding region.

When a mutation does occur in a protein coding region, the redundancy of the genetic code offers the cell a margin of safety. For example, a mutation that changes a UCU codon to UCC in an mRNA may not have further effects, because both codons specify serine. Mutations in a gene that have no effect on the gene's product are said to be silent. Other mutations are not silent: They may change an amino acid in a protein, or result in a premature stop codon that shortens it. Both outcomes can have severe consequences for the cell—and the organism.

The oxygen-binding properties of hemoglobin (Section 3.2) provide an example of how a mutation can change a protein. As red blood cells circulate through the lungs, the hemoglobin proteins inside of them bind to oxygen molecules. The cells then travel to other regions of the body, and the hemoglobin releases its oxygen cargo wherever the oxygen level is low. When the red blood cells return to the lungs, the hemoglobin binds to more oxygen.

Hemoglobin's structure allows it to bind and release oxygen. The protein consists of four polypeptides called globins (**Figure 9.13A**). Each globin folds around a heme, a cofactor with an iron atom at its center (Section 5.6). Oxygen molecules bind to hemoglobin at those iron atoms.

In adult humans, two alpha globin chains and two beta globin chains make up each hemoglobin molecule. Defects in either of the polypeptide chains can cause a condition called anemia, in which a person's blood is deficient in red blood cells or in hemoglobin.

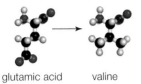

glutamic acid   valine

A A base-pair substitution results in the abnormal beta globin chain of sickle hemoglobin (HbS). The sixth amino acid in such chains is valine, not glutamic acid. The difference causes HbS molecules to form rod-shaped clumps that distort normally round blood cells into sickle shapes.

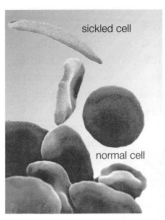

sickled cell

normal cell

B The sickled cells clog small blood vessels, causing circulatory problems that result in damage to many organs. Destruction of the cells by the body's immune system results in anemia.

Tionne "T-Boz" Watkins of the music group TLC is a celebrity spokesperson for the Sickle Cell Disease Association of America. She was diagnosed with sickle-cell anemia as a child.

Figure 9.14 Animated Sickle-cell anemia.

Both outcomes limit the blood's ability to carry oxygen, and the resulting symptoms range from mild to life-threatening.

Sickle-cell anemia, a type of anemia that is most common in people of African ancestry, arises because of a base-pair substitution in the beta globin gene. The substitution causes the body to produce a version of beta globin in which the sixth amino acid is valine instead of glutamic acid (**Figure 9.13B,C**). Hemoglobin with this mutation in its beta globin is called sickle hemoglobin, or HbS.

Unlike glutamic acid, which carries a negative charge, valine carries no charge. As a result of that one substitution, a tiny patch of the beta globin polypeptide changes from hydrophilic to hydrophobic, which in turn causes the hemoglobin's behavior to change slightly. HbS molecules stick together and form large, rodlike clumps under certain conditions. Red blood cells that contain the clumps become distorted into a crescent (sickle) shape (**Figure 9.14**). Sickled cells clog tiny blood vessels, thus disrupting blood circulation throughout the body. Over time, repeated episodes of sickling can damage organs and cause death.

A different type of anemia, beta thalassemia, is caused by the deletion of the twentieth base pair in the coding region of the beta globin gene (**Figure 9.13D**). Like many other deletions, this one causes a **frameshift**, in which the reading frame of the mRNA codons shifts. A frameshift usually has drastic consequences because it garbles the genetic message, just as incorrectly grouping a series of letters garbles the meaning of a sentence:

> The cat ate the rat.
> T hec ata tet her at.
> Th eca tat eth era t.

The frameshift caused by the beta globin deletion results in a polypeptide that differs drastically from normal beta globin. Hemoglobin molecules do not assemble correctly with the altered polypeptides, which are only 18 amino acids long (the normal beta globin chain has 147 amino acids). This outcome is the source of anemia in beta thalassemia.

Beta thalassemia can also be caused by insertion mutations, which, like deletions, often cause frameshifts (**Figure 9.13E**). Insertion mutations are often caused by the activity of **transposable elements**, which are segments of DNA that can move spontaneously within or between chromosomes. Transposable elements can be hundreds or thousands of base pairs long, so when one interrupts a gene it becomes a major insertion that changes the gene's product. Transposable elements are common in the DNA of all species; about 45 percent of human DNA consists of transposable elements or their remnants.

base-pair substitution Mutation in which a single base pair changes.
deletion Mutation in which one or more base pairs are lost.
frameshift Mutation that causes the reading frame of mRNA codons to shift.
insertion Mutation in which one or more base pairs become inserted into DNA.
transposable element Segment of chromosomal DNA that can spontaneously move to a new location.

Take-Home Message

What happens after a gene becomes mutated?

» Mutations that result in an altered protein can have drastic consequences.

» A base-pair substitution may change an amino acid in a protein, or shorten it by introducing a premature stop codon.

» Frameshifts that occur after an insertion or deletion change an mRNA's codon reading frame, so they garble its protein-building instructions.

The Aptly Acronymed RIPs (revisited)

The enzyme in ricin inactivates ribosomes by removing a particular adenine base from one of the rRNAs in the heavy subunit. That adenine is part of an RNA binding site for proteins that help with elongation. After the base has been removed, a ribosome can no longer bind to those proteins, and elongation stops.

However, the main function of RIPs may not be destroying ribosomes. Many function in immune signaling, but it is their antiviral and anti-

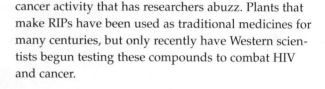

cancer activity that has researchers abuzz. Plants that make RIPs have been used as traditional medicines for many centuries, but only recently have Western scientists begun testing these compounds to combat HIV and cancer.

How would you vote? Exposure to ricin is unlikely, but extremists continue to attempt using it for terrorist activities. Researchers have developed a vaccine against ricin. Do you want to be vaccinated?

Summary

Section 9.1 The ability to make proteins is critical to all life processes. Molecules such as ricin and other proteins that inactivate ribosomes (RIPs) can be extremely toxic if they enter cells.

Section 9.2 DNA's genetic information is encoded within its base sequence. **Genes** are subunits of that sequence. A cell uses the information in a gene to make an RNA or protein product. The process of **gene expression** involves **transcription** of a DNA sequence to an RNA, and **translation** of the information in an **mRNA**, or **messenger RNA**, to a protein product. Translation requires the participation of **tRNA (transfer RNA)** and **rRNA (ribosomal RNA)**.

Section 9.3 During transcription, **RNA polymerase** binds to a **promoter** near a gene region of a chromosome. The polymerase assembles a strand of RNA by linking RNA nucleotides in the order dictated by the base sequence of the gene. Thus, the new RNA is complementary to the gene from which it was transcribed.

The RNA of eukaryotes is modified before it leaves the nucleus. **Introns** are removed. With **alternative splicing**, some **exons** may be removed also, and the remaining ones spliced in different combinations. A cap and a poly-A tail are also added to a new mRNA.

Section 9.4 mRNA carries DNA's protein-building information. The information consists of a series of **codons**, sets of three nucleotides. Sixty-four codons, most of which specify amino acids, constitute the **genetic code**. Each tRNA has an **anticodon** that can base-pair with a codon, and it binds to the amino acid specified by that codon. rRNA and proteins make up the two subunits of ribosomes.

Section 9.5 Genetic information carried by an mRNA directs the synthesis of a polypeptide during translation. First, an mRNA, an initiator tRNA, and two ribosomal subunits converge. The intact ribosome then catalyzes formation of a peptide bond between successive amino acids, which are delivered by tRNAs in the order specified by the codons in the mRNA. Translation ends when the ribosome encounters a stop codon.

Section 9.6 Insertions, **deletions**, and **base-pair substitutions** are mutations. The activity of **transposable elements** causes some mutations. A mutation that changes a gene's product may have harmful effects. Sickle-cell anemia, which is caused by a base-pair substitution in the gene for the beta globin chain of hemoglobin, is one example. Beta thalassemia is an outcome of a **frameshift** mutation in the beta globin gene.

Self-Quiz

Answers in Appendix III

1. A chromosome contains many different gene regions that are transcribed into different _____ .
   a. proteins   c. RNAs
   b. polypeptides   d. a and b

2. A binding site for RNA polymerase is called a _____ .
   a. gene   c. codon
   b. promoter   d. protein

3. Energy that drives transcription is provided mainly by _____ .
   a. ATP   c. GTP
   b. RNA nucleotides   d. all are correct

4. An RNA molecule is typically _____ ; a DNA molecule is typically _____ .
   a. single-stranded; double-stranded
   b. double-stranded; single-stranded
   c. both are single-stranded
   d. both are double-stranded

RIPs as Cancer Drugs Researchers are taking a page from the structure–function relationship of RIPs in their quest for cancer treatments. The most toxic RIPs, remember, have one domain that interferes with ribosomes, and another that carries them into cells. Melissa Cheung and her colleagues incorporated a peptide that binds to skin cancer cells into the enzymatic part of the *E. coli* entero-toxin RIP. The researchers created a new RIP that specifi-cally kills skin cancer cells, which are notoriously resistant to established therapies. Some of their results are shown in **Figure 9.15**.

1. Which cells had the greatest response to an increase in concentration of the engineered RIP?

2. At what concentration of RIP did all of the different kinds of cells survive?

3. Which cells survived best at 10^{-6} grams per liter RIP?

4. Why are some of the data points linked by straight lines?

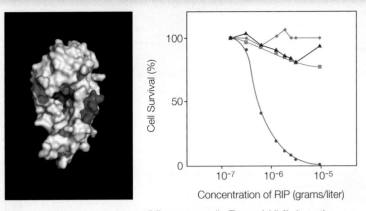

Figure 9.15 Effect of an engineered RIP on cancer cells. The model (*left*) shows the enzyme portion of *E. coli* enterotoxin that has been engineered to carry a small sequence of amino acids (in *blue*) that targets skin cancer cells. (*Red* indicates the active site.) The graph (*right*) shows the effect of this engineered RIP on human cancer cells of the skin (●); breast (♦); liver (▲); and prostate (■). *Source:* Cheung et al. *Molecular Cancer*, 9:28, 2010.

5. RNAs form by _____ ; proteins form by _____ .
   a. replication; translation
   b. translation; transcription
   c. transcription; translation
   d. replication; transcription

6. _____ different codons constitute the genetic code.
   a. 3   c. 20
   b. 64   d. 120

7. Most codons specify a(n) _____ .
   a. protein   c. amino acid
   b. polypeptide   d. mRNA

8. Anticodons pair with _____ .
   a. mRNA codons   c. RNA anticodons
   b. DNA codons   d. amino acids

9. What is the maximum length of a polypeptide encoded by an mRNA that is 45 nucleotides long?

10. _____ are removed from new mRNA transcripts.
   a. Introns   c. Telomeres
   b. Exons   d. Amino acids

11. Where does transcription take place in a eukaryotic cell?
   a. the nucleus   c. the cytoplasm
   b. extracellular fluid   d. b and c are correct

12. Where does translation take place in a typical eukaryotic cell?
   a. the nucleus   c. the cytoplasm
   b. extracellular fluid   d. b and c are correct

13. Energy that drives translation is provided mainly by _____ .
   a. ATP   c. GTP
   b. amino acids   d. all are correct

14. Each amino acid is specified by a set of _____ bases in an mRNA transcript.
   a. 3   b. 20   c. 64   d. 120

15. _____ are mutations.
   a. Transposable elements   d. Deletions
   b. Base-pair substitutions e. b, c, and d are correct
   c. Insertions   f. all of the above

16. Match the terms with the best description.
   ___ genetic message   a. coding region
   ___ promoter   b. gets around
   ___ polysome   c. read as base triplets
   ___ exon   d. removed before translation
   ___ genetic code   e. occurs only in groups
   ___ intron   f. complete set of 64 codons
   ___ transposable   g. binding site for RNA
       element   polymerase

Critical Thinking

1. Antisense drugs help us fight some types of cancer and viral diseases. The drugs consist of short mRNA strands that are complementary in base sequence to mRNAs linked to the diseases. Speculate on how antisense drugs work.

2. An anticodon has the sequence GCG. What amino acid does this tRNA carry? What would be the effect of a muta-tion that changed the C of the anticodon to a G?

3. Each position of a codon can be occupied by one of four nucleotides. What is the minimum number of nucleotides per codon necessary to specify all 20 of the amino acids that are typical of eukaryotic proteins?

4. Using **Figure 9.7**, translate this nucleotide sequence into an amino acid sequence, starting at the first base:

   5'—GGUUUCUUGAAGAGA—3'

5. Translate the sequence of bases in the previous question, starting at the second base.

6. The termination of prokaryotic DNA transcription often depends on the structure of a newly form-ing RNA. Transcription stops where the mRNA folds back on itself, forming a hairpin-looped structure like the one shown at *right*. How do you think that this structure stops transcription?

LEARNING ROADMAP

Where you have been This chapter explores metabolism (Sections 5.5, 5.6) in context of gene expression (9.2). You will be applying what you know about DNA: chromosomes (8.2), structure (8.4), replication (8.5), mutations and how they occur (9.6), and genetic reprogramming (8.7), as well as transcription (9.3) and translation (9.5). You will also revisit functional groups (3.3), carbohydrates (3.4), glycolysis (7.3), and fermentation (7.6).

Where you are now

Mechanisms of Gene Control
Gene expression changes in response to changing conditions both inside and outside the cell. Selective gene expression during development results in differentiation.

Master Genes
The orderly, localized expression of homeotic and other master genes gives rise to the body plan of complex multicelled organisms.

Examples in Eukaryotes
Genes that govern X chromosome inactivation and male sex determination in mammals, and flower formation in plants, offer examples of gene control in eukaryotes.

Examples in Prokaryotes
Most prokaryotic gene controls bring about fast adjustments in the rate of transcription in response to changes in external conditions.

Epigenetics
New research is revealing how gene expression changes that arise in response to environmental pressures can be passed to an individual's future offspring.

Where you are going Chapter 11 discusses the cell cycle and how controls over it go awry in cancer; Chapter 12 returns to the genetic basis of reproduction. Chapter 13 explores how inheritance works, and Chapter 14, inheritance patterns in humans. Genomics, the study of genomes, is explained in Section 15.5, and evolutionary processes involving changes in DNA sequence will be discussed more fully in Chapter 17. Section 19.3 discusses the hypothesis that RNA, not DNA, was genetic material in the distant past. Plant gene controls are discussed in more detail in Chapter 30.

10.1 Between You and Eternity

You are in college, your whole life ahead of you. Your risk of developing cancer is as remote as old age, an abstract statistic that is easy to forget. "There is a moment when everything changes—when the width of two fingers can suddenly be the total distance between you and eternity." Robin Shoulla wrote those words after being diagnosed with breast cancer. She was seventeen. At an age when most young women are thinking about school, friends, parties, and potential careers, Robin was dealing with radical mastectomy: the removal of a breast, all lymph nodes under the arm, and skeletal muscles in the chest wall under the breast. She was pleading with her oncologist not to use her jugular vein for chemotherapy and wondering if she would survive to see the next year (**Figure 10.1**).

Robin's ordeal became part of a statistic, one of more than 200,000 new cases of breast cancer diagnosed in the United States each year. About 5,700 of those cases occur in women and men under thirty-four years of age.

Every second, millions of cells in your skin, bone marrow, gut lining, liver, and elsewhere are dividing and replacing their worn-out, dead, and dying predecessors. They do not divide at random; in normal cells, growth and division is tightly regulated. When this control fails, cancer is the outcome.

Cancer is a multistep process in which abnormally growing and dividing cells disrupt body tissues. Mechanisms that normally keep cells from getting overcrowded in tissues are lost, so cancer cell populations may reach extremely high densities. Unless chemotherapy, surgery, or another procedure eradicates them, cancer cells can put an individual on a painful road to death. Each year, cancers cause 15 to 20 percent of all human deaths in developed countries alone.

Cancer typically begins with a mutation in a gene whose product is part of a system of stringent controls over cell growth and division. Such controls govern when and how fast specific genes are transcribed and translated. A cancer-causing mutation may be inherited, or it may be a new one, as when DNA becomes damaged by environmental agents. If the mutation alters the gene's protein product so that it no longer works properly, one level of control over the cell's growth and division has been lost. You will be considering the impact of gene controls in chapters throughout this book, and also in some chapters of your life.

Robin Shoulla survived. Radical mastectomy is rarely performed today, but it was her only option. Now, seventeen years later, she has what she calls a normal life: career, husband, children. Her goal as a cancer survivor: "To grow very old with gray hair and spreading hips, smiling."

normal cells in organized clusters

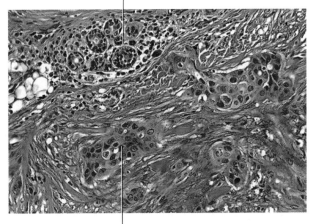

disorganized clusters of malignant cells

Figure 10.1 A case of breast cancer. *Top*, Robin Shoulla. *Bottom*, this light micrograph shows irregular clusters of cancer cells that have infiltrated milk ducts in human breast tissue. Diagnostic tests revealed abnormal cells such as these in Robin's body when she was just seventeen years old.

cancer Disease that occurs when the uncontrolled growth of body cells that can invade other tissues physically and metabolically disrupts normal function.

10.2 Switching Genes On and Off

■ Gene controls govern the kinds and amounts of substances that are present in a cell at any given time.
■ Links to Functional groups 3.3, Chromatin 4.6, Phosphorylation 5.6, Glycolysis 7.3, Histones 8.2, Gene expression 9.2, Transcription 9.3, Translation 9.5, Globin 9.6

All of the cells in your body are descended from the same fertilized egg, so they all contain the same DNA with the same genes. However, each cell rarely uses more than 10 percent of its genes at one time. Some of those genes affect structural features and metabolic pathways common to all cells. Others are expressed only by certain subsets of your cells. For example,

Figure 10.3 Hypothetical part of a chromosome that contains a gene. Molecules that affect the rate of transcription of the gene bind at promoter (*yellow*) or enhancer (*green*) sequences.

most of your body cells express genes that encode the enzymes of glycolysis, but only your immature red blood cells express genes that encode globin.

Differentiation, the process by which cells of a multicelled organism become specialized, occurs as different cell lineages begin to express different subsets of their genes during development. Which genes a cell uses determines the molecules it will produce, which in turn determines what kind of cell it will be.

Gene Controls

Control over which genes are expressed at a particular time is crucial for proper development of complex, multicelled bodies. It also allows individual cells to respond appropriately to changes in their environment. The "switches" that turn gene expression on or off are called gene controls—molecules or structures that start, enhance, slow, or stop the individual steps of gene expression (**Figure 10.2**).

Transcription Many gene controls affect whether and how fast a gene is transcribed into RNA ❶. Proteins called **transcription factors** affect the rate of transcription by binding directly to DNA at promoters or other special nucleotide sequences. Whether and how fast a gene is transcribed depends on which transcription factors are bound to the DNA. A **repressor** slows or stops transcription by interfering with RNA polymerase binding to a promoter. An **activator** speeds up transcription when it binds. Many activators work by helping RNA polymerase attach to the promoter. Some speed transcription by binding to DNA sequences called **enhancers** (**Figure 10.3**). An enhancer is not necessarily close to the gene it affects, and may even be on a different chromosome.

Chromatin structure also affects transcription: RNA polymerase can only attach to DNA that has been unwound from histones (Section 8.2). Certain modifications to histone proteins change the way they interact with the DNA that wraps around them. Some modifications make them release their grip on the DNA; others make them tighten it. For example, adding an acetyl group to a histone loosens the DNA wrapped around it, so enzymes that acetylate histones encourage transcription. Conversely, adding a methyl

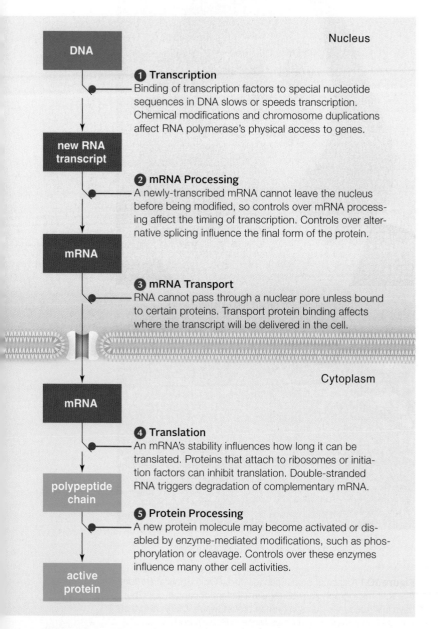

Figure 10.2 Animated Points of control over eukaryotic gene expression.

Nucleus

❶ **Transcription**
Binding of transcription factors to special nucleotide sequences in DNA slows or speeds transcription. Chemical modifications and chromosome duplications affect RNA polymerase's physical access to genes.

❷ **mRNA Processing**
A newly-transcribed mRNA cannot leave the nucleus before being modified, so controls over mRNA processing affect the timing of transcription. Controls over alternative splicing influence the final form of the protein.

❸ **mRNA Transport**
RNA cannot pass through a nuclear pore unless bound to certain proteins. Transport protein binding affects where the transcript will be delivered in the cell.

Cytoplasm

❹ **Translation**
An mRNA's stability influences how long it can be translated. Proteins that attach to ribosomes or initiation factors can inhibit translation. Double-stranded RNA triggers degradation of complementary mRNA.

❺ **Protein Processing**
A new protein molecule may become activated or disabled by enzyme-mediated modifications, such as phosphorylation or cleavage. Controls over these enzymes influence many other cell activities.

transcription start site →   ← transcription end

group to a histone can make the DNA wind more tightly around it, so enzymes that methylate histones discourage transcription.

In a few types of cells, the number of copies of a gene affects how fast its product is made. DNA in these cells is copied repeatedly. The result is a set of

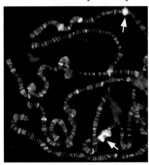

polytene chromosomes, each of which consists of hundreds or thousands of side-by-side copies of the same DNA molecule. Transcription of one gene occurs simultaneously on all of the identical DNA strands, and is visible as a puff on the chromosome (*above*). Transcription produces a lot of mRNA that can be translated quickly into a lot of protein. Polytene chromosomes are common in immature amphibian eggs, storage tissues of some plants, and saliva gland cells of some insect larvae.

mRNA Processing As you know, before eukaryotic mRNAs leave the nucleus, they are modified—spliced, capped, and finished with a poly-A tail (Section 9.3). Controls over these modifications can affect the form of a protein product and when it will appear in the cell ❷. For example, controls that determine which exons are spliced out of an mRNA affect which form of a protein will be translated from it.

mRNA Transport In eukaryotes, transcription occurs in the nucleus, and translation in the cytoplasm. Thus, mRNA transport out of the nucleus is another potential point of control ❸. A new RNA can pass through pores of the nuclear envelope only after it has been processed appropriately. Mechanisms that delay processing of an mRNA also delay its appearance in the cytoplasm, and thereby delay its translation.

Transport of mRNA outside the nucleus also influences gene expression. A short base sequence near an mRNA's poly-A tail is like a zip code. Proteins that attach to the zip code drag the mRNA along cytoskeletal elements to the organelle or area of the cytoplasm specified by the code. Other proteins that attach to the zip code region prevent translation of the mRNA before it reaches its destination. Localization of mRNA to a particular area of a cell allows growth or move-

ment in a specific direction. In an egg, it is crucial for proper development of a forthcoming embryo.

Translational Control Most controls over eukaryotic gene expression affect translation ❹. Many of these controls govern the production or function of the various molecules that carry out the process. Others affect mRNA stability: The longer an mRNA lasts, the more protein can be made from it. Enzymes begin to disassemble a new mRNA as soon as it arrives in the cytoplasm. The fast turnover allows cells to adjust their protein synthesis quickly in response to changing needs. How long an mRNA persists depends on its base sequence, the length of its poly-A tail, and which proteins are attached to it.

MicroRNAs inhibit translation. Part of a microRNA folds back on itself and forms a small double-stranded region. By a process called RNA interference, any double-stranded RNA (including a microRNA) is cut up into small bits that are taken up by special enzyme complexes. These complexes destroy every mRNA in a cell that can base-pair with the bits. So, expression of a microRNA complementary in sequence to a gene inhibits expression of that gene.

Post-Translational Modification Many newly synthesized polypeptide chains must be modified before they become functional ❺. For example, some enzymes become active only after they have been phosphorylated (Section 5.6). Such post-translational modifications inhibit, activate, or stabilize many molecules, including the enzymes that participate in transcription and translation.

activator Regulatory protein that increases the rate of transcription when it binds to a promoter or enhancer.
differentiation The process by which cells become specialized.
enhancer Binding site in DNA for proteins that enhance the rate of transcription.
repressor Regulatory protein that blocks transcription.
transcription factor Regulatory protein that influences transcription; e.g., an activator or repressor.

Take-Home Message

What is gene control?

» Gene controls consist of molecules and structures that can start, enhance, slow, or stop individual steps of gene expression.

» Most cells of multicelled organisms differentiate as they start expressing a unique subset of their genes. Which genes a cell expresses depends on the type of organism, its stage of development, and environmental conditions.

10.3 Master Genes

■ Cascades of gene expression govern the development of a complex, multicelled body.

As an animal embryo develops, its differentiating cells form tissues, organs, and body parts. Some cells that alternately migrate and stick to other cells develop into nerves, blood vessels, and other structures that weave through the tissues. Events like these fill in the body's details, and all are driven by cascades of master gene expression. **Master genes** encode products that affect the expression of many other genes. Expression of a master gene causes other genes to be expressed, with the final outcome being the completion of an intricate task such as the formation of an eye.

Pattern formation is the process by which a complex body forms from local processes in an embryo. Pattern formation begins when maternal mRNAs are delivered to opposite ends of an unfertilized egg as it forms. These mRNAs are translated only after the egg is fertilized, and then their protein products diffuse away in gradients that span the entire developing embryo. Depending on where a nucleus falls within the gradients, one or another set of master genes will be transcribed inside of it. The products of those master genes also form in gradients that span the embryo. Still other master genes are transcribed depending on where a nucleus falls within these gradients, and so on.

Such regional gene expression during development results in a three-dimensional map that consists of overlapping concentration gradients of master gene products. Which master genes are transcribed at any given time changes, so the map changes too.

Figure 10.5
A homeodomain. The protein product of a homeotic gene called *antennapedia* (*gold*) is shown attached to a promoter sequence in a fragment of DNA. The region of the protein that recognizes and binds directly to regulatory sequences in DNA is the homeodomain.

Eventually, the products of some master genes cause undifferentiated cells to differentiate, and specialized tissues and structures are the outcome (**Figure 10.4**).

Homeotic Genes

Master genes called **homeotic genes** control the formation of specific body parts (eyes, legs, and so on) during development. All homeotic genes encode transcription factors with a homeodomain, which is a region of about sixty amino acids that can bind to a promoter or some other sequence of nucleotides in a chromosome (**Figure 10.5**).

Long before body parts develop, certain master genes trigger the transcription of homeotic genes in particular parts of the embryo. Products of the different homeotic genes cause cells in those parts to differentiate into tissues that form specific structures such as a wing, a leg, or an eye.

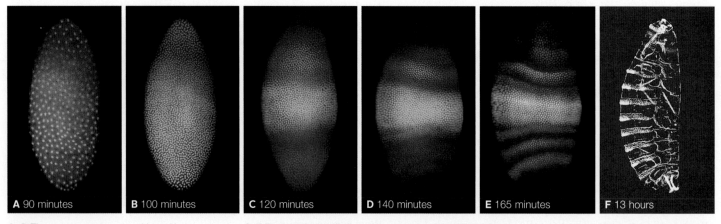

A 90 minutes   **B** 100 minutes   **C** 120 minutes   **D** 140 minutes   **E** 165 minutes   **F** 13 hours

A, B The master gene *even-skipped* is expressed (in *red*) only where two maternal gene products (*blue* and *green*) overlap.

C–E The products of several master genes, including the two shown here in *green* and *blue*, confine the expression of *even-skipped* (*red*) to seven stripes.

F Seven segments that develop later correspond to the position of the *even-skipped* stripes.

Figure 10.4 How gene expression control makes a fly, as illuminated by segmentation. Expression of different master genes is shown by different colors in fluorescence microscopy images of whole *Drosophila* embryos at successive stages of development. Bright dots are individual nuclei.

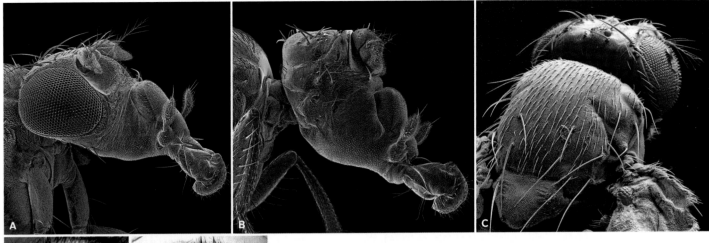

Figure 10.6 *Eyeless*: the eyes have it. (**A**) A normal fruit fly has large, round eyes. (**B**) A fruit fly with a mutation in its *eyeless* gene develops with no eyes. (**C**) Eyes form wherever the *eyeless* gene is expressed in fly embryos—here, on the head and wing.

Humans, mice, squids, and other animals have a gene called *PAX6*. In humans, *PAX6* mutations result in missing irises, a condition called aniridia (**D**). Compare a normal iris (**E**). *PAX6* is so similar to *eyeless* that it triggers eye development when expressed in fly embryos.

The function of many homeotic genes has been discovered by manipulating their expression, one at a time. Researchers inactivate a homeotic gene by introducing a mutation or deleting it entirely, an experiment called a **knockout**. An organism that carries the knocked-out gene may differ from normal individuals, and the differences are clues to the function of the missing gene product.

normal fly head

groucho mutation

Researchers often name homeotic genes based on what happens in their absence. For instance, fruit flies with a mutated *eyeless* gene develop with no eyes (**Figure 10.6A,B**). *Dunce* is required for learning and memory. *Wingless*, *wrinkled*, and *minibrain* are self-explanatory. *Tinman* is necessary for development of a heart. Flies with a mutated *groucho* gene have extra bristles above their eyes (*inset*). One gene was named *toll*, after what its German discoverer exclaimed upon seeing the disastrous effects of the mutation (*toll* means "cool!" in German slang).

Homeotic genes control development by the same mechanisms in all multicelled eukaryotes, and many are interchangeable among different species. Thus, we can infer that they evolved in the most ancient eukaryotic cells. Homeodomains often differ among species only in conservative substitutions (one amino acid has replaced another with similar chemical properties).

Consider the *eyeless* gene. Eyes form in embryonic fruit flies wherever this gene is expressed, which is typically in tissues of the head. If the *eyeless* gene is expressed in another part of the developing embryo, eyes form there too (**Figure 10.6C**). Humans, squids, mice, fish, and many other animals have a gene called *PAX6*, which is very similar to the *eyeless* gene. In humans, mutations in *PAX6* cause eye disorders such as aniridia, in which a person's irises are underdeveloped or missing (**Figure 10.6D,E**). *PAX6* works the same way in animals of different phyla. For example, if the *PAX6* gene from a human or mouse is inserted into an *eyeless* mutant fly, it has the same effect as the *eyeless* gene: An eye forms wherever it is expressed. Such studies are evidence of shared ancestry among these evolutionarily distant animals.

homeotic gene Type of master gene; its expression controls formation of specific body parts during development.
knockout An experiment in which a gene is deliberately inactivated in a living organism.
master gene Gene encoding a product that affects the expression of many other genes.
pattern formation Process by which a complex body forms from local processes during embryonic development.

Take-Home Message

How do genes control development?

» Development is orchestrated by cascades of master gene expression in embryos.

» The expression of homeotic genes during development governs the formation of specific body parts. Homeotic genes that function in similar ways across taxa are evidence of shared ancestry.

10.4 Examples of Gene Control in Eukaryotes

- Selective gene expression gives rise to many traits.
- Links to Chromosomes and sex determination 8.2, Mutations 9.6

Gene controls influence many traits that are characteristic of humans and other eukaryotic organisms, as the following examples illustrate.

X Chromosome Inactivation

In humans and other mammals, a female's cells each contain two X chromosomes, one inherited from her mother, the other one from her father. One X chromosome is always tightly condensed (**Figure 10.7A**). We call the condensed X chromosomes "Barr bodies," after Murray Barr, who discovered them. Condensation prevents transcription, so most of the genes on a Barr body are not expressed. This **X chromosome inactivation** ensures that only one of the two X chromosomes in a female's cells is active. According to a theory called **dosage compensation**, X chromosome inactivation equalizes expression of X chromosome genes between the sexes. The body cells of male mammals (XY) have one set of X chromosome genes. Body cells of female mammals have two sets, but female embryos do not develop properly when both sets are expressed.

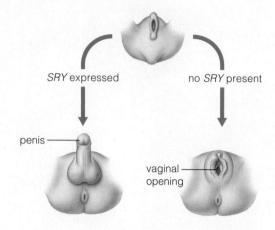

Figure 10.8 Development of reproductive organs in human embryos. An early embryo appears neither male nor female. *SRY* gene expression determines what reproductive organs form.

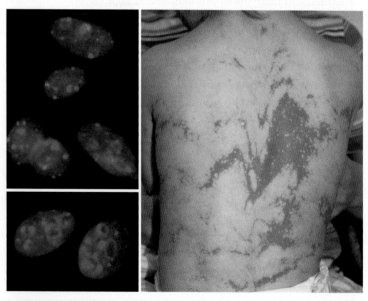

A Inactivated X chromosomes—Barr bodies—show up as *red* spots in the nucleus of these XX cells (*top*). Compare the nucleus of two XY cells (*bottom*).

B Visible evidence of mosaic tissues in a human female. One of this girl's X chromosomes carries a mutation that causes incontinentia pigmenti, a disorder that affects the skin, hair, nails, and teeth. This chromosome is active in darker patches of skin. Her other X chromosome, which does not carry the mutation, is active in the lighter patches of skin.

Figure 10.7 Animated X chromosome inactivation.

X chromosome inactivation occurs when an embryo is a ball of about 200 cells. In humans and many other mammals, it occurs independently in every cell of a female embryo. The maternal X chromosome may get inactivated in one cell, and the paternal or maternal X chromosome may get inactivated in a cell next to it. Once the selection is made in a cell, all of that cell's descendants make the same selection as they continue dividing and forming tissues. As a result of the inactivation, a female mammal is a "mosaic" for the expression of genes on the X chromosome. She has patches of tissue in which genes on the maternal X chromosome are expressed, and patches in which genes on the paternal X chromosome are expressed (**Figure 10.7B**).

How does just one of two X chromosomes get inactivated? A gene called *XIST* is transcribed on only one of the two X chromosomes. The gene's product, a long, noncoding RNA, sticks to the chromosome that expresses the gene. The RNA coats the chromosome, and by an unknown mechanism causes it to condense into a Barr body. Thus, transcription of the *XIST* gene keeps a chromosome from transcribing other genes. The other chromosome does not express *XIST*, so it does not get coated with RNA; its genes remain available for transcription. How a cell "chooses" which X chromosome will express *XIST* remains unknown.

Male Sex Determination in Humans

The human X chromosome carries 1,336 genes. Some of those genes are associated with sexual traits, such as the distribution of body fat and hair. However, most of the genes on the X chromosome govern nonsexual traits such as blood clotting and color perception. Such

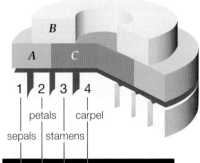

A The pattern in which the floral identity genes *A*, *B*, and *C* are expressed affects differentiation of cells growing in whorls in the plant's tips. Their gene products guide expression of other genes in cells of each whorl; a flower results.

B Mutations in *ABC* genes result in malformed flowers.

Top left, right: Flowers of plants with *A* gene mutations have no petals, and no structures in place of the missing petals.

Bottom left: Flowers of plants with *B* gene mutations have sepals instead of petals.

Bottom right: Flowers of plants with *C* gene mutations have petals instead of sepals and carpels. Compare the normal flower in **A**.

Figure 10.9 **Animated** Control of flower formation, as revealed by mutations in *Arabidopsis thaliana*.

genes are expressed in both males and females. Males, remember, also inherit one X chromosome.

The human Y chromosome carries only 307 genes, but one of them is *SRY*—the master gene for male sex determination in mammals. Its expression in XY embryos triggers the formation of testes, which are male gonads (**Figure 10.8**). Some of the cells in these primary male reproductive organs make testosterone, a sex hormone that controls the emergence of male secondary sexual traits such as facial hair, increased musculature, and a deep voice. We know that *SRY* is the master gene that controls emergence of male sexual traits because mutations in this gene cause XY individuals to develop external genitalia that appear female. An XX embryo has no Y chromosome, no *SRY* gene, and much less testosterone, so primary female reproductive organs (ovaries) form instead of testes. Ovaries make estrogens and other sex hormones that will govern the development of female secondary sexual traits, such as enlarged, functional breasts, and fat deposits around the hips and thighs.

Flower Formation

When it is time for a plant to flower, populations of cells that would otherwise give rise to leaves instead differentiate into floral parts—sepals, petals, stamens, and carpels. How does the switch happen? Three sets of master genes called *A*, *B*, and *C* guide the development of specialized parts of a flower. These genes are switched on by environmental cues such as seasonal changes in the length of night, as you will see in Section 30.9.

At the tip of a shoot, cells form whorls of tissue, one over the other like layers of an onion. Cells in each whorl give rise to different tissues depending on which of their *ABC* genes are activated (**Figure 10.9**). Studies of the phenotypic effects of mutations elucidated the function of these genes. In the outer whorl, only the *A* genes are switched on, and their products trigger events that cause sepals to form. Cells in the next whorl express both *A* and *B* genes; they give rise to petals. Cells farther in express *B* and *C* genes; they give rise to stamens, the structures that produce male reproductive cells. The cells of the innermost whorl express only the *C* genes; they give rise to carpels, the structures that produce female reproductive cells.

dosage compensation Theory that X chromosome inactivation equalizes gene expression between males and females.
X chromosome inactivation Shutdown of one of the two X chromosomes in the cells of female mammals.

10.5 Examples of Gene Control in Prokaryotes

■ Bacteria control gene expression mainly by adjusting the rate of transcription.
■ Links to Carbohydrates 3.4, Controls over metabolism 5.5, Lactate fermentation 7.6

Bacteria and archaea do not undergo development, so these cells have no need of master genes. However, they do respond to environmental fluctuations by adjusting gene expression. For example, when a preferred nutrient becomes available, a bacterium begins transcribing genes whose products allow the cell to use that nutrient. When the nutrient is no longer available, transcription of those genes stops. Thus, the cell does not waste energy and resources producing gene products that are not needed at a particular moment.

Bacteria control gene expression mainly by adjusting the rate of transcription. Genes that are used together often occur together on the chromosome, one after the other. A single promoter precedes the genes, so all are transcribed together into a single RNA strand. Thus, their transcription is controllable in a single step. Transcription control may occur at an **operator**, a region of DNA that serves as a binding site for a repressor. (Repressors, remember, stop transcription.) A promoter and one or more operators that together control transcription of multiple genes are collectively called an **operon**. Operons occur in bacteria, archaea, and eukaryotes.

The Lac Operon

Escherichia coli lives in the gut of mammals, where it dines on dissolved nutrients traveling past. Its carbohydrate of choice is glucose, but it can also use other sugars such as the lactose in milk. The operon that controls lactose metabolism in *E. coli* is called the lac operon (**Figure 10.10**). This operon includes three genes and a promoter flanked by two operators ❶.

Lactose consists of two monosaccharide monomers: glucose and galactose. The three genes in the

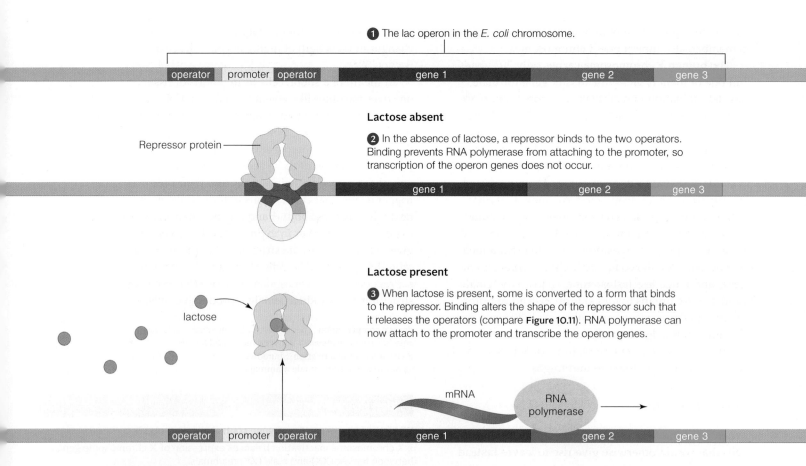

❶ The lac operon in the *E. coli* chromosome.

| operator | promoter | operator | gene 1 | gene 2 | gene 3 |

Lactose absent

❷ In the absence of lactose, a repressor binds to the two operators. Binding prevents RNA polymerase from attaching to the promoter, so transcription of the operon genes does not occur.

Repressor protein

| gene 1 | gene 2 | gene 3 |

Lactose present

❸ When lactose is present, some is converted to a form that binds to the repressor. Binding alters the shape of the repressor such that it releases the operators (compare **Figure 10.11**). RNA polymerase can now attach to the promoter and transcribe the operon genes.

lactose

mRNA

RNA polymerase

| operator | promoter | operator | gene 1 | gene 2 | gene 3 |

Figure 10.10 **Animated** Example of gene control in bacteria: the lactose operon on a bacterial chromosome. The operon consists of a promoter flanked by two operators, and three genes for lactose-metabolizing enzymes. **Figure It Out: What portion of the operon binds RNA polymerase when lactose is present?** Answer: The promoter

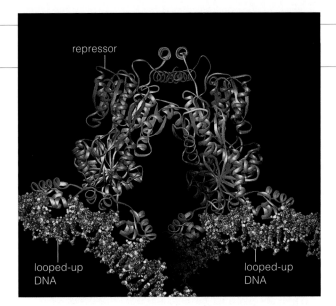

Figure 10.11 Model of the lactose operon repressor, shown here bound to operators. Binding twists the bacterial chromosome into a loop, which in turn prevents RNA polymerase from binding to the lac operon promoter.

repressor

looped-up DNA

looped-up DNA

lac operon encode proteins that allow *E. coli* to harvest glucose monomers from lactose. Bacteria conserve resources by making these proteins only when lactose is present. When lactose is not present, a repressor binds to the two operators, and lactose-metabolizing genes stay switched off. A repressor molecule that binds to the operators twists the region of DNA with the promoter into a loop (**Figure 10.11**). RNA polymerase cannot bind to the twisted-up promoter, so it cannot transcribe the operon genes ❷.

When lactose is present in the gut, some of it is converted to another sugar that binds to the repressor and changes its shape. The altered repressor can no longer stay attached to the DNA; it releases the operators and the looped DNA unwinds. The promoter is now accessible to RNA polymerase, and transcription begins ❸.

In bacteria, glucose metabolism requires fewer enzymes than lactose metabolism does, so it requires less energy and fewer resources. Accordingly, when both lactose and glucose are present, the cells will use up all of the available glucose before switching to lactose metabolism. How does a cell ignore the presence of one sugar while it uses another? Another level of gene control over the lac operon regulates this metabolic switch, but how it works is still being debated even after decades of research.

Lactose Intolerance

Like *E. coli*, humans and other mammals break down lactose into monosaccharide subunits, but most do so only when they are young. An individual's ability to digest lactose declines at a species-specific age. In most humans, the switch occurs at about age five, with

a programmed shutdown of the gene encoding lactase. The resulting decline in production of this lactose-metabolizing enzyme results in a common condition known as lactose intolerance.

Cells in the intestinal lining secrete lactase into the small intestine, where the enzyme cleaves lactose into its glucose and galactose monomers. Monosaccharides are absorbed directly by the small intestine, but lactose and other disaccharides are not. Thus, when lactase secretion slows, lactose passes undigested through the small intestine. Lactose ends up in the large intestine, which hosts huge numbers of *E. coli* and a variety of other bacteria. These resident organisms respond to the abundant sugar supply by switching on their lac operons. Carbon dioxide, methane, hydrogen, and other gaseous products of their various fermentation reactions accumulate quickly in the large intestine, distending its wall and causing pain. The other products of their metabolism (undigested carbohydrates) disrupt the solute–water balance inside the large intestine, and diarrhea results.

Not everybody is lactose intolerant. Many people of northern and central European ancestry carry a mutation in one of the genes responsible for programmed lactase shutdown. These people make enough lactase to continue drinking milk without problems into adulthood.

Riboswitches

Some bacterial mRNAs regulate their own translation with riboswitches, which are small sequences of RNA nucleotides that bind to a target molecule. The binding affects translation of the mRNA. Consider what happens when bacteria make vitamin B_{12}. The enzymes involved in synthesis of this vitamin are produced from mRNAs that have riboswitches. These particular riboswitches bind to vitamin B_{12}. Binding changes the shape of the mRNA so that ribosomes can no longer attach to it, and translation stops. This example also illustrates feedback inhibition, in which an end product inhibits its own production.

operator Part of an operon; a DNA binding site for a repressor.
operon Group of genes together with a promoter–operator DNA sequence that controls their transcription.

Take-Home Message

Do bacteria control gene expression?

» In bacteria, the main gene expression controls regulate gene expression in response to shifts in nutrient availability and other environmental conditions.

» Prokaryotes can regulate gene expression using operons and riboswitches.

10.6 Epigenetics

- Methylations and other modifications that accumulate in DNA during an individual's lifetime can be passed to offspring.
- Links to DNA structure 8.4, DNA replication 8.5, Environmental causes of mutations 8.6, Reprogramming DNA 8.7, Transposable elements 9.6

Heritable Methylations

You learned in Section 10.2 that methylation of histone proteins silences DNA transcription. Direct methylation of DNA also suppresses gene expression, but in an often more permanent manner than histone modification. The *XIST* gene offers an example. The one active X chromosome in cells of female mammals does not express this gene because its promoter is heavily methylated. Methylation also suppresses the activity of transposable elements, so it is important for the stability of a cell's chromosomes. Cancer is often associated with the loss of methylation.

Between 3 and 6 percent of the DNA in a normal, differentiated body cell is methylated. Methyl groups are most often attached to a cytosine that is followed by a guanine (**Figure 10.12A**), but which of these cytosines are methylated and which are not varies by the individual. The pattern starts early. For instance, individuals conceived during a famine end up with an unusually low number of methyl groups attached to certain genes. One of those genes encodes a hormone that fosters prenatal growth and development. The resulting increase in expression of this gene may offer a survival advantage in a poor nutritional environment.

Figure 10.13 A cause of epigenetic changes. In 1944, a supply blockade followed by an unusually harsh winter caused a severe famine in the Nazi-occupied Netherlands. Grandsons of boys who endured the famine (such as the one pictured in this photo) lived about 32 years longer than grandsons of boys who ate well during the same winter.

An individual's DNA also acquires methyl groups on an ongoing basis. Once a particular base has become methylated in a cell's DNA, it will usually stay methylated in all of the cell's descendants (**Figure 10.12B**). Genes actively expressed in early development become silenced as their promoters get methylated during differentiation. Environmental factors, including the chemicals in cigarette smoke, add more methyl groups. Methylation also occurs by chance during DNA replication, so cells that divide a lot tend to have more methyl groups in their DNA than inactive cells. Thus, environmental factors that result in a gene being expressed can have long-term effects.

When an organism reproduces, it passes its chromosomes to offspring. The methylation of parental chromosomes is normally "reset" in the first cell of the new individual, with new methyl groups being added and old ones being removed. This reprogramming does not remove all of the parental methyl groups, however, so methylations acquired during an individual's lifetime can be passed to future offspring.

Any heritable changes in gene expression that are not due to changes in the underlying DNA sequence are said to be **epigenetic**. Epigenetic inheritance can adapt offspring to an environmental stressor much more quickly than evolutionary processes. Epigenetic

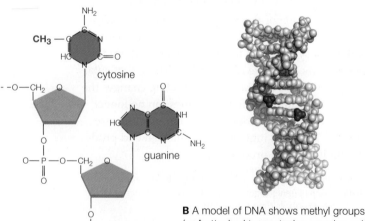

A In the DNA of differentiated cells, a methyl group (*red*) is most often attached to a cytosine that is followed by a guanine.

B A model of DNA shows methyl groups (*red*) attached to a cytosine–guanine pair on complementary DNA strands. When the cytosine on one strand is methylated, enzymes methylate the cytosine on the other strand. This is why a methylation tends to persist in a cell's descendants.

Figure 10.12 DNA methylation.

Between You and Eternity (revisited)

An effective cancer treatment must eliminate cancerous cells from a person's body. However, cancer cells are body cells, so drugs and other therapies that kill them also kill normal body cells. There is often a fine line between eliminating cancer from a patient's body, and killing the patient.

A normal cell has layers upon layers of gene expression controls and fail-safe mechanisms that determine when the division occurs and when it does not. This finely tuned system becomes unbalanced in a cancer cell, so that some genes are expressed at a higher level than they should be, and some are expressed at a lower level. For three decades, researchers have been studying the system and how it goes awry in cancer, because understanding how a cancer cell differs from a normal cell at a molecular level offers our best chance at finding a cure.

In the early 1980s, researchers discovered that mutations in some genes predispose individuals to develop certain kinds of cancer. Some of the genes are tumor suppressors, so named because tumors are more likely to occur when these genes mutate. Two examples are *BRCA1* and *BRCA2*: A mutated version of one or both of these genes is often found in breast and ovarian cancer cells. Because mutations in genes such as *BRCA* can be inherited, cancer is not only a disease of the elderly, as Robin Shoulla's story illustrates. Robin is one of the unlucky people who carry mutations in both *BRCA1* and *BRCA2*.

If a *BRCA* gene mutates in one of three especially dangerous ways, a woman has an 80 percent chance of developing breast cancer before the age of seventy. *BRCA* genes are master genes whose protein products help maintain the structure and number of chromosomes in a dividing cell. The multiple functions of these proteins are still being unraveled. We do know they participate directly in DNA repair (Section 8.6),

so any mutations that alter this function also alter the cell's capacity to repair damaged DNA. Other mutations are likely to accumulate, and that sets the stage for cancer.

The products of *BRCA* genes also bind to receptors for the hormones estrogen and progesterone, which are abundant on cells of breast and ovarian tissues. Binding suppresses transcription of growth factor genes in these cells. Among other things, growth factors stimulate cells to divide during normal, cyclic renewals of breast and ovarian tissues. When a mutation alters a *BRCA* gene so that its product cannot bind to hormone receptors, the cells overproduce growth factors. Cell division goes out of control, and tissue growth becomes disorganized. In other words, cancer develops.

Researchers recently found that the RNA product of the *XIST* gene localizes abnormally in breast cancer cells. In those cells, both X chromosomes are active. It makes sense that two active X chromosomes would have something to do with abnormal gene expression, but why the RNA product of an unmutated *XIST* gene does not localize properly in cancer cells remains a mystery.

Mutations in the *BRCA1* gene may be part of the answer. The researchers found that the protein product of the *BRCA1* gene physically associates with the RNA product of the *XIST* gene. They were able to restore proper *XIST* RNA localization—and proper X chromosome inactivation—by restoring the function of the *BRCA1* gene product in breast cancer cells.

How would you vote? Some women at high risk of developing breast cancer opt for preventive breast removal. Many of them never would have developed cancer. Should the surgery be restricted to cancer treatment?

marks such as methylation patterns are not evolutionary because the underlying DNA sequence does not change, but these marks may persist for generations after an environmental stressor has faded. For example, even when corrected for social factors, grandsons of boys who endured a winter of famine when they were 6 years old (**Figure 10.13**) lived about 32 years longer than the grandsons of boys who overate at the same age. Nine-year-old boys whose fathers smoked cigarettes before age 11 are very overweight compared with boys whose fathers did not smoke in childhood. In these and other studies, the effect was sex-limited:

boys were affected by lifestyle of individuals in the paternal line; girls, by individuals in the maternal line. Animal studies have confirmed the epigenetic effects seen in human historical data.

epigenetic Refers to heritable changes in gene expression that are not the result of changes in DNA sequence.

Take-Home Message

Can gene expression be inherited?

» Epigenetic marks in chromosomal DNA, including DNA methylations acquired during an individual's lifetime, can be passed to offspring.

Summary

Section 10.1 Controls over gene expression are required for normal functioning of cells. When gene controls fail, typically as a consequence of some mutations, **cancer** may be the outcome.

Section 10.2 Which genes a cell uses depends on the type of organism, the type of cell, conditions inside and outside the cell, and, in complex multicelled species, the organism's stage of development.

Controls over gene expression help sustain all organisms. They also drive development in multicelled eukaryotes. All cells of an embryo share the same genes. As different cell lineages use different subsets of those genes during development, they become specialized, a process called **differentiation**. Specialized cells form tissues and organs in the adult.

Different molecules and processes govern every step between transcription of a gene and delivery of the gene's product to its final destination. Most controls operate at transcription; **transcription factors** such as **activators** and **repressors** influence transcription by binding to promoters, **enhancers**, or other sequences in chromosomal DNA.

Section 10.3 Local controls over gene expression govern embryonic development of complex, multicelled bodies, a process called **pattern formation**. Various **master genes** are expressed locally in different parts of an embryo as it develops. Their products diffuse through the embryo and affect expression of other master genes, which affect the expression of others, and so on. These cascades of master gene products form a dynamic spatial map of overlapping gradients that spans the entire embryo. Cells differentiate according to their location on the map. Eventually, master gene expression induces local expression of **homeotic genes** that govern development of particular body parts, as revealed by **knockout** experiments in fruit flies (*Drosophila melanogaster*).

Section 10.4 In female mammals, most genes on one of the two X chromosomes are permanently inaccessible. **X chromosome inactivation** balances gene expression between the sexes. Such **dosage compensation** arises because the RNA product of the *XIST* gene shuts down the chromosome that transcribes it. The *SRY* gene determines male sex in humans.

Studies of mutations in *Arabidopsis thaliana* showed that three sets of master genes (*A*, *B*, and *C*) guide cell differentiation in the whorls of a floral shoot so that sepals, petals, stamens, and carpels form.

Section 10.5 Bacterial cells do not have great structural complexity and do not undergo development. Most of their gene controls reversibly adjust transcription rates in response to environmental conditions, especially

nutrient availability. **Operons** are examples of bacterial gene controls. The lactose operon governs expression of three genes, the three products of which allow the bacterial cell to metabolize lactose. Two **operators** that flank the promoter are binding sites for a repressor that blocks transcription.

Section 10.6 Epigenetic refers to heritable changes in gene expression that do not involve changes to the DNA sequence. DNA methylations and other epigenetic marks acquired during an individual's lifetime can persist for generations.

Self-Quiz

Answers in Appendix III

1. The expression of a gene may depend on _____ .
   a. the type of organism   c. the type of cell
   b. environmental conditions d. all of the above

2. Gene expression in multicelled eukaryotic cells changes in response to _____ .
   a. extracellular conditions   c. operons
   b. master gene products   d. a and b

3. Binding of _____ to _____ in DNA can increase the rate of transcription of specific genes.
   a. activators; promoters   c. repressors; operators
   b. activators; enhancers   d. both a and b

4. Proteins that influence gene expression by binding to DNA are called _____ .
   a. promoters   c. operators
   b. transcription factors   d. all of the above

5. Polytene chromosomes form in some types of cells that _____ .
   a. have a lot of chromosomes c. are differentiating
   b. are making a lot of protein d. b and c are correct

6. Eukaryotic gene controls govern _____ .
   a. transcription   e. translation
   b. RNA processing   f. protein modification
   c. RNA transport   g. all of the above
   d. mRNA degradation

7. Controls over eukaryotic gene expression guide _____ .
   a. natural selection   c. development
   b. nutrient availability   d. all of the above

8. The expression of *ABC* genes _____ .
   a. occurs in layers
   b. controls flower formation
   c. causes mutations in flowers
   d. both a and b

9. Cell differentiation _____ .
   a. occurs in all complex multicelled organisms
   b. requires unique genes in different cells
   c. involves selective gene expression
   d. both a and c
   e. all of the above

10. Homeotic gene products _____ .
    a. flank a bacterial operon
    b. map out the overall body plan in embryos
    c. control the formation of specific body parts

Effect of Paternal Grandmother's Food Supply on Infant Mortality Researchers are investigating long-reaching epigenetic effects of starvation, in part because historical data on periods of famine are widely available. Before the industrial revolution, a failed harvest in one autumn typically led to severe food shortages the following winter.

A retrospective study has correlated female infant mortality at certain ages with the abundance of food during the paternal grandmother's childhood. **Figure 10.14** shows some of the results of this study.

1. Compare the mortality of girls whose paternal grandmothers ate well at age 2 with that of those who experienced famine at the same age. Which girl was more likely to die early? How much more likely was she to die?

2. Children have a period of slow growth around age 9. What trend in this data can you see around that age?

3. There was no correlation between early death of a male child and eating habits of his paternal grandmother, but there was a strong correlation with the eating habits of his paternal grand*father*. What does this tell you about the probable location of epigenetic changes that gave rise to this data?

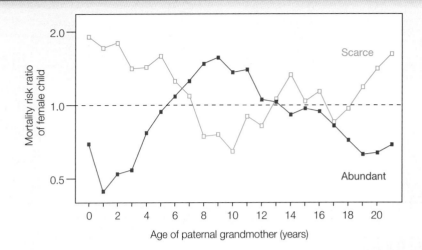

Figure 10.14 Graph showing the relative risk of early death of a female child, correlated with the age at which her paternal grandmother experienced a winter with a food supply that was scarce (*blue*) or abundant (*red*) during childhood. The dotted line represents no difference in risk of mortality. A value above the line means an increased risk; one below the line indicates a reduced risk.

---

11. A gene that is knocked out is _____ .
    a. deleted   c. expressed
    b. inactivated   d. either a or b

12. During X chromosome inactivation, _____ .
    a. female cells shut down   c. pigments form
    b. RNA coats chromosomes d. both a and b

13. A cell with a Barr body is _____ .
    a. a bacterium   c. from a female mammal
    b. a sex cell   d. infected by Barr virus

14. A promoter and a set of operators that control access to two or more prokaryotic genes _____ .
    a. occurs in all complex multicelled organisms
    b. is called an operon
    c. involves selective gene expression
    d. regulates gene expression in all organisms

15. True or false? Gene expression patterns are heritable.

16. Match the terms with the most suitable description.
    ____ repressor   a. mRNA self-regulation
    ____ riboswitch   b. binding site for repressor
    ____ operator   c. cells become specialized
    ____ Barr body   d. may cause epigenetic effects
    ____ differentiation   e. inactivated X chromosome
    ____ methylation   f. binds to promoter

Critical Thinking

1. Why are some genes expressed and some not expressed?

2. Children conceived during times of famine tend to be overweight. Suggest a possible reason for this effect.

3. Do the same gene controls operate in bacterial cells and eukaryotic cells? Why or why not?

BRCA Mutations in Women Diagnosed With Breast Cancer				
	BRCA1	*BRCA2*	No *BRCA* Mutation	Total
Total number of patients	89	35	318	442
Avg. age at diagnosis	43.9	46.2	50.4	
Preventive mastectomy	6	3	14	23
Preventive oophorectomy	38	7	22	67
Number of deaths	16	1	21	38
Percent died	18.0	2.8	6.9	8.6

4. Investigating a correlation between specific cancer-causing mutations and risk of mortality in humans is challenging, in part because each cancer patient is given the best treatment available at the time. There are no "untreated control" cancer patients, and the idea of what treatments are the best changes quickly as new drugs become available and new discoveries are made.

The table *above* shows some results of a 2007 study in which 442 women who had been diagnosed with breast cancer were checked for *BRCA* mutations, and their treatments and progress were followed over several years. All of the women in the study had at least two affected close relatives, so their risk of developing breast cancer due to an inherited factor was estimated to be greater than that of the general population. Some of the women underwent preventive mastectomy (removal of the noncancerous breast) during their course of treatment. Others had preventive oophorectomy (surgical removal of the ovaries) to prevent the possibility of getting ovarian cancer.

According to this study, is a *BRCA1* or *BRCA2* mutation more dangerous in breast cancer cases? What other data would you have to see in order to make a conclusion about the effectiveness of preventive surgeries?

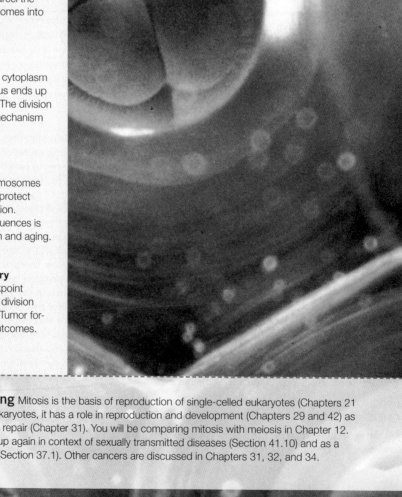

LEARNING ROADMAP

Where you have been Be sure you understand cell structure (Sections 4.2, 4.6, 4.10, 4.11), chromosomes (8.2), and DNA replication (8.5) before beginning this chapter. What you know about phosphorylation (5.6), membrane proteins (5.7), fermentation (7.6), mutations (8.6), and gene control (10.2) will help you understand how cancer (10.1) develops. You will also revisit animal cloning (8.7) and knockout experiments (10.3).

Where you are now

The Cell Cycle
A cell cycle starts when a new cell forms by division of a parent cell, and ends when the cell completes its own division. Built-in checkpoints control the timing and rate of the cycle.

Mitosis
Mitosis divides the nucleus and maintains the chromosome number. Four sequential stages parcel the cell's duplicated chromosomes into two new nuclei.

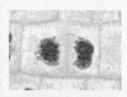

Cytoplasmic Division
After nuclear division, the cytoplasm may divide, so one nucleus ends up in each of two new cells. The division proceeds by a different mechanism in animal and plant cells.

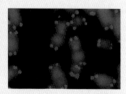

Mitotic Clocks
Built into eukaryotic chromosomes are DNA sequences that protect the cell's genetic information. Degradation of these sequences is associated with cell death and aging.

The Cell Cycle Gone Awry
On rare occasions, checkpoint mechanisms fail, and cell division becomes uncontrollable. Tumor formation and cancer are outcomes.

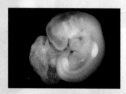

Where you are going Mitosis is the basis of reproduction of single-celled eukaryotes (Chapters 21 and 23). In multicelled eukaryotes, it has a role in reproduction and development (Chapters 29 and 42) as well as growth and tissue repair (Chapter 31). You will be comparing mitosis with meiosis in Chapter 12. The HPV virus will come up again in context of sexually transmitted diseases (Section 41.10) and as a cause of cervical cancer (Section 37.1). Other cancers are discussed in Chapters 31, 32, and 34.

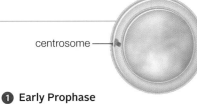

centrosome

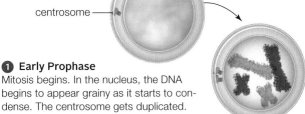

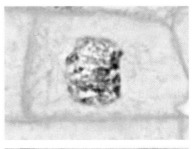

❶ Early Prophase
Mitosis begins. In the nucleus, the DNA begins to appear grainy as it starts to condense. The centrosome gets duplicated.

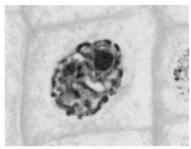

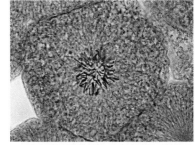

❷ Prophase
The duplicated chromosomes become visible as they condense. One of the two centrosomes moves to the opposite side of the nucleus. The nuclear envelope breaks up.

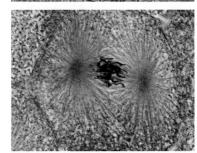

❸ Transition to Metaphase
The nuclear envelope is gone, and the chromosomes are at their most condensed. Spindle microtubules assemble and bind to chromosomes at the centromere. Sister chromatids are attached to opposite spindle poles.

microtubule of spindle

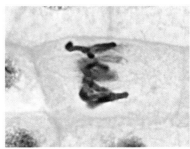

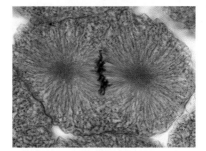

❹ Metaphase
All of the chromosomes are aligned midway between the spindle poles.

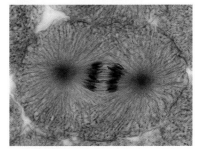

❺ Anaphase
Spindle microtubules separate the sister chromatids and move them toward opposite spindle poles. Each sister chromatid has now become an individual, unduplicated chromosome.

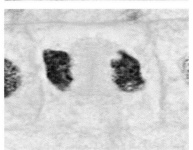

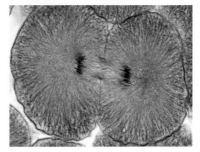

❻ Telophase
The chromosomes reach the spindle poles and decondense. A nuclear envelope forms around each cluster, and mitosis ends.

11.4 Cytokinesis: Division of Cytoplasm

■ In most eukaryotes, the cell cytoplasm divides between late anaphase and the end of telophase. The mechanism of division differs between plants and animals.

■ Links to Cytoskeleton 4.10, Primary wall of plant cells 4.11

A cell's cytoplasm usually divides after mitosis, so two cells form. The process of cytoplasmic division, which is called **cytokinesis**, differs among eukaryotes.

Typical animal cells pinch themselves in two after nuclear division (**Figure 11.6**). How? The spindle begins to disassemble during telophase ❶. The cell cortex, which is the mesh of cytoskeletal elements just under the plasma membrane, includes a band of actin and myosin filaments that wraps around the cell's midsection. The band is called a contractile ring because it contracts when its component proteins are energized by phosphate-group transfers from ATP. When the ring contracts, it drags the attached plasma membrane inward ❷. The sinking plasma membrane is visible on the outside of the cell as an indentation between the former spindle poles ❸. The indentation, a **cleavage furrow**, advances around the cell and deepens until the cytoplasm (and the cell) is pinched in two ❹. Each of the two cells formed by this division has its own nucleus and some of the parent cell's cytoplasm; each is enclosed by a plasma membrane.

Dividing plant cells face a particular challenge because they have stiff cell walls on the outside of their plasma membrane (Section 4.11). Accordingly, plant cells have a different mechanism of cytokinesis (**Figure 11.7**). By the end of anaphase, a set of short microtubules has formed on either side of the future plane of division. These microtubules now guide vesicles from Golgi bodies and the cell surface to the division plane ❺. There, the vesicles and their wall-building contents start to fuse into a disk-shaped **cell plate** ❻. The plate expands at its edges until it reaches the plasma membrane ❼. When the cell plate attaches to the membrane, it partitions the cytoplasm. In time, the cell plate will develop into a primary cell wall that merges with the parent cell's wall. Thus, by the end of division, each of the descendant cells will be enclosed by its own plasma membrane and its own cell wall ❽.

cell plate After nuclear division in a plant cell, a disk-shaped structure that forms a cross-wall between the two new nuclei.
cleavage furrow In a dividing animal cell, the indentation where cytoplasmic division will occur.
cytokinesis Cytoplasmic division.

Take-Home Message

How do eukaryotic cells divide?

» After mitosis, the cytoplasm of the parent cell typically is partitioned into two descendant cells, each with its own nucleus.

» In animal cells, a contractile ring pinches the cytoplasm in two. In plant cells, a cell plate that forms midway between the spindle poles partitions the cytoplasm when it reaches and connects to the parent cell wall.

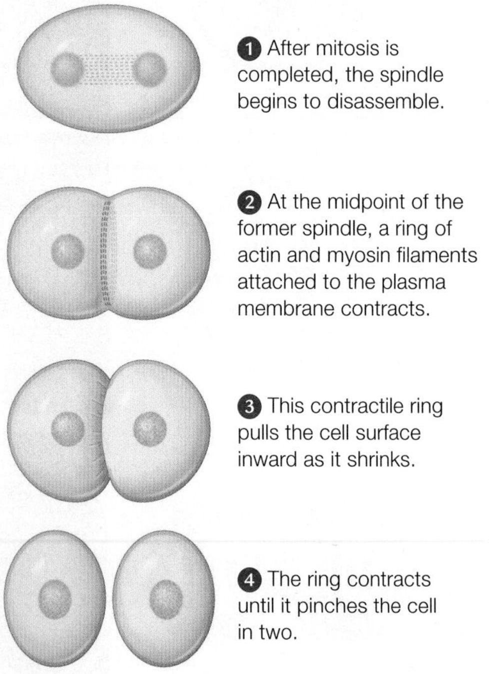

❶ After mitosis is completed, the spindle begins to disassemble.

❷ At the midpoint of the former spindle, a ring of actin and myosin filaments attached to the plasma membrane contracts.

❸ This contractile ring pulls the cell surface inward as it shrinks.

❹ The ring contracts until it pinches the cell in two.

Figure 11.6 Animated Cytoplasmic division of an animal cell.

❺ The future plane of division was established before mitosis began. Vesicles cluster here when mitosis ends.

❻ As the vesicles fuse with each other, they form a cell plate along the plane of division.

❼ The cell plate expands outward along the plane of division. When it reaches the plasma membrane, it attaches to the membrane and partitions the cytoplasm.

❽ The cell plate matures as two new cell walls. These walls join with the parent cell wall, so each descendant cell becomes enclosed by its own cell wall.

Figure 11.7 Animated Cytoplasmic division of a plant cell.

11.5 Marking Time With Telomeres

- Telomeres protect eukaryotic chromosomes from losing genetic information at their ends.
- Shortening telomeres are associated with aging.
- Links to DNA replication 8.5, Animal cloning 8.7, Knockout experiments 10.3

The very ends of the DNA strands composing eukaryotic chromosomes consist of noncoding sequences called **telomeres (Figure 11.8)**. Vertebrate telomeres have a short DNA sequence, 5'-TTAGGG-3', repeated perhaps thousands of times—"junk" sequences that provide a buffer against the loss of more valuable DNA internal to the chromosomes. This is particularly important because the chromosomes of a normal eukaryotic body cell shorten by about 100 bases with each replication. The shortening occurs in part because DNA polymerase can synthesize DNA only in the 3' to 5' direction, and it can start synthesis only at the 3' hydroxyl group of a primer. Thus, a polymerase cannot copy the last hundred bases or so of the 3' end of a chromosome. Each new DNA strand ends up a bit shorter than its parent.

When telomeres get too short, checkpoint gene products come into play to halt the cell cycle. The cell stops dividing, and it dies shortly thereafter. Thus, body cells can divide only a certain number of times. Scientists speculate that this limitation is a fail-safe mechanism because cells sometimes lose control over the cell cycle and begin to divide over and over. A limit on the number of divisions keeps such cells from overrunning the body, an outcome that, as you will see in the next section, has dangerous consequences.

In an adult, a few normal cells—stem cells—retain the ability to divide indefinitely. Descendants of these cells replace cell lineages that eventually die out. Stem cells are immortal because they continue to make telomerase, a molecule that consists of a noncoding RNA and a replication enzyme. The enzyme uses the RNA as a template to extend telomere repeat sequences at the 3' end of a chromosome. Telomerase reverses telomere shortening that normally occurs after DNA replication.

Mice that have had their telomerase enzyme knocked out age prematurely. Their tissues degenerate much more quickly than those of normal mice, and their life expectancy declines to about half that of a normal mouse. When one of these knockout mice is close to the end of its shortened life span, rescuing the function of its telomerase enzyme results in lengthened telomeres. The rescued mouse also regains vitality: Decrepit tissue in its brain and other organs repairs itself and begins to function normally, and the

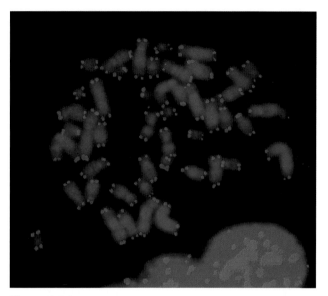

Figure 11.8 Telomeres. The *pink* dots at the end of each DNA strand in these duplicated chromosomes are telomere sequences.

once-geriatric individual even begins to reproduce again. This and other research in animals suggests that the built-in limit on cell divisions may be part of the mechanism that sets an organism's life span.

The very first cloned animal, Dolly the sheep, had telomeres that were abnormally short. When Dolly was only two years old, her telomeres were as short as those of a six-year-old sheep—the exact age of the adult animal that had been her genetic donor. By the time Dolly was five, she was as fat and arthritic as a twelve-year-old sheep. The following year, she contracted a lung disease that is typical of geriatric sheep, and had to be euthanized.

Dolly's early demise may well have been the result of short telomeres, but scientists are careful to point out that short telomeres could be an effect of aging rather than a cause. They also caution that while telomerase holds therapeutic promise for rejuvenation of aged tissues, it can also be dangerous: Cancer cells characteristically express high levels of the molecule.

telomere Noncoding, repetitive DNA sequence at the end of chromosomes; protects the coding sequences from degradation.

Take-Home Message

What is the function of telomeres?

» Telomeres at the ends of chromosomes provide a buffer against loss of genetic information.

» Telomeres shorten with every cell division in normal body cells. When they get too short, the cell stops dividing and dies.

11.6 When Mitosis Becomes Pathological

- On rare occasions, controls over cell division are lost and a neoplasm forms.
- Cancer develops as cells of a neoplasm become malignant.
- Links to Phosphorylations and energy 5.6, Membrane proteins 5.7, Environmental causes of mutation 8.6, Fermentation 7.6, Cancer 10.1

Sometimes a checkpoint gene mutates so that its protein product no longer works properly. In other cases, the controls that regulate its production fail, and a cell makes too much or too little of its product. When enough checkpoint mechanisms fail, a cell loses control over its cell cycle. The cell may skip interphase, so division occurs over and over with no resting period. Signaling mechanisms that make an abnormal cell die may stop working. The problem is compounded

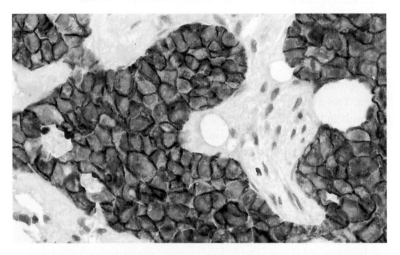

Figure 11.9 An oncogene. In this section of human breast tissue, phosphorylated EGF receptor is stained *brown*. Normal cells are the ones with lighter staining. The heavily stained cells have formed a neoplasm; the abnormal overabundance of the phosphorylated EGF receptor means that mitosis is being continually stimulated in these cells. The EGF receptor is overproduced or overactive in most neoplasms.

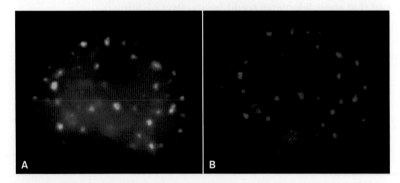

Figure 11.10 Checkpoint genes in action. Radiation damaged the DNA inside this nucleus. (**A**) *Green* dots pinpoint the location of the product of a gene called *53BP1*; (**B**) *red* dots show the location of the *BRCA1* gene product. Both proteins have clustered around the same chromosome breaks in the same nucleus; both function to recruit DNA repair enzymes. The integrated action of these and other checkpoint gene products blocks mitosis until the DNA breaks are fixed.

because checkpoint malfunctions are passed along to the cell's descendants, which form a **neoplasm**, an accumulation of cells that lost control over how they grow and divide.

A neoplasm that forms a lump in the body is called a **tumor**, but the two terms are sometimes used interchangeably. Once a tumor-causing mutation has occurred, the gene it affects is called an oncogene. An **oncogene** is any gene that helps transform a normal cell into a tumor cell (Greek *onkos*, or bulging mass). Some oncogene mutations can be passed to offspring, which is a reason that some types of tumors tend to run in families.

Genes encoding proteins that promote mitosis are called **proto-oncogenes** because mutations can turn them into oncogenes. A gene that encodes the epidermal growth factor (EGF) receptor is an example of a proto-oncogene. **Growth factors** are molecules that stimulate a cell to divide and differentiate. The EGF receptor is a plasma membrane protein that phosphorylates itself upon binding EGF. When phosphorylated, the receptor initiates a series of intracellular events that moves the cell cycle out of interphase and into mitosis. Mutations can result in an EGF receptor that always phosphorylates itself, so it stimulates mitosis even when EGF is not present. Most neoplasms carry mutations resulting in an overactivity or overabundance of this particular receptor (**Figure 11.9**).

Checkpoint gene products that inhibit mitosis are called tumor suppressors because tumors form when they are missing. The products of the *BRCA1* and *BRCA2* genes (Chapter 10) are examples of tumor suppressors. These proteins regulate, among other things, the expression of DNA repair enzymes (**Figure 11.10**). Mutations in *BRCA* genes are often found in cells of neoplasms. As another example, viruses such as HPV (human papillomavirus) cause a cell to make proteins that interfere with its own tumor suppressors. Infection with HPV causes skin growths called warts, and some kinds are associated with neoplasms that form on the cervix.

Cancer

Benign neoplasms such as ordinary skin moles are not dangerous (**Figure 11.11**). They grow very slowly, and their cells retain the plasma membrane adhesion proteins that keep them properly anchored to the other cells in their home tissue ❶.

A malignant neoplasm is one that gets progressively worse, and is dangerous to health. The disease called cancer occurs when the abnormally dividing cells of a malignant neoplasm disrupt body tissues, both physically and metabolically. Malignant cells typically display the following three characteristics:

First, like cells of all neoplasms, malignant cells grow and divide abnormally. Controls that usually keep cells from getting overcrowded in tissues are lost, so their populations may reach extremely high densities with cell division occurring very rapidly. The number of small blood vessels that transport blood to the growing cell mass also increases abnormally.

Second, the cytoplasm and plasma membrane of malignant cells are altered. The cytoskeleton may be shrunken, disorganized, or both. Malignant cells typically have an abnormal chromosome number, with some chromosomes present in multiple copies, and others missing or damaged. The balance of metabolism is often shifted, as in an amplified reliance on ATP formation by fermentation rather than aerobic respiration.

Altered or missing proteins impair the function of the plasma membrane of malignant cells. For example, these cells do not stay anchored properly in tissues because their plasma membrane adhesion proteins are defective or missing ❷. Malignant cells can slip easily into and out of vessels of the circulatory and lymphatic systems ❸. By migrating through these vessels, the cells establish neoplasms elsewhere in the body ❹. The process in which malignant cells break loose from their home tissue and invade other parts of the body is called **metastasis**. Metastasis is the third hallmark of malignant cells.

Unless chemotherapy, surgery, or another procedure eliminates malignant cells from the body, they can put an individual on a painful road to death. Each year, cancer causes 15 to 20 percent of all human deaths in developed countries. The good news is that mutations in multiple checkpoint genes are required to transform a normal cell into a malignant one, and these mutations may take a lifetime to accumulate. Lifestyle choices such as not smoking and avoiding exposure of unprotected skin to sunlight can reduce one's risk of acquiring mutations in the first place. Some neoplasms can be detected with periodic screening procedures such as Pap tests or dermatology exams (**Figure 11.12**). If detected early enough, many types of malignant neoplasms can be removed before metastasis occurs.

growth factor Molecule that stimulates mitosis and differentiation.
metastasis The process in which cancer cells spread from one part of the body to another.
neoplasm An accumulation of abnormally dividing cells.
oncogene Gene that helps transform a normal cell into a tumor cell.
proto-oncogene Gene that, by mutation, can become an oncogene.
tumor A neoplasm that forms a lump.

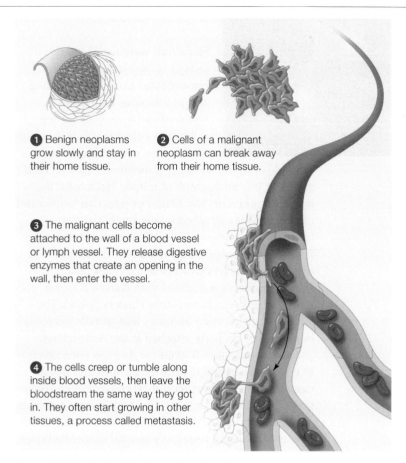

❶ Benign neoplasms grow slowly and stay in their home tissue.

❷ Cells of a malignant neoplasm can break away from their home tissue.

❸ The malignant cells become attached to the wall of a blood vessel or lymph vessel. They release digestive enzymes that create an opening in the wall, then enter the vessel.

❹ The cells creep or tumble along inside blood vessels, then leave the bloodstream the same way they got in. They often start growing in other tissues, a process called metastasis.

Figure 11.11 Animated Benign and malignant neoplasms.

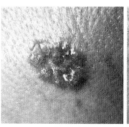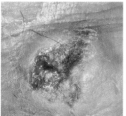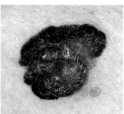

A Basal cell carcinoma is the most common type of skin cancer. This slow-growing, raised lump may be uncolored, reddish-brown, or black.

B The second most common form of skin cancer is a squamous cell carcinoma. This pink growth, firm to the touch, grows under the surface of skin.

C Melanoma, a malignant neoplasm of skin cells, spreads fastest. Cells form dark, encrusted lumps that may itch or bleed easily.

Figure 11.12 Skin cancer can be detected early with periodic screening.

Take-Home Message

What is cancer?

» Cancer is a disease that occurs when the abnormally dividing cells of a neoplasm physically and metabolically disrupt body tissues.

» A malignant neoplasm results from mutations in multiple checkpoint genes.

» Although some mutations are inherited, lifestyle choices and early intervention can reduce one's risk of cancer.

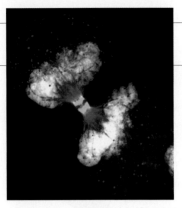

Henrietta's Immortal Cells (revisited)

HeLa cells were used in early tests of taxol, a drug that keeps microtubules from disassembling and so interferes with mitosis. Frequent divisions make cancer cells more vulnerable to this poison than normal cells. A more recent example of cancer research is shown in **Figure 11.1** (and *above*). In this micrograph of mitotic HeLa cells, the *blue* stain pinpoints INCENP, a protein that helps sister chromatids stay attached to one another at the centromere. The *green* stain identifies an enzyme, Aurora B kinase, that helps attach spindle microtubules (*red*) to centromeres. At this stage of telophase, INCENP should be closely associated with Aurora B kinase midway between chromosome clusters (*white*). The abnormal distribution indicates that spindle microtubules are not properly attached to the centromeres.

Defects in Aurora B kinase or its expression result in unequal segregation of chromosomes into descendant cells. Researchers recently correlated overexpression of Aurora B in cancer cells with shortened patient survival rates. Thus, drugs that inhibit Aurora B function are now being tested as potential cancer therapies.

Despite the invaluable cellular legacy of Henrietta Lacks, her body rests in an unmarked grave in an unmarked cemetery. These days, physicians and researchers are required to obtain a signed consent form before they take tissue samples from a patient. No such requirement existed in the 1950s, when it was common for doctors to experiment on patients without their knowledge or consent. Thus, the young resident who was treating Henrietta Lacks's cancerous cervix probably never even thought about asking permission before he took a sample of it. That sample was the one that the Geys used to establish the HeLa cell line. No one in Henrietta's family knew about the cells until 25 years after she died. HeLa cells are still being sold worldwide, but her family has not received any compensation to date.

How would you vote? You can legally donate—but not sell—your own organs and tissues. However, companies can profit from research on donated organs or tissues, and also from cell lines derived from these materials. Companies that do so are not obligated to share their profits with the donors. Should profits derived from donated tissues or cells be shared with donors or their families?

Summary

Section 11.1 An immortal line of cancer cells is a legacy of cancer victim Henrietta Lacks. Researchers trying to unravel the mechanisms of cancer continue to work with these cells.

Section 11.2 A **cell cycle** includes all the stages through which a eukaryotic cell passes during its lifetime. Most of a cell's activities, including replication of a diploid cell's **homologous chromosomes**, occur in **interphase**. A cell reproduces by dividing: nucleus first, then cytoplasm. Nuclear division partitions duplicated chromosomes into new nuclei. **Mitosis** maintains the chromosome number. It is the basis of growth, cell replacements, and tissue repair in multicelled species, and **asexual reproduction** in many species.

Section 11.3 Mitosis proceeds in four stages. In **prophase**, the duplicated chromosomes condense. Microtubules form a **spindle**, and the nuclear envelope breaks up. Spindle microtubules drag each chromosome toward the center of the cell. At **metaphase**, all chromosomes are aligned at the spindle's midpoint. During **anaphase**, sister chromatids of each chromosome detach from each other, and spindle microtubules start moving them toward opposite spindle poles. During **telophase**, a complete set of chromosomes reaches each spindle pole. A nuclear envelope forms around each cluster. Two new nuclei, each with the parental chromosome number, are the result.

Section 11.4 In most cases, **cytokinesis** follows nuclear division. A contractile ring pulls the plasma membrane of an animal cell inward, forming a **cleavage furrow** that eventually pinches the cytoplasm in two. In a plant cell, vesicles guided by microtubules to the future plane of division merge as a **cell plate**. The plate expands until it fuses with the parent cell wall, thus becoming a cross-wall that partitions the cytoplasm.

Section 11.5 Telomeres protecting the ends of eukaryotic chromosomes shorten with every cell division. When they become too short, the cell dies. Telomerase can restore the length of a cell's telomeres.

Section 11.6 The products of checkpoint genes are part of the controls governing the cell cycle. When checkpoint mechanisms fail, a cell loses control over its cell cycle, and the cell's descendants form a **neoplasm**. Mutations can turn **proto-oncogenes**, includ-

HeLa Cells Are a Genetic Mess HeLa cells continue to be an extremely useful tool in cancer research. One early finding was that HeLa cells can vary in chromosome number. The panel of chromosomes on the *right*, originally published in 1989 by Nicholas Popescu and Joseph DiPaolo, shows all of the chromosomes in a single metaphase HeLa cell.

1. What is the chromosome number of this HeLa cell?

2. How many extra chromosomes does this cell have, compared to a normal human body cell?

3. Can you tell that this cell came from a female? How?

ing **growth factor** receptor genes, into **oncogenes**. Such mutations typically disrupt checkpoint gene products or their expression, and can result in neoplasms. Neoplasms may form lumps called **tumors**. Malignant cells break loose from their home tissues and colonize other parts of the body, a process called **metastasis**.

Self-Quiz

Answers in Appendix III

1. Mitosis and cytoplasmic division function in _____ .
   a. asexual reproduction of single-celled eukaryotes
   b. growth and tissue repair in multicelled species
   c. gamete formation in bacteria and archaea
   d. both a and b

2. A duplicated chromosome has _____ chromatid(s).
   a. one   c. three
   b. two   d. four

3. Except for a pairing of sex chromosomes, homologous chromosomes _____ .
   a. carry the same genes   c. are the same length
   b. are the same shape   d. all of the above

4. Most cells spend the majority of their lives in _____ .
   a. prophase   d. telophase
   b. metaphase   e. interphase
   c. anaphase   f. a and c

5. A plant cell divides by the process of _____ .
   a. prophase   d. telophase
   b. metaphase   e. interphase
   c. anaphase   f. cytokinesis

6. The spindle attaches to chromosomes at the _____ .
   a. centriole   c. centromere
   b. contractile ring d. centrosome

7. Interphase is the part of the cell cycle when _____ .
   a. a cell ceases to function   c. mitosis proceeds
   b. the spindle forms   d. a cell grows and
                                     duplicates its DNA

8. After mitosis, the chromosome number of a descendant cell is _____ the parent cell's.
   a. the same as   c. rearranged compared to
   b. one-half of   d. doubled compared to

9. Stem cells are immortal as long as they express _____ .
   a. checkpoint genes   c. telomerase
   b. oncogenes   d. growth factors

10. Only _____ is not a stage of mitosis.
    a. prophase   b. interphase   c. metaphase   d. anaphase

11. In intervals of interphase, G stands for _____ .
    a. gap   b. growth   c. Gey   d. gene

12. *BRCA1*, *BRCA2*, and *53BP1* are examples of _____ .
    a. checkpoint genes   c. tumor suppressors
    b. proto-oncogenes   d. all of the above

13. Which one of the following encompasses the other two?
    a. cancer   b. neoplasm   c. tumor

14. Match each term with its best description.
    ___ cell plate   a. lump of cells
    ___ spindle   b. made of microfilaments
    ___ tumor   c. divides plant cells
    ___ cleavage furrow   d. organize(s) the spindle
    ___ contractile ring   e. migrating, metastatic cells
    ___ cancer   f. made of microtubules
    ___ telomere   g. indentation
    ___ centrosomes   h. shortens with age

15. Match each stage with the events listed.
    ___ metaphase   a. sister chromatids move apart
    ___ prophase   b. chromosomes start to condense
    ___ telophase   c. new nuclei form
    ___ anaphase   d. all duplicated chromosomes are
                        aligned at the spindle equator

Critical Thinking

1. When a cell reproduces by mitosis and cytoplasmic division, does its life end?

 2. The eukaryotic cell in the photo on the *left* is in the process of cytoplasmic division. Is this cell from a plant or an animal? How do you know?

3. Exposure to radioisotopes or other sources of radiation can damage DNA. Humans exposed to high levels of radiation face a condition called radiation poisoning. Why do you think that hair loss and damage to the lining of the gut are early symptoms of radiation poisoning? Speculate about why exposure to radiation is used as a therapy to treat some kinds of cancers.

4. Suppose you have a way to measure the amount of DNA in one cell during the cell cycle. You first measure the amount at the G1 phase. At what points in the rest of the cycle will you see a change in the amount of DNA per cell?

LEARNING ROADMAP

Where you have been Be sure you understand how eukaryotic chromosomes are organized (Sections 8.2 and 11.2) and how genes work (9.2). You will draw on your knowledge of DNA replication (8.5), cytoplasmic division (11.4), and cell cycle controls (11.6) as we compare meiosis with mitosis (11.3). This chapter also revisits clones (8.1, 8.7), the effects of mutation (9.6), and telomeres (11.5).

Where you are now

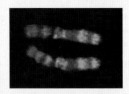

Sexual or Asexual Reproduction
In asexual reproduction, one parent transmits its genes to offspring. In sexual reproduction, offspring inherit genes from two parents who usually differ in some number of alleles.

Stages of Meiosis
Meiosis is a nuclear division process that occurs only in cells set aside for sexual reproduction. Meiosis reduces the chromosome number by sorting chromosomes into four new nuclei.

Recombinations and Shufflings
During meiosis, homologous chromosomes swap segments, then are randomly sorted into separate nuclei. Both processes lead to novel combinations of alleles among offspring.

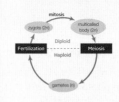

Sexual Reproduction in Life Cycles
Gametes form by different mechanisms in males and females, but meiosis is part of both processes. In most plants, spores occur between meiosis and gamete formation.

Mitosis and Meiosis Compared
Similarities suggest that meiosis originated by evolutionary remodeling of mechanisms that already existed for mitosis and, before that, for repairing damaged DNA.

Where you are going Meiosis is the basis of sexual reproduction, a topic covered in detail in Chapters 29 (plants) and 41 (animals). The variation in traits among individuals of sexually reproducing species arises from differences in alleles, a concept we revisit in context of inheritance in Chapters 13 and 14, and natural selection in Chapter 17.

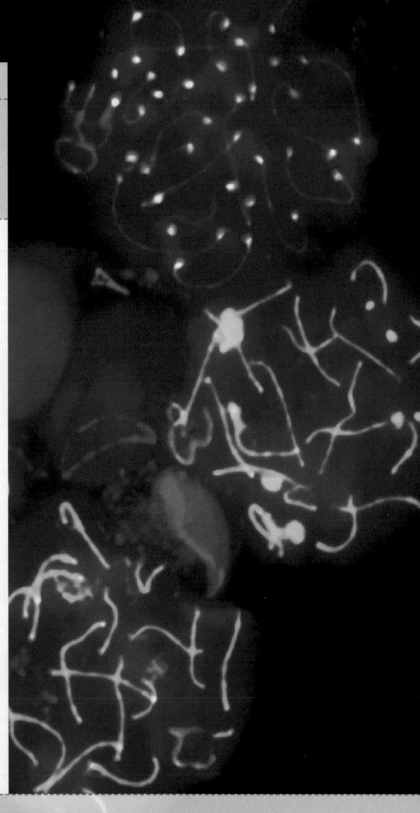

12.1 Why Sex?

If the function of reproduction is the perpetuation of one's genes, then an asexual reproducer would seem to win the evolutionary race. In asexual reproduction, a single individual gives rise to offspring. All of the individual's genes are passed to all of its offspring. **Sexual reproduction** mixes up the genes of two parents (**Figure 12.1**), so only about half of each parent's genetic information is passed to offspring.

So why sex? Consider that all offspring of asexual reproducers are clones of their parent, so all are adapted the same way to an environment—and all are equally vulnerable to changes in it. By contrast, the offspring of sexual reproducers have different combinations of many traits. As a group, their diversity offers them flexibility: a better chance of surviving environmental change than clones. Some of them may have a particular combination of traits that suits them perfectly to a change. The variation among offspring of sexual reproducers is part of the reason that beneficial mutations tend to spread through sexually reproducing populations more quickly than asexual ones.

Other organisms are part of the environment, and they, too, can change. Think of predator and prey, say, foxes and rabbits. If one rabbit is better than others at outrunning the foxes, it has a better chance of escaping, surviving, and passing to its offspring the genes that help it evade foxes. Thus, over many generations, rabbits may get faster. If one fox is better than others at outrunning faster rabbits, it has a better chance of eating, surviving, and passing to its offspring genes that help it catch faster rabbits. Thus, over many generations, the foxes may tend to get faster. As one species changes, so does the other—an idea called the Red Queen hypothesis, after Lewis Carroll's book *Through the Looking Glass*. In the book, the Queen of Hearts tells Alice, "It takes all the running you can do, to keep in the same place."

Perhaps the most important advantage of sexual reproduction involves harmful mutations that inevitably occur. Collectively, the offspring of sexual reproducers have a better chance of surviving the effects of a harmful mutation that arises in the population. This is because, with asexual reproduction, individuals bearing a harmful mutation necessarily pass it to all of their offspring. By contrast, sexual reproduction mixes up the genetic material of two individuals. If an individual bearing a harmful mutation reproduces, it is very unlikely that all of the individual's offspring will inherit the mutation. Thus, all else being equal, harmful mutations accumulate in an asexually reproducing population much more quickly than in a sexually reproducing one.

Sexual reproduction generates new combinations of traits in fewer generations than does asexual reproduction. The process inherent to sexual reproduction that gives rise to this variation is **meiosis**, a nuclear division mechanism that halves the chromosome number.

meiosis Nuclear division process that halves the chromosome number. Basis of sexual reproduction.
sexual reproduction Reproductive mode by which offspring arise from two parents and inherit genes from both.

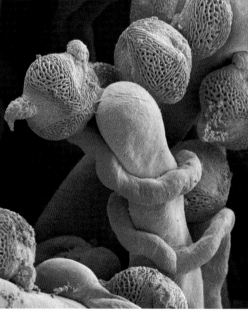

Figure 12.1 Moments in the stages of sexual reproduction of humans (*left*) and plants (*right*). Sexual reproduction mixes up the genetic material of two organisms.

In flowering plants, pollen grains (*orange*) germinate on flower carpels (*yellow*). Pollen tubes with male gametes inside grow from the grains down into tissues of the ovary, which house the flower's female gametes.

12.2 Meiosis Halves the Chromosome Number

- Sexual reproduction mixes up alleles from two parents.
- Meiosis, the basis of sexual reproduction, is a nuclear division mechanism that occurs in immature reproductive cells of eukaryotes.
- Links to Clones 8.1 and 8.7, Eukaryotic chromosomes 8.2, DNA replication 8.5, Genes 9.2, Mutations 9.6, Homologous chromosomes and asexual reproduction and mitosis 11.2

Introducing Alleles

The **somatic** (body) cells of humans and many other sexually reproducing organisms are diploid—they contain pairs of chromosomes (Section 8.2). One chromosome of each pair is typically maternal, and the other is paternal. Except for a pairing of nonidentical sex chromosomes, homologous chromosomes carry the same set of genes (**Figure 12.2A**).

If the DNA sequences of homologous chromosomes were identical, then sexual reproduction would produce clones, just like asexual reproduction does. But the two chromosomes of a homologous pair are typically not identical. Why not? Mutations that inevitably occur in chromosomes change their DNA sequence. Over time, unique mutations accumulate in separate lines of descent, and some of those mutations occur in gene regions. Thus, any gene may differ a bit in sequence from the corresponding gene on the homologous chromosome (**Figure 12.2B**). Different forms of the same gene are called **alleles**.

Alleles may encode slightly different forms of the gene's product. Such differences influence thousands of traits. For example, the beta globin gene you encountered in Section 9.6 has more than 700 alleles: one that causes sickle-cell anemia, one that causes beta thalassemia, and so on. The beta globin gene is one of

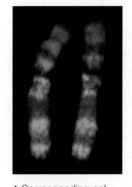

Genes occur in pairs on homologous chromosomes.

The members of each pair of genes may be identical, or they may differ slightly, as alleles.

A Corresponding colored patches in this fluorescence micrograph indicate corresponding DNA sequences in a homologous chromosome pair. These chromosomes carry the same set of genes.

B Homologous chromosomes carry the same series of genes, but the DNA sequence of any one of those genes might differ just a bit from that of its partner on the homologous chromosome.

Figure 12.2 Animated Genes on chromosomes. Different forms of a gene are called alleles.

about 20,000 human genes, and most genes have multiple alleles. Allele differences among individuals are one reason that the members of a sexually reproducing species do not all look identical. The offspring of sexual reproducers inherit new combinations of alleles, which is the basis of new combinations of traits.

What Meiosis Does

Sexual reproduction involves the fusion of haploid reproductive cells—**gametes**—from two parents. In both plants and animals, gametes form inside special reproductive structures or organs (**Figure 12.3**). Division

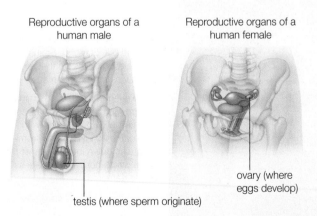

Reproductive organs of a human male

Reproductive organs of a human female

testis (where sperm originate)

ovary (where eggs develop)

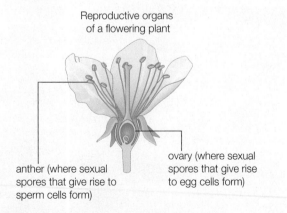

Reproductive organs of a flowering plant

anther (where sexual spores that give rise to sperm cells form)

ovary (where sexual spores that give rise to egg cells form)

Figure 12.3 Animated Examples of reproductive organs. In human ovaries or testes, meiosis in germ cells produces gametes called eggs and sperm. In the anthers or ovaries of flowering plants, meiosis produces haploid spores that give rise to gametes by mitosis.

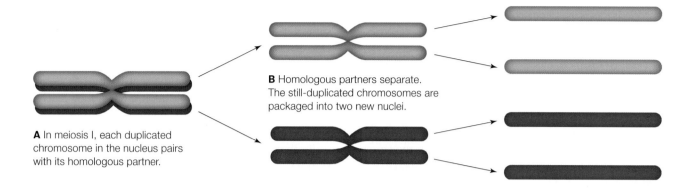

A In meiosis I, each duplicated chromosome in the nucleus pairs with its homologous partner.

B Homologous partners separate. The still-duplicated chromosomes are packaged into two new nuclei.

C Sister chromatids separate in meiosis II. The now unduplicated chromosomes are packaged into four new nuclei.

Figure 12.4 How meiosis halves the chromosome number.

of immature reproductive cells called **germ cells** gives rise to gametes. Meiosis in animal germ cells gives rise to eggs (female gametes) or sperm (male gametes). In plants, haploid germ cells form by meiosis. Gametes form when these cells divide by mitosis.

Gametes have a single set of chromosomes, so they are **haploid** (*n*): Their chromosome number is half of the diploid (2*n*) number (Section 8.2). Human body cells are diploid, with 23 pairs of homologous chromosomes. Meiosis of a human germ cell (2*n*) normally produces gametes with 23 chromosomes: one of each pair (*n*). The diploid chromosome number is restored at **fertilization**, when two haploid gametes (one egg and one sperm, for example) fuse to form a **zygote**, the first cell of a new individual.

The first part of meiosis is similar to mitosis. A cell duplicates its DNA before either nuclear division process starts. As in mitosis, the microtubules of a spindle move the duplicated chromosomes to opposite spindle poles. However, meiosis sorts the chromosomes into new nuclei not once, but twice, so it results in the formation of four haploid nuclei. The two consecutive nuclear divisions are called meiosis I and meiosis II:

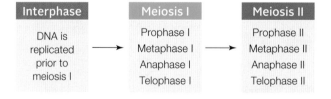

Interphase	Meiosis I	Meiosis II
DNA is replicated prior to meiosis I	Prophase I Metaphase I Anaphase I Telophase I	Prophase II Metaphase II Anaphase II Telophase II

In some cells meiosis II occurs immediately after meiosis I. In others, a period of protein synthesis—but no DNA synthesis—intervenes between the divisions.

During meiosis I, every duplicated chromosome aligns with its homologous partner (**Figure 12.4A**). Then the homologous chromosomes are pulled away from

one another (**Figure 12.4B**). After homologous chromosomes separate, each ends up in one of two new nuclei. At this stage, the chromosomes are still duplicated (the sister chromatids are still attached to one another).

During meiosis II, the sister chromatids of each chromosome are pulled apart, so each becomes an individual, unduplicated chromosome (**Figure 12.4C**). The chromosomes are sorted into four new nuclei. With one unduplicated version of each chromosome, the new nuclei are all haploid (*n*).

Thus, meiosis partitions the chromosomes of one diploid nucleus (2*n*) into four haploid (*n*) nuclei. The next section zooms in on the details of this process.

alleles Forms of a gene with slightly different DNA sequences; may encode slightly different versions of the gene's product.
fertilization Fusion of two gametes to form a zygote.
gamete Mature, haploid reproductive cell; e.g., an egg or a sperm.
germ cell Immature reproductive cell that gives rise to haploid gametes when it divides.
haploid Having one of each type of chromosome characteristic of the species.
somatic Relating to the body.
zygote Diploid cell formed by fusion of two gametes; the first cell of a new individual.

Take-Home Message

Why do populations that reproduce sexually tend to have the most variation in heritable traits?

» The nuclear division process of meiosis is the basis of sexual reproduction in eukaryotes. It precedes the formation of gametes or spores.

» Meiosis halves the diploid (2*n*) chromosome number, to the haploid number (*n*). When two gametes fuse at fertilization, the chromosome number is restored. The resulting zygote has one set of chromosomes from each parent.

» Corresponding genes on homologous chromosomes may vary in sequence as alleles.

» Alleles are the basis of traits. Sexual reproduction mixes up alleles from two parents.

12.3 Visual Tour of Meiosis

- Meiosis halves the chromosome number.
- During meiosis, chromosomes of a diploid nucleus become distributed into four haploid nuclei.
- Links to Diploid chromosome number 8.2, DNA replication 8.5, Mitosis 11.2

The nucleus of a diploid (2n) cell contains two sets of chromosomes, one from each parent. DNA replication occurs before meiosis I begins, so each one of the chromosomes has two sister chromatids. The first stage of meiosis I is prophase I (**Figure 12.5**). During this phase, the chromosomes condense, and homologous chromosomes align tightly and swap segments (more about

segment-swapping in the next section). The centrosome gets duplicated along with its two centrioles. One centriole pair moves to the opposite side of the cell as the nuclear envelope breaks up. Spindle microtubules begin to extend from the centrosomes ❶. Some of the microtubules stop lengthening when they reach the middle of the cell. Others lengthen until they reach a chromosome and attach to it at the centromere.

By the end of prophase I, microtubules of the spindle connect all of the chromosomes to the spindle poles. The maternal chromosome of each homologous pair is attached to one spindle pole, and the paternal chromosome to the other. The opposing sets of microtubules then begin a tug-of-war by adding and losing tubulin subunits. As the microtubules extend and shrink, they push and pull the chromosomes. At metaphase I, the microtubules are the same length, and the chromosomes are aligned midway between the spindle

Figure 12.5 Animated Meiosis. Two pairs of chromosomes are illustrated in a diploid (2n) animal cell. Homologous chromosomes are indicated in *blue* and *pink*. Micrographs show meiosis in a lily plant cell (*Lilium regale*).

Figure It Out: During which phase of meiosis does the chromosome number become reduced?

Answer: Anaphase I

Meiosis I One diploid nucleus to two haploid nuclei

 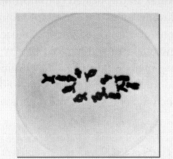 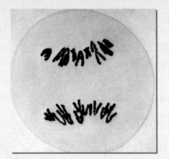 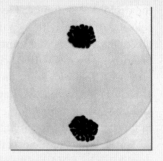

❶ Prophase I. Homologous chromosomes condense, pair up, and swap segments. Spindle microtubules attach to them as the nuclear envelope breaks up.

❷ Metaphase I. The homologous chromosome pairs are aligned midway between spindle poles. The two chromosomes of each pair are attached to opposite spindle poles.

❸ Anaphase I. Homologous chromosomes separate and begin heading toward the spindle poles.

❹ Telophase I. A complete set of chromosomes reaches each spindle pole. A nuclear envelope forms around each set, so two haploid (n) nuclei form.

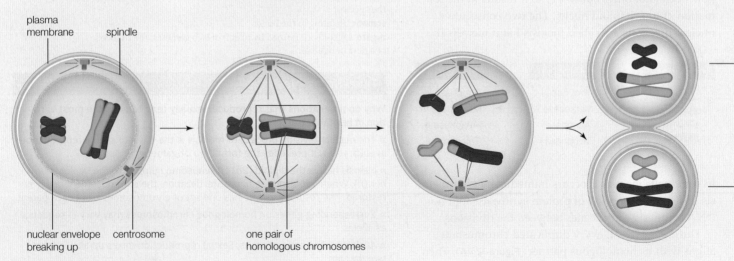

plasma membrane   spindle

nuclear envelope   centrosome
breaking up

one pair of
homologous chromosomes

poles ❷. In anaphase I, the spindle microtubules pull the homologous chromosomes of each pair apart and toward opposite spindle poles ❸. The two sets of chromosomes reach the spindle poles during telophase I ❹, and a new nuclear envelope forms around each cluster of chromosomes as the DNA loosens up. The two new nuclei are haploid (*n*); each contains one set of (duplicated) chromosomes. The cytoplasm may divide at this point, producing two haploid cells.

DNA replication does not occur before meiosis II begins. During prophase II, the chromosomes condense as a new spindle forms ❺. One centriole moves to the opposite side of each new nucleus, and the nuclear envelopes break up. By the end of prophase II, microtubules connect the chromosomes to the spindle poles. Each chromatid is now attached to one spindle pole; its sister is attached to the other. The microtubules lengthen and shorten, pushing and pulling the

chromosomes as they do. At metaphase II, the microtubules are the same length, and the chromosomes are aligned midway between the spindle poles ❻. In anaphase II, the spindle microtubules pull the sister chromatids apart ❼. Each chromosome now consists of one molecule of DNA. During telophase II, the chromosomes (now unduplicated) reach the spindle poles ❽. New nuclear envelopes form around the four clusters of chromosomes as the DNA loosens up. Each of the four haploid (*n*) nuclei that form contains one set of unduplicated chromosomes. The cytoplasm may divide at this point, producing four haploid cells.

Take-Home Message

What happens to a cell during meiosis?

» During meiosis, the nucleus of a diploid (2*n*) cell divides twice. Four haploid (*n*) nuclei form, each with a full set of chromosomes—one of each type.

Meiosis II Two haploid nuclei to four haploid nuclei

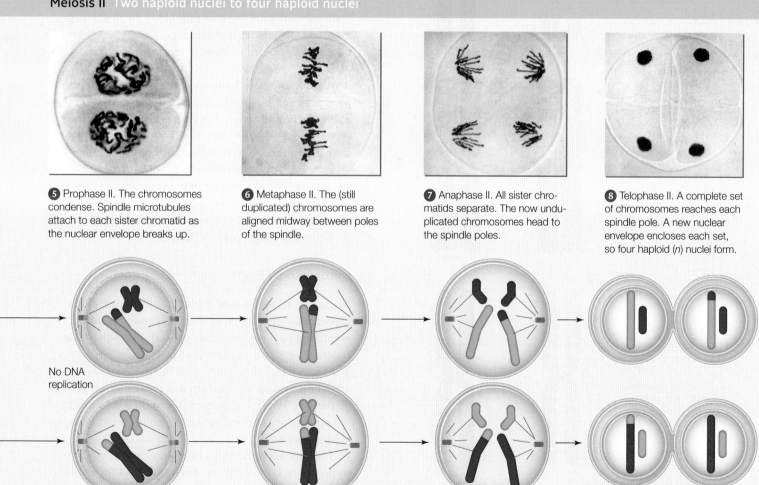

❺ Prophase II. The chromosomes condense. Spindle microtubules attach to each sister chromatid as the nuclear envelope breaks up.

❻ Metaphase II. The (still duplicated) chromosomes are aligned midway between poles of the spindle.

❼ Anaphase II. All sister chromatids separate. The now unduplicated chromosomes head to the spindle poles.

❽ Telophase II. A complete set of chromosomes reaches each spindle pole. A new nuclear envelope encloses each set, so four haploid (*n*) nuclei form.

No DNA replication

12.4 How Meiosis Introduces Variations in Traits

- Crossovers and the random sorting of chromosomes into gametes result in new combinations of traits among offspring of sexual reproducers.
- Link to Chromosome structure 8.2

The previous section mentioned briefly that duplicated chromosomes swap segments with their homologous partners during prophase I. It also showed how spindle microtubules align and then separate homologous chromosomes during anaphase I. Both events introduce novel combinations of alleles into gametes.

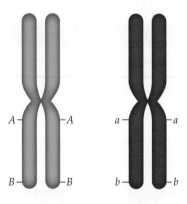

A Here, we focus on only two of the many genes on a chromosome. In this example, one gene has alleles *A* and *a*; the other has alleles *B* and *b*.

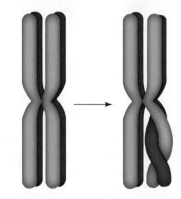

B Close contact between homologous chromosomes promotes crossing over between nonsister chromatids. Paternal and maternal chromatids exchange corresponding pieces.

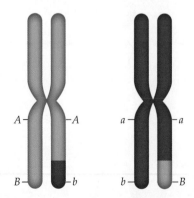

C Crossing over mixes up paternal and maternal alleles on homologous chromosomes.

Figure 12.6 Animated Crossing over. *Blue* signifies a paternal chromosome, and *pink*, its maternal homologue. For clarity, we show only one pair of homologous chromosomes and one crossover, but more than one crossover may occur in each chromosome pair.

Along with fertilization, these events contribute to the variation in combinations of traits among the offspring of sexually reproducing species.

Crossing Over in Prophase I

Early in prophase I of meiosis, all chromosomes in a germ cell condense. When they do, each is drawn close to its homologue. The chromatids of one homologous chromosome become tightly aligned with the chromatids of the other along their length:

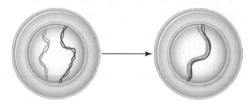

This tight, parallel orientation favors **crossing over**, a process in which a chromosome and its homologous partner exchange corresponding pieces of DNA (**Figure 12.6**). Homologous chromosomes may swap any segment of DNA along their length, although crossovers tend to occur more frequently in certain regions.

Swapping segments of DNA shuffles alleles between homologous chromosomes. It breaks up the particular combinations of alleles that occurred on the parental chromosomes, and makes new ones on the chromosomes that end up in gametes. Thus, crossing over introduces novel combinations of traits among offspring. It is a normal and frequent process in meiosis, but the rate of crossing over varies among species and among chromosomes. In humans, between 46 and 95 crossovers occur per meiosis, so on average each chromosome crosses over at least once.

Chromosome Segregation

Normally, all of the new nuclei that form in meiosis I receive a complete set of chromosomes. However, whether a new nucleus ends up with the maternal or paternal version of a chromosome is entirely random. The chance that the maternal or the paternal version of any chromosome will end up in a particular nucleus is 50 percent. Why? The answer has to do with the way the spindle segregates the homologous chromosomes during meiosis I.

The process of chromosome segregation begins in prophase I. Imagine one of your own germ cells undergoing meiosis. Crossovers have already made genetic mosaics of its chromosomes, but for simplicity let's put crossing over aside for a moment. Just call the

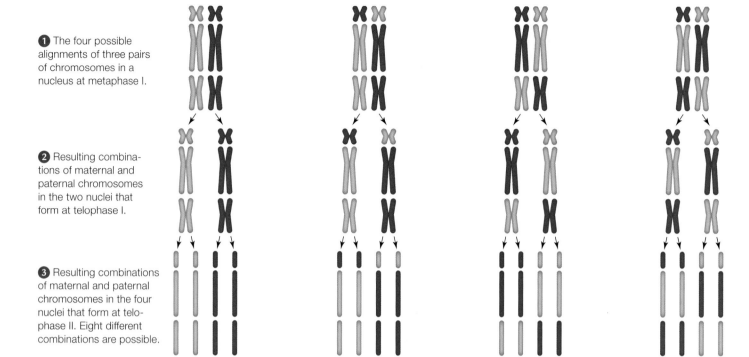

① The four possible alignments of three pairs of chromosomes in a nucleus at metaphase I.

② Resulting combinations of maternal and paternal chromosomes in the two nuclei that form at telophase I.

③ Resulting combinations of maternal and paternal chromosomes in the four nuclei that form at telophase II. Eight different combinations are possible.

Figure 12.7 **Animated** Hypothetical segregation of three pairs of chromosomes in meiosis I. Maternal chromosomes are *pink*; paternal, *blue*. Which chromosome of each pair gets packaged into which of the two new nuclei that form at telophase I is random. For simplicity, no crossing over occurs in this example, so all sister chromatids are identical.

twenty-three chromosomes you inherited from your mother the maternal ones, and the twenty-three from your father the paternal ones.

During prophase I, microtubules fasten your cell's chromosomes to the spindle poles. Chances are fairly slim that all of the maternal chromosomes get attached to one pole and all of the paternal chromosomes get attached to the other. Microtubules extending from a spindle pole bind to the centromere of the first chromosome they contact, regardless of whether it is maternal or paternal. Each homologous partner gets attached to the opposite spindle pole. Thus, there is no pattern to the attachment of the maternal or paternal chromosomes to a particular pole.

Now imagine that your germ cell has just three pairs of chromosomes (**Figure 12.7**). By metaphase I, those three pairs of maternal and paternal chromosomes are divvied up between the two spindle poles in one of four ways ①. In anaphase I, homologous chromosomes separate and are pulled toward opposite spindle poles. In telophase I, a new nucleus forms around the chromosomes that cluster at each spindle pole. Each nucleus contains one of eight possible combinations of maternal and paternal chromosomes ②.

In telophase II, each of the two nuclei divides and gives rise to two new haploid nuclei. The two new nuclei are identical because no crossing over occurred

in our hypothetical example, so all of the sister chromatids were identical. Thus, at the end of meiosis in this cell, two (2) spindle poles have divvied up three (3) chromosome pairs. The resulting four nuclei have one of eight (2^3) possible combinations of maternal and paternal chromosomes ③.

Cells that give rise to human gametes have twenty-three pairs of homologous chromosomes, not three. Each time a human germ cell undergoes meiosis, the four gametes that form end up with one of 8,388,608 (or 2^{23}) possible combinations of homologous chromosomes. That number does not even take into account crossing over, which mixes up the alleles on maternal and paternal chromosomes, or fusion with another gamete at fertilization. Are you getting an idea of why such fascinating combinations of traits show up among the generations of your own family tree?

crossing over Process in which homologous chromosomes exchange corresponding segments during prophase I of meiosis.

12.5 From Gametes to Offspring

■ Aside from meiosis, the details of gamete formation and fertilization differ among plants and animals.

■ Link to Cytoplasmic division 11.4

Gametes are the specialized reproductive cells of sexual reproducers. All are haploid, but they differ in other details. For example, human male gametes—sperm—have one flagellum (Section 4.10). Opossum sperm have two, and roundworm sperm have none. A flowering plant's male gamete consists simply of a nucleus. We leave details of reproduction for later chapters, but you will need to know a few concepts before you get there.

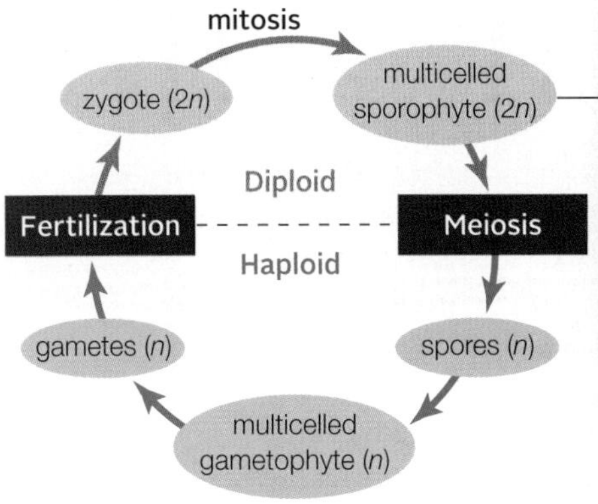

A Generalized life cycle for most plants. A sequoia tree is a sporophyte.

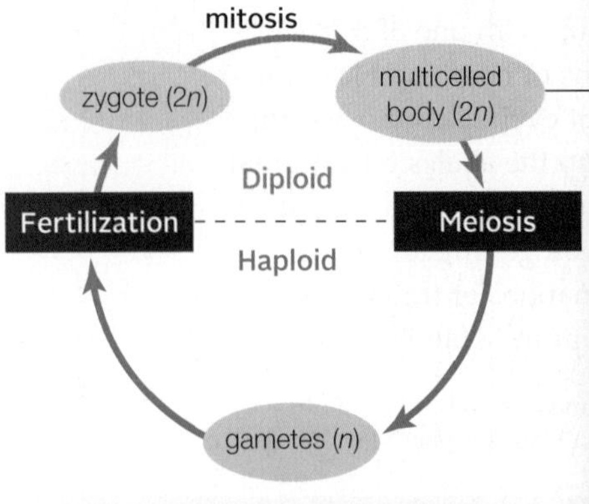

B Generalized life cycle for animals. The zygote is the first cell to form when the nuclei of two gametes, such as a sperm and an egg, fuse at fertilization.

Figure 12.8 Animated Comparing the life cycles of animals and plants.

Gamete Formation in Plants Two kinds of multicelled bodies form during the life cycle of a plant: sporophytes and gametophytes. A **sporophyte** is typically diploid, and spores form by meiosis in its specialized parts. Spores consist of one or a few haploid cells. These cells undergo mitosis and give rise to **gametophytes**, multicelled haploid bodies inside which one or more gametes form. A sequoia tree is an example of a sporophyte (**Figure 12.8A**). Male and female gametophytes develop inside different types of cones that form on each tree. In flowering plants, gametophytes form in flowers, which are specialized reproductive shoots of the sporophyte body.

Gamete Formation in Animals In a typical animal life cycle, a zygote develops into a multicelled individual that produces gametes (**Figure 12.8B**). Animal gametes arise by meiosis of diploid germ cells. The germ cell of a male animal (**Figure 12.9**) develops into a primary spermatocyte ❶. Meiosis I in a primary spermatocyte results in two haploid secondary spermatocytes ❷, which undergo meiosis II and become four spermatids ❸. Each spermatid then matures as a **sperm** ❹.

The germ cell of a female animal (**Figure 12.10**) develops into a primary oocyte, which is an immature egg ❺. This cell undergoes meiosis and division, as occurs with a primary spermatocyte. Two haploid cells form when the primary oocyte divides after meiosis I. However, the cytoplasm of a primary oocyte divides unequally, so these haploid cells differ in size and function. One of the cells is called a first polar body. The other cell, the secondary oocyte, is much larger because it gets nearly all of the parent cell's cytoplasm ❻. The larger cell undergoes meiosis II and cytoplasmic division, which again is unequal ❼. One of the two cells that forms is a second polar body. The other cell gets most of the cytoplasm and matures into a female gamete, which is called an ovum (plural, ova), or **egg**.

Polar bodies are not nutrient-rich or plump with cytoplasm, and generally do not function as gametes. In time they degenerate. Their formation ensures that the egg will have a haploid chromosome number, and also will get enough metabolic machinery to support early divisions of the new individual.

Fertilization Meiosis produces haploid gametes. When two gametes fuse at fertilization, the resulting zygote is diploid. Thus, meiosis halves the chromosome number, and fertilization restores it. If meiosis did not precede fertilization, the chromosome number would double with every generation. An individual's

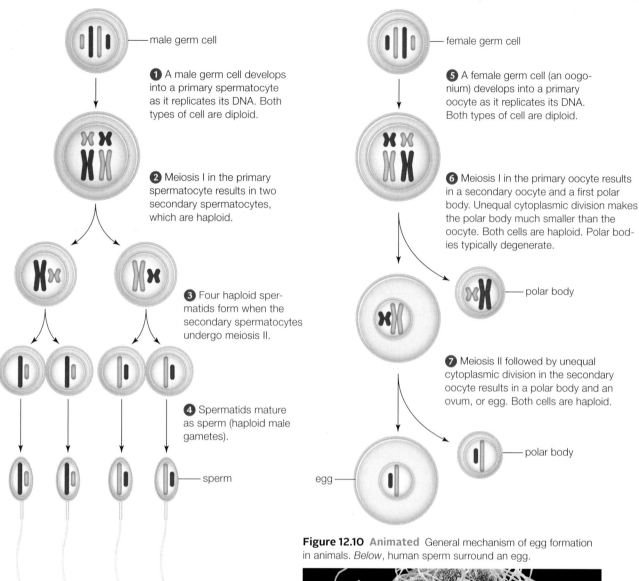

male germ cell

1 A male germ cell develops into a primary spermatocyte as it replicates its DNA. Both types of cell are diploid.

2 Meiosis I in the primary spermatocyte results in two secondary spermatocytes, which are haploid.

3 Four haploid spermatids form when the secondary spermatocytes undergo meiosis II.

4 Spermatids mature as sperm (haploid male gametes).

sperm

Figure 12.9 Animated General mechanism of sperm formation in animals.

female germ cell

5 A female germ cell (an oogonium) develops into a primary oocyte as it replicates its DNA. Both types of cell are diploid.

6 Meiosis I in the primary oocyte results in a secondary oocyte and a first polar body. Unequal cytoplasmic division makes the polar body much smaller than the oocyte. Both cells are haploid. Polar bodies typically degenerate.

polar body

7 Meiosis II followed by unequal cytoplasmic division in the secondary oocyte results in a polar body and an ovum, or egg. Both cells are haploid.

polar body

egg

Figure 12.10 Animated General mechanism of egg formation in animals. *Below*, human sperm surround an egg.

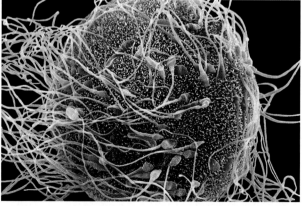

set of chromosomes is like a fine-tuned blueprint that must be followed exactly, page by page, in order to build a body that functions normally. As you will see in Chapter 14, chromosome number changes can have drastic consequences, particularly in animals.

egg Mature female gamete, or ovum.
gametophyte A haploid, multicelled body in which gametes form during the life cycle of land plants and some algae.
sperm Mature male gamete.
sporophyte Diploid, spore-producing stage of a plant life cycle.

Take-Home Message

How does meiosis fit into the life cycle of plants and animals?

» Meiosis and cytoplasmic division precede the development of haploid gametes in animals and spores in plants.

» The union of two haploid gametes at fertilization results in a diploid zygote.

12.6 Mitosis and Meiosis—An Ancestral Connection?

■ Though they have different results, mitosis and meiosis are fundamentally similar processes.

■ Links to Mitosis 11.3, Telomeres 11.5, Cell cycle controls 11.6

This chapter opened with hypotheses about the evolutionary advantages of asexual and sexual reproduction. It seems like a giant evolutionary step from producing clones to producing genetically varied offspring, but was it really?

By mitosis and cytoplasmic division, one cell becomes two new cells that have the parental chromosomes. Mitotic (asexual) reproduction results in clones of the parent. Meiosis results in the formation of haploid gametes. Gametes of two parents fuse to form a zygote, which is a cell of mixed parentage. Meiotic (sexual) reproduction results in offspring that differ genetically from the parent, and from one another.

Though their end results differ, there are striking parallels between the four stages of mitosis and meiosis II (**Figure 12.11**). As one example, a spindle forms and separates chromosomes during both processes. There are many more similarities at the molecular level.

Figure 12.11 Animated Comparative summary of key features of mitosis and meiosis, starting with a diploid cell. Only two paternal and two maternal chromosomes are shown, for clarity. Mitosis maintains the parental chromosome number. Meiosis halves it, to the haploid number.

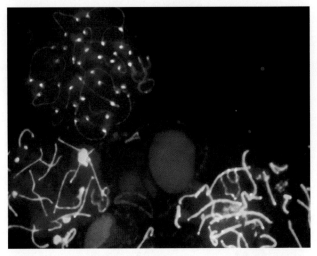

Figure 12.12 Fluorescence micrograph of mouse cell nuclei during meiosis. *Blue* stain: DNA; *green:* paired-up chromosomes before crossing over; *red:* BRCA1, which is clustered around telomeres and the sex chromosomes. This checkpoint protein is involved in DNA repair in mitosis (DNA damage) and meiosis (sealing crossover breaks), and in X chromosome inactivation.

Long ago, some of the molecular machinery of mitosis may have been remodeled into meiosis. Evidence for this hypothesis includes a host of molecules, including the products of the *BRCA* genes (Chapters 10 and 11) that are made by all modern eukaryotes. These molecules monitor and repair breaks in DNA, for example during DNA replication prior to mitosis. They actively maintain the integrity of

Meiosis I One diploid nucleus to two haploid nuclei

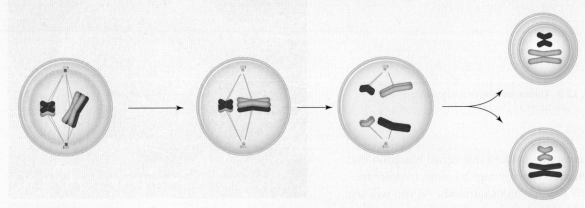

Prophase I
- Chromosomes condense.
- Homologous chromosomes pair.
- Crossovers occur (not shown).
- Spindle forms and attaches chromosomes to spindle poles.
- Nuclear envelope breaks up.

Metaphase I
- Chromosomes align midway between spindle poles.

Anaphase I
- Homologous chromosomes separate and move toward opposite spindle poles.

Telophase I
- Chromosome clusters arrive at spindle poles.
- New nuclear envelopes form.
- Chromosomes decondense.

a cell's chromosomes. The very same set of molecules monitor and fix the breaks in homologous chromosomes during crossing over in prophase I of meiosis (Figure 12.12). Some of them function as checkpoint proteins in both mitosis and meiosis, so mutations that affect them or the rate at which they are made can affect the outcomes of both nuclear division processes.

In anaphase of mitosis, sister chromatids are pulled apart. What would happen if the connections between the sisters did not break? Each duplicated chromosome would be pulled to one or the other spindle pole—which is exactly what happens in anaphase I of meiosis. Sexual reproduction probably originated by mutations that affected mitosis. As you will see in later chapters, the remodeling of existing processes into new ones is a common evolutionary theme.

Take-Home Message

Are the processes of mitosis and meiosis related?

» Meiosis may have evolved by the remodeling of existing mechanisms of mitosis.

Mitosis One diploid nucleus to two diploid nuclei

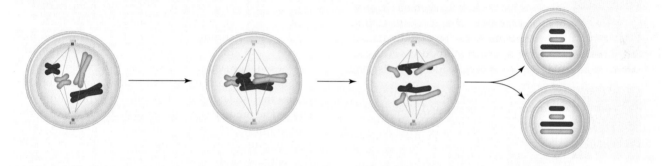

Prophase
- Chromosomes condense.
- Spindle forms and attaches chromosomes to spindle poles.
- Nuclear envelope breaks up.

Metaphase
- Chromosomes align midway between spindle poles.

Anaphase
- Sister chromatids separate and move toward opposite spindle poles.

Telophase
- Chromosome clusters arrive at spindle poles.
- New nuclear envelopes form.
- Chromosomes decondense.

Meiosis II Two haploid nuclei to four haploid nuclei

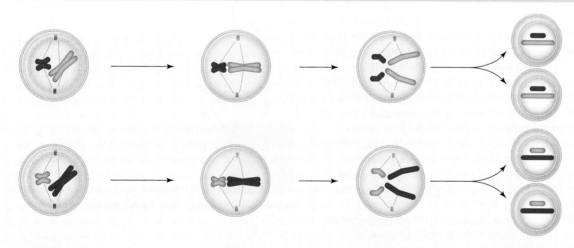

Prophase II
- Chromosomes condense.
- Spindle forms and attaches chromosomes to spindle poles.
- Nuclear envelope breaks up.

Metaphase II
- Chromosomes align midway between spindle poles.

Anaphase II
- Sister chromatids separate and move toward opposite spindle poles.

Telophase II
- Chromosome clusters arrive at spindle poles.
- New nuclear envelopes form.
- Chromosomes decondense.

Why Sex? (revisited)

There are a few all-female species of fishes, reptiles, and birds in nature, but not mammals. In 2004, researchers fused two mouse eggs in a test tube and made an embryo using no DNA from a male. The embryo developed into Kaguya, the world's first fatherless mammal (*right*). The mouse grew up healthy,

engaged in sex with a male mouse, and gave birth to offspring. The researchers wanted to find out if sperm was required for normal development.

How would you vote? Researchers made a "fatherless" mouse by fusing two eggs. Should they be prevented from trying the process with human eggs?

Summary

Section 12.1 Sexual reproduction, which involves the nuclear division mechanism of **meiosis**, mixes up genetic information of two parents. Variation in traits among offspring of sexual reproducers is an evolutionary advantage over genetically identical offspring.

Section 12.2 Offspring of sexual reproducers differ from parents, and often from one another, in details of shared traits. These individuals inherit pairs of chromosomes, one of each pair from the mother and the other from the father. Paired genes on homologous chromosomes in **somatic** cells may vary in DNA sequence, in which case they are called **alleles**. In animals, haploid **gametes** arise from meiosis in **germ cells**. Meiosis shuffles parental DNA, so offspring inherit new combinations of alleles. The fusion of two haploid **gametes** during **fertilization** restores the parental chromosome number in the **zygote**.

Section 12.3 Two divisions, meiosis I and II, halve the parental chromosome number. Chromosomes are duplicated during interphase. In prophase I, chromosomes condense and align tightly with their homologous partners. Each pair of homologues typically undergoes crossing over. Microtubules extending from the spindle poles penetrate the nuclear region and attach to one or the other chromosome of each homologous pair. At metaphase I, all chromosomes are lined up at the spindle equator. During anaphase I, homologous chromosomes separate and move to opposite spindle poles. Two nuclei form during telophase I. The cytoplasm may divide at this point.

Meiosis II occurs in both nuclei that formed in meiosis I. The chromosomes are still duplicated; each still consists of two sister chromatids. The chromosomes condense in prophase II, and align in metaphase II. Sister chromatids of each chromosome are pulled apart from each other in anaphase II, so each becomes an individual chromosome. By the end of telophase II, four haploid nuclei have formed.

Section 12.4 Events in prophase I and metaphase I produce nonparental combinations of alleles. The nonsister chromatids of homologous chromosomes undergo **crossing over** during prophase I: They exchange segments at the same place along their length, so each can end up with a new combination of alleles that was not present in either parental chromosome. The random segregation of maternal and paternal chromosomes into new nuclei also contributes to variation in traits among offspring. Microtubules can attach the maternal or the paternal chromosome of each pair to one or the other spindle pole. Either chromosome may end up in any new nucleus, and in any gamete. Such chromosome shufflings, along with crossovers during prophase I of meiosis, are the basis of variation in traits among individuals of sexually reproducing species.

Section 12.5 Multicelled diploid bodies are typical in life cycles of plants and animals. A diploid **sporophyte** is a multicelled plant body that makes haploid spores. Spores give rise to haploid **gametophytes**, which are multicelled plant bodies in which haploid gametes form. Germ cells in the reproductive organs of most animals give rise to **sperm** or **eggs**. The fusion of haploid gametes at fertilization results in a diploid zygote.

Section 12.6 Like mitosis, meiosis requires a spindle to move and sort duplicated chromosomes, but meiosis occurs only in cells that are set aside for sexual reproduction. Some parts of meiosis resemble those of mitosis, and may have evolved from them.

Self-Quiz   *Answers in Appendix III*

1. The main evolutionary advantage of sexual over asexual reproduction is that it produces _____ .
   a. more offspring per individual
   b. more variation among offspring
   c. healthier offspring

BPA and Abnormal Meiosis In 1998, researchers at Case Western University were studying meiosis in mouse oocytes when they saw an unexpected and dramatic increase of abnormal meiosis events (**Figure 12.13**). The improper segregation of chromosomes during meiosis is one of the main causes of human genetic disorders, which we will discuss in Chapter 14.

The researchers discovered that the spike in meiotic abnormalities began immediately after the mouse facility started washing the animals' plastic cages and water bottles in a new, alkaline detergent. The detergent had damaged the plastic, which began to leach bisphenol A (BPA). BPA is a synthetic chemical that mimics estrogen, the main female sex hormone in animals. BPA is used to manufacture polycarbonate plastic items (including baby bottles and water bottles) and epoxies (including the coating on the inside of metal cans of food).

1. What percentage of mouse oocytes displayed abnormalities of meiosis with no exposure to damaged caging?

2. Which group of mice showed the most meiotic abnormalities in their oocytes?

3. What is abnormal about metaphase I as it is occurring in the oocytes shown in **Figure 12.13B, C,** and **D**?

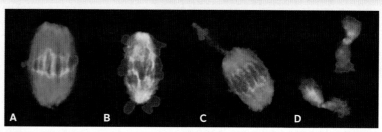

Caging materials	Total number of oocytes	Abnormalities
Control: New cages with glass bottles	271	5 (1.8%)
Damaged cages with glass bottles		
Mild damage	401	35 (8.7%)
Severe damage	149	30 (20.1%)
Damaged bottles	197	53 (26.9%)
Damaged cages with damaged bottles	58	24 (41.4%)

Figure 12.13 Meiotic abnormalities associated with exposure to damaged plastic caging. Fluorescent micrographs show nuclei of single mouse oocytes in metaphase I. (**A**) Normal metaphase; (**B–D**) examples of abnormal metaphase. Chromosomes are stained *red;* spindle fibers, *green.*

---

2. Meiosis functions in _____ .
   a. asexual reproduction of single-celled eukaryotes
   b. growth and tissue repair in multicelled species
   c. sexual reproduction
   d. both a and b

3. Sexual reproduction in animals requires _____ .
   a. meiosis   c. spore formation
   b. fertilization   d. a and b

4. Meiosis _____ the parental chromosome number.
   a. doubles   c. maintains
   b. halves   d. mixes up

5. Crossing over mixes up _____ .
   a. chromosomes   c. zygotes
   b. alleles   d. gametes

6. Crossing over happens during which phase of meiosis?

7. The stage of meiosis that makes descendant cells haploid is _____ .
   a. prophase I   d. anaphase II
   b. prophase II   e. metaphase I
   c. anaphase I   f. metaphase II

8. Dogs have a diploid chromosome number of 78. How many chromosomes do their gametes have?
   a. 39   c. 156
   b. 78   d. 234

9. The cell in the diagram to the *right* is in anaphase I, not anaphase II. I know this because _____ .

10. _____ contributes to variation in traits among the offspring of sexual reproducers.
    a. Crossing over   c. Fertilization
    b. Random attachment of   d. both a and b
       chromosomes to spindle poles   e. all are factors

11. Which of the following is one of the most important differences between mitosis and meiosis?
    a. Chromosomes align midway between spindle poles only in meiosis.
    b. Homologous chromosomes pair up only in meiosis.
    c. DNA is replicated only in mitosis.
    d. Sister chromatids separate only in meiosis.

12. Match each term with its description.
    ___ DNA replication   a. different molecular form
    ___ metaphase I   of a gene
    ___ allele   b. none between meiosis I
    ___ sporophyte   and meiosis II
    ___ gamete   c. all chromosomes are aligned
                           at spindle equator
                           d. haploid
                           e. does not occur in animals

Critical Thinking

1. Make a simple sketch of meiosis in a cell with a diploid chromosome number of 4. Now try it when the chromosome number is 3.

2. The diploid chromosome number for the body cells of a frog is 26. What would that number be after three generations if meiosis did not occur before gamete formation?

3. Assume you can measure the amount of DNA in the nucleus of a primary oocyte, and then in the nucleus of a primary spermatocyte. Each gives you a mass m. What mass of DNA would you expect to find in the nucleus of each mature gamete (each egg and sperm) that forms after meiosis? What mass of DNA will be (1) in the nucleus of a zygote that forms at fertilization and (2) in that zygote's nucleus after the first DNA duplication?

LEARNING ROADMAP

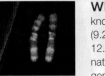

Where you have been You may want to review what you know about traits (Chapter 1), chromosomes (Section 8.2), genes (9.2), sexual reproduction (12.1), alleles (12.2), and meiosis (12.3, 12.4). You will revisit probability and sampling error (1.8), laws of nature (1.9), protein structure (3.6), pigments (6.2), clones (8.7), gene control (10.2, 11.6), and epigenetics (10.6).

Where you are now

Where Modern Genetics Started
Gregor Mendel discovered that inherited traits are specified in units. The units, which are distributed into gametes in predictable patterns, were later identified as genes.

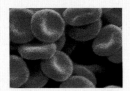

Insights from Monohybrid Crosses
Tracing inheritance patterns of single traits led to the discovery that during meiosis, pairs of genes on homologous chromosomes separate and end up in different gametes.

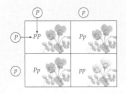

Insights from Dihybrid Crosses
Tracing inheritance patterns of two unrelated traits led to the discovery that pairs of genes often segregate into gametes independently of how other gene pairs are distributed.

Variations on Mendel's Theme
An allele may be partly dominant over a nonidentical partner, or codominant with it. Multiple genes may influence a trait; some genes influence many traits.

Complex Variations in Traits
Environmental factors can alter the expression of genes that influence a trait. Many traits appear in a continuous range of forms.

Where you are going We return to melanin and human skin color in Section 14.1, and human genetic disorders in the rest of Chapter 14. The complex interplay between genes and the environment is further explored in an evolutionary context in Chapter 17. Signaling pathways in the body that lead to changes in gene expression comprise stimulus, perception, and response (Sections 31.9 and 33.2). Chapter 29 details reproduction in flowering plants. Neurons and how they work are the topic of Chapter 32, with neurological disorders in Section 32.6.

13.1 Menacing Mucus

In 1988, researchers discovered a gene that, when mutated, causes cystic fibrosis (CF). Cystic fibrosis is the most common fatal genetic disorder in the United States. The gene in question, *CFTR*, encodes a protein that moves chloride ions out of epithelial cells. Sheets of these cells line the passageways and ducts of the lungs, liver, pancreas, intestines, reproductive system, and skin. When the CFTR protein pumps chloride ions out of these cells, water follows the ions by osmosis. The two-step process maintains a thin film of water on the surface of the epithelial sheets. Mucus slides easily over the wet sheets of cells.

The most common mutation in CF is a deletion of three base pairs—one codon that specifies phenylalanine as the 508th amino acid of the CFTR protein. This deletion, which is called Δ*F508*, prevents proper membrane trafficking of CFTR so that newly assembled polypeptides are stranded in the endoplasmic reticulum. The altered protein still functions properly, but it never reaches the cell surface to do its job.

One outcome is that the transport of chloride ions out of epithelial cells is disrupted. If not enough chloride ions leave the cells, not enough water leaves them either, so the surfaces of epithelial cell sheets are not as wet as they should be. Mucus that normally slips and slides through the body's tubes sticks to the walls of the tubes instead. Thick globs of mucus accumulate and clog passageways and ducts throughout the body. Breathing becomes difficult as the mucus obstructs the smaller airways of the lungs.

The CFTR protein also functions as a receptor that alerts the body to the presence of bacteria. Bacteria bind to CFTR. The binding triggers endocytosis, which speeds the immune system's defensive responses. Without the CFTR protein on the surface of epithelial cell linings, disease-causing bacteria that enter the ducts and passageways of the body can persist there. Thus, chronic bacterial infections of the intestine and lungs are hallmarks of cystic fibrosis. Daily routines of posture changes and thumps on the chest and back help clear the lungs of some of the thick mucus, and antibiotics help control infections, but there is no cure. Even with a lung transplant, most cystic fibrosis patients live no longer than thirty years, at which time their tormented lungs usually fail (**Figure 13.1**).

About 1 in 25 people carry the Δ*F508* mutation in one of their two copies of the *CFTR* gene, but most of them do not realize it because they do not have the symptoms of cystic fibrosis. The disease occurs only when a person inherits two mutated genes, one from each parent. This unlucky event occurs in about 1 of 3,300 births worldwide.

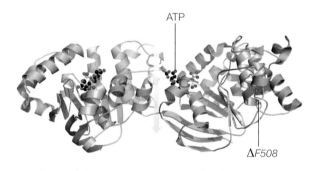

ATP

Δ*F508*

Figure 13.1 Cystic fibrosis. *Left*, model of the CFTR protein. The parts shown here are a pair of ATP-driven motors that widen or narrow a channel (*gray* arrow) across the plasma membrane. The tiny part of the protein that is deleted in most people with cystic fibrosis is shown in *green*.

Below, a few of the many young victims of cystic fibrosis, which occurs most often in people of northern European ancestry. At least one young person dies every day in the United States from complications of this disease.

Lindsay, 22　Savannah, 19　Cody, 23　Ben, 23　Jeff, 21　Brandon, 18

13.2 Mendel, Pea Plants, and Inheritance Patterns

■ Recurring patterns of inheritance offer observable evidence of how heredity works.

■ Links to Traits Chapter 1, Diploid 8.2, Genes 9.2, Alleles and haploid 12.2

In the nineteenth century, people thought that hereditary material must be some type of fluid, with fluids from both parents blending at fertilization like milk into coffee. However, the idea of "blending inheritance" failed to explain what people could see with their own eyes. Children sometimes have traits such as freckles that do not appear in either parent. A cross between a black horse and a white one does not produce gray offspring.

The naturalist Charles Darwin did not accept the idea of blending inheritance, but he could not come up with an alternative even though inheritance was cen-

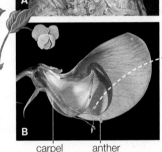

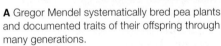

carpel   anther

A Gregor Mendel systematically bred pea plants and documented traits of their offspring through many generations.

B Garden pea flower, cut in half. Male gametes form in pollen grains produced by anthers, and female gametes form in carpels. Experimenters can control the transfer of hereditary material from one flower to another by snipping off a flower's anthers (to prevent the flower from self-fertilizing), and then brushing pollen from another flower onto its carpel.

C In this example, pollen from a plant that has purple flowers is brushed onto the carpel of a white-flowered plant.

D Later, seeds develop inside pods of the cross-fertilized plant. An embryo in each seed develops into a mature pea plant.

E Every plant that arises from the cross has purple flowers. Predictable patterns such as this are evidence of how inheritance works.

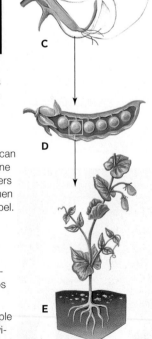

Figure 13.2 Animated Breeding garden pea plants (*Pisum sativum*), which can self-fertilize or cross-fertilize.

tral to his theory of natural selection (we return to natural selection in Chapter 16). Neither Darwin nor anyone else at the time knew that hereditary information (DNA) is divided into discrete units (genes), an insight that is critical to understanding how heredity really works. However, even before Darwin presented his theory, someone had been gathering evidence that would support it. Gregor Mendel, an Austrian monk, had been carefully breeding thousands of pea plants (**Figure 13.2A**). By meticulously documenting the passage of traits from one generation to the next, Mendel had been collecting evidence of how inheritance works.

Mendel's Experimental Approach

Mendel studied variation in the traits of the garden pea plant, *Pisum sativum*. This species is naturally self-fertilizing, which means that its flowers produce male gametes (in pollen) and female gametes (in carpels) that form viable embryos when they meet up.

In order to study inheritance, Mendel had to breed particular individuals together, then observe and document the traits of their offspring. Control over the reproduction of an individual pea plant begins with preventing it from self-fertilizing. Mendel did this by removing a flower's pollen-bearing anthers, then brushing its carpel with pollen from another plant (**Figure 13.2B,C**). He collected the seeds from the cross-fertilized individual, and recorded the traits of the new pea plants that grew from them (**Figure 13.2D,E**).

Many of Mendel's experiments started with plants that "breed true" for a particular trait. Breeding true for a trait means that, new mutations aside, all offspring have the same form of the trait as the parent(s), generation after generation. For example, all offspring of pea plants that breed true for white flowers also have white flowers.

Breeders cross-fertilize plants when they transfer pollen among individuals that have different traits. As you will see in the next section, Mendel discovered that the traits of the offspring of cross-fertilized pea plants often appear in predictable patterns. Mendel's meticulous work tracking pea plant traits led him to conclude (correctly) that hereditary information passes from one generation to the next in discrete units.

Inheritance in Modern Terms

DNA was not proven to be hereditary material until the 1950s (Section 8.3), but Mendel discovered its units, which we now call genes, almost a century before then. Today, we know that individuals of a

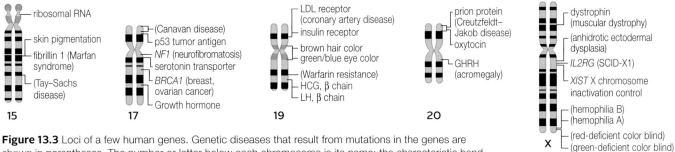

Figure 13.3 Loci of a few human genes. Genetic diseases that result from mutations in the genes are shown in parentheses. The number or letter below each chromosome is its name; the characteristic banding patterns appear after staining. Appendix VI has a similar map of all 23 human chromosomes.

species share certain traits because their chromosomes carry the same genes. Offspring tend to look like their parents because they inherited their parents' genes.

Each gene occurs at a specific location, or **locus** (plural, loci), on a particular chromosome (**Figure 13.3**). Somatic cells of humans and other animals are diploid: They have pairs of genes, on pairs of homologous chromosomes. In most cases, both genes of a pair are expressed (Section 9.2).

The two genes of a pair may be identical, or they may vary as alleles (Section 12.2). Organisms that breed true for a specific trait probably have identical alleles of the gene governing that trait. An individual with identical alleles of a gene is said to be **homozygous** for the allele. The particular set of alleles that an individual carries is called the individual's **genotype**.

Alleles are the major source of variation in a trait. New alleles arise by mutation (Section 8.6). A mutation may cause a trait to change, as when a gene that causes flowers to be purple mutates so the resulting flowers are white. "White-flowered" is an example of a **phenotype**, which refers to one or more aspects of an individual's observable traits. Any mutated gene is an allele, whether or not it affects phenotype.

Hybrids are offspring of a cross, or mating, between individuals that breed true for different forms of a trait. A hybrid has two different alleles of a gene, so it is said to be **heterozygous** for the alleles (*hetero–*, mixed). In many cases, the effect of one allele influences the effect of the other, and the outcome of this interaction is visible in the hybrid phenotype. An allele is **dominant** when its effect masks that of a **recessive** allele paired with it. Usually, a dominant allele is represented by an italic capital letter such as *A*; a recessive allele, with a lowercase italic letter such as *a*.

Mendel crossed plants that breed true for purple flowers with plants that breed true for white flowers. All offspring of these crosses have purple flowers, but Mendel did not know why. We now understand that one gene governs purple and white flower color in pea plants. The allele that specifies purple (let's call it *P*) is dominant over the allele that specifies white (*p*). Thus, a pea plant homozygous for two dominant alleles (*PP*) has purple flowers; one homozygous for two recessive alleles (*pp*) has white flowers. A plant heterozygous at this gene locus (*Pp*) has purple flowers (**Figure 13.4**).

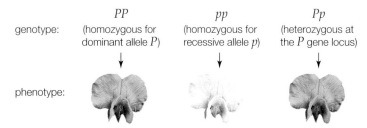

Figure 13.4 Genotype gives rise to phenotype. In this example, the dominant allele *P* specifies purple flowers; the recessive allele *p*, white flowers.

Figure It Out Which individual is a hybrid? *Answer: The heterozygous one*

Take-Home Message

How do alleles contribute to traits?

» Gregor Mendel indirectly discovered the role of alleles in inheritance by carefully breeding pea plants and tracking traits of their offspring.

» Genotype refers to the particular set of alleles carried by an individual's somatic cells. Phenotype refers to the individual's observable traits. Genotype is the basis of phenotype.

» A homozygous individual has two identical alleles at a particular locus. A heterozygous individual has nonidentical alleles at the locus.

» Dominant alleles mask the effects of recessive ones in heterozygous individuals.

dominant Refers to an allele that masks the effect of a recessive allele paired with it.
genotype The particular set of alleles carried by an individual.
heterozygous Having two different alleles of a gene.
homozygous Having identical alleles of a gene.
hybrid The offspring of a cross between two individuals that breed true for different forms of a trait; a heterozygous individual.
locus Location of a gene on a chromosome.
phenotype An individual's observable traits.
recessive Refers to an allele with an effect that is masked by a dominant allele on the homologous chromosome.

■ Pairs of genes on homologous chromosomes separate during meiosis, so they end up in different gametes.
■ Links to Probability and sampling error 1.8, Laws of nature 1.9, Meiosis 12.3, Chromosome segregation 12.4

When homologous chromosomes separate during meiosis, the gene pairs on those chromosomes separate too. Each gamete that forms carries only one of the two genes of a pair (**Figure 13.5**). Thus, plants homozygous for the dominant allele (*PP*) can only make gametes that carry the allele *P* ❶. Plants homozygous for the recessive allele (*pp*) can only make gametes that carry the allele *p* ❷. If these homozygous plants are crossed (*PP* × *pp*), only one outcome is possible: A gamete carrying a *P* allele meets up with a gamete carrying a *p* allele ❸. All of the offspring of this cross have one of each allele, so their genotype is *Pp*. A grid called a **Punnett square** is helpful for predicting the genetic and phenotypic outcomes of crosses ❹. In this example, all offspring of the cross carry the dominant allele *P*, so all have purple flowers.

This pattern is so predictable that it can be used as evidence of a dominance relationship between alleles.

Breeding experiments use such patterns to reveal genotype. In a **testcross**, an individual that has a dominant trait (but an unknown genotype) is crossed with an individual known to be homozygous recessive. The pattern of traits among the offspring of the cross can reveal whether the tested individual is heterozygous or homozygous.

For example, we may do a testcross between a purple-flowered pea plant (which could have a genotype of either *PP* or *Pp*) and a white-flowered pea plant (*pp*). If all of the offspring of this cross had purple flowers, we could be reasonably certain that the genotype of the purple-flowered parent was *PP*.

Dominance relationships between alleles determine the phenotypic outcome of a **monohybrid cross**, in which individuals identically heterozygous for one gene—*Pp*, for example—are bred together or self-fertilized. The frequency at which the two traits appear among the offspring of this cross depends on whether one of the alleles is dominant over the other.

To produce identically heterozygous individuals for a monohybrid cross, we would start with two individuals that breed true for two different forms of a trait. In pea plants, flower color (purple and white) is one example of a trait with two distinct forms, but there are many others. Mendel investigated seven of them: stem length (tall and short), seed color (yellow and green), pod texture (smooth and wrinkled), and so on (**Table 13.1**). A cross between the two true-breeding individuals yields offspring that are identically heterozygous for the alleles that govern the trait. When these F₁ (first generation) hybrids are crossed, the frequency at which the two traits appear in the F₂ (second generation) offspring offers information about a dominance relationship between the two alleles. F is an abbreviation for filial, which means offspring.

DNA replication

meiosis I

meiosis II

gametes (*P*)

gametes (*p*)

zygote (*Pp*)

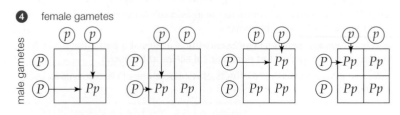

Figure 13.5 Gene segregation. Homologous chromosomes separate during meiosis, so the pairs of genes they carry separate too. Each of the resulting gametes carries one of the two members of each gene pair. For clarity, only one set of chromosomes is shown.

❶ All gametes made by a parent homozygous for a dominant allele carry that allele.

❷ All gametes made by a parent homozygous for a recessive allele carry that allele.

❸ If these two parents are crossed, the union of any of their gametes at fertilization produces a zygote with both alleles. All offspring of this cross will be heterozygous.

❹ This outcome is easy to see with a Punnett square. Parental gametes are listed in circles on the top and left sides of a grid. Each square is filled with the combination of alleles that would result if the gametes in the corresponding row and column met up.

Table 13.1 Mendel's Seven Pea Plant Traits

Trait:	Recessive Form	Dominant Form
Seed Shape:	Round	Wrinkled
Seed Color:	Yellow	Green
Pod Texture:	Smooth	Wrinkled
Pod Color:	Green	Yellow
Flower Color:	Purple	White
Flower Position:	Along Stem	At Tip
Stem Length:	Tall	Short

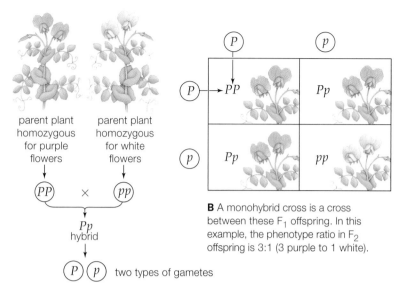

parent plant homozygous for purple flowers

parent plant homozygous for white flowers

PP × pp

Pp hybrid

P p two types of gametes

A All of the F_1 offspring of a cross between two plants that breed true for different forms of a trait are identically heterozygous (Pp). These offspring make two types of gametes: P and p.

B A monohybrid cross is a cross between these F_1 offspring. In this example, the phenotype ratio in F_2 offspring is 3:1 (3 purple to 1 white).

Figure 13.6 **Animated** A monohybrid cross. **Figure It Out:** In this example, how many possible genotypes are there in the F_2 generation?

Answer: Three: PP, Pp, and pp

A cross between purple-flowered heterozygous plants (Pp) offers an example. Each plant can make two types of gametes: ones that carry a P allele, and ones that carry a p allele (**Figure 13.6A**). So, in a monohybrid cross between Pp plants ($Pp \times Pp$), the two types of gametes can meet up in four possible ways at fertilization:

Possible Event	Probable Outcome
Sperm P meets egg P ⟶	zygote genotype is PP
Sperm P meets egg p ⟶	zygote genotype is Pp
Sperm p meets egg P ⟶	zygote genotype is Pp
Sperm p meets egg p ⟶	zygote genotype is pp

Three of four possible outcomes of this cross include at least one copy of the dominant allele P. Each time fertilization occurs, there are 3 chances in 4 that the resulting offspring will inherit a P allele, and have purple flowers. There is 1 chance in 4 that it will

inherit two recessive p alleles, and have white flowers. Thus, the probability that a particular offspring of this cross will have purple or white flowers is 3 purple to 1 white, represented as a ratio of 3:1 (**Figure 13.6B**).

If the probability of one individual inheriting a particular genotype is difficult to imagine, think about probability in terms of the phenotypes of many offspring. In this example, there will be roughly three purple-flowered plants for every white-flowered one. The 3:1 pattern is an indication that the purple and white flower colors are specified by alleles with a clear dominant–recessive relationship: Purple is dominant, and white is recessive.

The phenotype ratios in F_2 offspring of Mendel's monohybrid crosses were all close to 3:1. These results became the basis of his **law of segregation**, which we state here in modern terms: Diploid cells carry pairs of genes, on pairs of homologous chromosomes. The two genes of each pair separate from one another during meiosis, so they end up in different gametes.

law of segregation The members of each pair of genes on homologous chromosomes end up in different gametes during meiosis.
monohybrid cross Cross between two individuals identically heterozygous for one gene; for example $Aa \times Aa$.
Punnett square Diagram used to predict the genetic and phenotypic outcome of a cross.
testcross Method of determining genotype by tracking a trait in the offspring of a cross between an individual of unknown genotype and an individual known to be homozygous recessive.

■ Many gene pairs tend to sort into gametes independently of one another.

■ Links to Meiosis 12.3, Crossing over and chromosome segregation 12.4

A monohybrid cross allows us to track alleles of one gene pair. What about alleles of two gene pairs? How two gene pairs get sorted into gametes depends partly on whether the two genes are on the same chromosome. When homologous chromosomes separate during meiosis, either member of the pair can end up in a particular nucleus. Thus, gene pairs on one chromosome tend to sort into gametes independently of gene pairs on other chromosomes (**Figure 13.7**).

Punnett squares are particularly useful for predicting inheritance patterns of two or more genes simultaneously, such as with a dihybrid cross. In a **dihybrid cross**, individuals identically heterozygous for alleles of two genes (dihybrids) are crossed. As with a monohybrid cross, the pattern of traits seen in the offspring of the cross depends on the dominance relationships between alleles of the genes.

To make a dihybrid cross, we would start with individuals that breed true for two different traits. Let's use a gene for flower color (*P*, purple; *p*, white) and one for plant height (*T*, tall; *t*, short). Assume that

these two genes are on separate chromosomes. **Figure 13.8** shows a dihybrid cross starting with a parent plant that breeds true for purple flowers and tall stems (*PPTT*), and one that breeds true for white flowers and short stems (*pptt*). Each homozygous plant makes only one type of gamete ❶. So, all offspring from a cross between these parent plants (*PPTT* × *pptt*) will be dihybrids (*PpTt*) with purple flowers and tall stems ❷.

Four combinations of *P* and *T* alleles are possible in the gametes of *PpTt* dihybrids ❸. If two *PpTt* plants are crossed (a dihybrid cross, *PpTt* × *PpTt*), the four types of gametes can combine in sixteen possible ways at fertilization ❹. Those sixteen genotypes would result in four different phenotypes. Nine would be tall with purple flowers, three would be short with purple flowers, three would be tall with white flowers, and one would be short with white flowers. Thus, the ratio of these phenotypes is 9:3:3:1.

Mendel discovered the 9:3:3:1 ratio of phenotypes among the offspring of his dihybrid crosses, but he had no idea what it meant. He could only say that "units" specifying one trait (such as flower color) are inherited independently of "units" specifying other traits (such as plant height). In time, Mendel's hypothesis became known as the **law of independent assortment**, which we state here in modern terms: During meiosis, the two

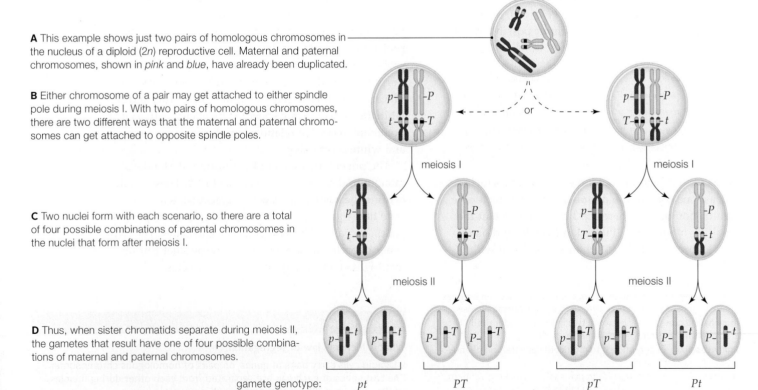

A This example shows just two pairs of homologous chromosomes in the nucleus of a diploid (*2n*) reproductive cell. Maternal and paternal chromosomes, shown in *pink* and *blue*, have already been duplicated.

B Either chromosome of a pair may get attached to either spindle pole during meiosis I. With two pairs of homologous chromosomes, there are two different ways that the maternal and paternal chromosomes can get attached to opposite spindle poles.

C Two nuclei form with each scenario, so there are a total of four possible combinations of parental chromosomes in the nuclei that form after meiosis I.

D Thus, when sister chromatids separate during meiosis II, the gametes that result have one of four possible combinations of maternal and paternal chromosomes.

meiosis I   meiosis I

meiosis II   meiosis II

gamete genotype: *pt*   *PT*   *pT*   *Pt*

Figure 13.7 Animated Independent assortment.

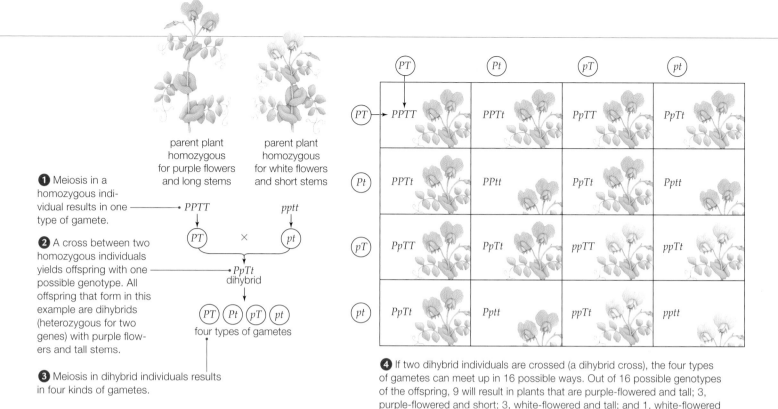

① Meiosis in a homozygous individual results in one type of gamete.

② A cross between two homozygous individuals yields offspring with one possible genotype. All offspring that form in this example are dihybrids (heterozygous for two genes) with purple flowers and tall stems.

③ Meiosis in dihybrid individuals results in four kinds of gametes.

parent plant homozygous for purple flowers and long stems

parent plant homozygous for white flowers and short stems

PPTT   pptt

PT × pt

PpTt
dihybrid

PT Pt pT pt
four types of gametes

④ If two dihybrid individuals are crossed (a dihybrid cross), the four types of gametes can meet up in 16 possible ways. Out of 16 possible genotypes of the offspring, 9 will result in plants that are purple-flowered and tall; 3, purple-flowered and short; 3, white-flowered and tall; and 1, white-flowered and short. Thus, the ratio of phenotypes in a dihybrid cross is 9:3:3:1.

Figure 13.8 **Animated** A dihybrid cross between plants that differ in flower color and plant height. *P* and *p* stand for dominant and recessive alleles for flower color; *T* and *t*, dominant and recessive alleles for height.
Figure It Out: What do the flowers inside the boxes represent?   Answer: Phenotypes of F₂ offspring

genes of a pair tend to be sorted into gametes independently of how other gene pairs are sorted into gametes.

Mendel published his results in 1866, but apparently his work was read by few and understood by no one at the time. In 1871 he became abbot of his monastery, and his pioneering experiments ended. He died in 1884, never to know that they would be the starting point for modern genetics.

The Contribution of Crossovers

It makes sense that gene pairs on different chromosomes would assort independently into gametes, but what about gene pairs on the same chromosome? Mendel studied seven genes in pea plants, which have seven chromosomes. Was he lucky enough to choose one gene on each of those chromosomes? As it turns out, some of the genes Mendel studied *are* on the same chromosome. The genes are far enough apart that crossing over occurs between them very frequently—so frequently that they tend to assort into gametes independently, just as if they were on different chromosomes.

By contrast, genes that are very close together on a chromosome do not assort independently, because crossing over does not happen very often between

them. Such genes are said to be linked. Alleles of some linked genes stay together during meiosis more frequently than others, an effect due to the relative distance between the genes. Genes that are closer together get separated less frequently by crossovers. Thus, the closer together any two genes are on a chromosome, the more likely gametes will be to receive parental combinations of alleles of those genes. Genes are said to be tightly linked if the distance between them is relatively small.

All of the genes on a single chromosome are called a **linkage group**. Peas have 7 different chromosomes, so they have 7 linkage groups. Humans have 23 different chromosomes, so they have 23 linkage groups.

dihybrid cross Cross between two individuals identically heterozygous for two genes; for example *AaBb* × *AaBb*.
law of independent assortment During meiosis, members of a pair of genes on homologous chromosomes get distributed into gametes independently of other gene pairs.
linkage group All genes on a chromosome.

13.5 Beyond Simple Dominance

■ Mendel focused on traits that are based on clearly dominant and recessive alleles. However, many other traits are influenced by alleles with less straightforward relationships.

■ Links to Fibrous proteins 3.6, Pigments 6.2

The inheritance patterns in the last two sections offer examples of simple dominance, in which the effect of a recessive allele is fully masked by that of a dominant one. This is not the only way in which alleles influence traits. In some cases, two alleles affect a trait equally; in others, one is incompletely dominant over the other. Many traits are influenced by multiple genes, and many single genes influence multiple traits.

Codominance

With **codominance**, both alleles are fully expressed in heterozygotes, and neither is dominant or recessive. Codominance may occur in **multiple allele systems**, in which three or more alleles persist in a population. The three alleles of the *ABO* gene are an example. An enzyme encoded by this gene modifies a carbohydrate on the surface of human red blood cell membranes. The *A* and *B* alleles encode slightly different versions of the enzyme, which in turn modify the carbohydrate differently. The *O* allele has a mutation that prevents its enzyme product from becoming active at all.

The two alleles you carry for the *ABO* gene determine the form of the carbohydrate on your blood cells, and that carbohydrate is the basis of your blood type (**Figure 13.9**). The *A* and the *B* alleles are codominant when paired. If your genotype is *AB*, then you have both versions of the enzyme, and your blood is type AB. The *O* allele is recessive when paired with either the *A* or *B* allele. If your genotype is *AA* or *AO*, your blood is type A. If your genotype is *BB* or *BO*, it is type B. If you are *OO*, it is type O.

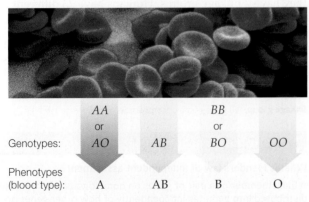

Figure 13.9 Combinations of alleles that are the basis of human blood type.

homozygous (*RR*) × homozygous (*rr*) ⟶ heterozygous (*Rr*)

A Cross a red-flowered with a white-flowered snapdragon, and all of the offspring will be pink-flowered heterozygotes.

B If two of the pink heterozygotes are crossed, the phenotypes of the resulting offspring will occur in a 1:2:1 ratio.

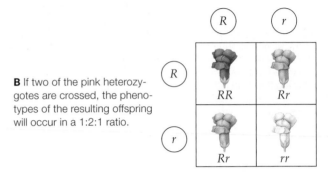

Figure 13.10 Animated Incomplete dominance in heterozygous (*pink*) snapdragons. An allele that affects red pigment is paired with a "white" allele. **Figure It Out: Is the experiment in (B) a monohybrid or dihybrid cross?** Answer: A monohybrid cross

Receiving incompatible blood cells in a transfusion is very dangerous, because the immune system usually attacks red blood cells bearing molecules that do not occur in one's own body. The attack can cause the blood cells to clump or burst, a transfusion reaction with potentially fatal consequences. People with type O blood can donate blood to anyone else, so they are called universal donors. However, they can receive transfusions of type O blood only. People with type AB blood can receive a transfusion of any blood type, so they are called universal recipients.

Incomplete Dominance

With **incomplete dominance**, one allele is not fully dominant over the other, so the heterozygous phenotype is somewhere between the two homozygous phenotypes. A gene that influences flower color in snapdragon plants is an example. There is one copy of this gene on each homologous chromosome; both copies are expressed. One allele (*R*) encodes an enzyme that makes a red pigment. The enzyme encoded by a mutated allele (*r*) cannot make any pigment. Plants homozygous for the *R* allele (*RR*) make a lot of red pigment, so they have red flowers. Plants homozygous for the *r* allele (*rr*) do not make any pigment at all, so their flowers are white. Heterozygous plants (*Rr*) make only enough red pigment to color their flowers pink (**Figure 13.10A**). A cross between two pink-flowered heterozygous plants yields red-, pink-, and white-flowered offspring in a 1:2:1 ratio (**Figure 13.10B**).

Epistasis and Pleiotropy

Some traits are affected by multiple gene products, an effect called polygenic inheritance or **epistasis**. For example, several gene products affect the coat color of a Labrador retriever, which can be black, yellow, or brown (**Figure 13.11**). One gene is involved in the synthesis of the pigment melanin. A dominant allele of the gene specifies black fur, and its recessive partner specifies brown fur. A dominant allele of a different gene causes melanin to be deposited in fur, and its recessive partner reduces melanin deposition.

A **pleiotropic** gene is one that influences multiple traits. Mutations in such genes are associated with complex genetic disorders such as sickle-cell anemia (Section 9.6) and cystic fibrosis. For example, thickened mucus in cystic fibrosis patients affects the entire body, not just the respiratory tract. The mucus clogs ducts that lead to the gut, which results in digestive problems. Male CF patients are typically infertile because their sperm flow is hampered by the thickened secretions.

Marfan syndrome is another example of a genetic disorder caused by mutation in a pleiotropic gene.

	EB	Eb	eB	eb
EB	EEBB	EEBb	EeBB	EeBb
Eb	EEBb	EEbb	EeBb	Eebb
eB	EeBB	EeBb	eeBB	eeBb
eb	EeBb	Eebb	eeBb	eebb

Figure 13.11 Epistasis in dogs. Epistatic interactions among products of two gene pairs affect coat color in Laborador retrievers. All dogs with an *E* and *B* allele have black fur. Those with an *E* and two recessive *b* alleles have brown fur. All dogs homozygous for the recessive *e* allele have yellow fur.

In this case, the gene encodes fibrillin. Long fibers of this protein impart elasticity to tissues of the heart, skin, blood vessels, tendons, and other body parts. Mutations in the fibrillin gene result in tissues that form with defective fibrillin or none at all. The largest blood vessel leading from the heart, the aorta, is particularly affected. Muscle cells in the aorta's thick wall do not work very well, and the wall itself is not as elastic as it should be. The aorta expands under pressure, so the lack of elasticity eventually makes it thin and leaky. Calcium deposits accumulate inside. Inflamed, thinned, and weakened, the aorta can rupture abruptly during exercise.

Marfan syndrome is particularly difficult to diagnose. Affected people are often tall, thin, and loose-jointed, but there are plenty of tall, thin, loose-jointed people that do not have the syndrome. Symptoms may not be apparent, so many people die suddenly and early without ever knowing they had the disorder (**Figure 13.12**).

Figure 13.12 Marfan syndrome. Basketball star Haris Charalambous died suddenly in 2006 when his aorta burst during warm-up exercises. He was 21.

Charalambous was very tall and lanky, with long arms and legs—traits that are valued in professional athletes such as basketball players. These traits are also associated with Marfan syndrome.

About 1 in 5,000 people are affected by Marfan syndrome worldwide. Like many of them, Charalambous did not realize he had the syndrome.

TOLEDO

codominant Refers to two alleles that are both fully expressed in heterozygotes and neither is dominant over the other.
epistasis Effect in which a trait is influenced by the products of multiple genes.
incomplete dominance Effect in which one allele is not fully dominant over another, so the heterozygous phenotype is between the two homozygous phenotypes.
multiple allele system Gene for which three or more alleles persist in a population.
pleiotropic Refers to a gene that influences multiple traits.

Take-Home Message

Are all alleles dominant or recessive?

» An allele may be fully dominant, incompletely dominant, or codominant with its partner on a homologous chromosome.

» In epistasis, two or more gene products influence a trait.

» The product of a pleiotropic gene influences two or more traits.

13.6 Nature and Nurture

■ Variations in traits are not always the result of differences in alleles. Many traits are also influenced by environmental factors.

■ Links to Gene control 10.2, Epigenetics 10.6, Tumor suppressors and HPV 11.6, Advantages of sexual over asexual reproduction 12.1

The phrase "nature vs. nurture" refers to a centuries-old debate about whether human behavioral traits arise from one's genetics (nature) or from environmental factors (nurture). It turns out that both play a role. The environment affects the expression of many genes, which in turn affects phenotype—including behavioral traits. We can summarize this thinking with the following equation:

$$\text{genotype} + \text{environment} \longrightarrow \text{phenotype}$$

The science of epigenetics (Section 10.6) is revealing that the environment has an even greater contribution to this equation than most biologists had suspected. Environmentally driven changes in gene expression patterns can be permanent and heritable. Such changes are implemented by gene controls such as chromatin modifications and RNA interference that act on the DNA itself (Section 10.2).

As an example of how the environment influences gene controls, consider DNA methylation.

Figure 13.13 Example of environmental effects on animal phenotype. The color of the snowshoe hare's fur varies by season. In summer, the fur is brown (*left*); in winter, white (*right*).

Environmental cues can initiate cell-signaling pathways (you will learn more about such pathways in later chapters). Some cell-signaling pathways end with methyl groups being removed from or added to particular regions of DNA. The change in methylation enhances or suppresses gene expression in those regions. Diet, stress, exercise, drugs, and exposure to toxins such as tobacco, alcohol, arsenic, and asbestos affect DNA methylation patterns.

Examples of Environmental Effects on Phenotype

Most of the research into environmental effects on phenotype concerns diseases and disorders. However, mechanisms that adjust phenotype in response to external cues are part of an individual's normal ability to adapt to environmental change, as the following examples illustrate.

Seasonal Changes in Coat Color Seasonal changes in temperature and the length of day affect the production of melanin and other pigments that color the skin and fur of many animals. These animals have different color phases in different seasons (**Figure 13.13**). Hormonal signals triggered by changes in daylength cause the fur to be shed, and different types and amounts of pigments to be deposited in fur that grows back. The resulting change in phenotype offers these animals seasonally appropriate camouflage from predators.

Effect of Altitude on Yarrow Yarrow is a type of plant useful for genetics experiments because it grows easily from cuttings. All cuttings of a plant have the same genotype, so experimenters know that genes are not

A Mature cutting at high elevation (3,060 meters above sea level)

B Mature cutting at mid-elevation (1,400 meters above sea level)

C Mature cutting at low elevation (30 meters above sea level)

Figure 13.14 Experiment showing environmental effects on phenotype in yarrow (*Achillea millefolium*). Cuttings from the same parent plant were grown in the same kind of soil at three different elevations.

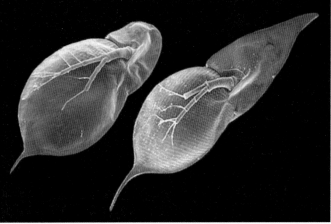

A Light micrograph of a living water flea.

B Electron micrographs comparing *Daphnia* body form that develops in the presence of few predators (*left*) with the form that develops in the presence of many predators (*right*). Note the difference in the length of the tail spine and the pointiness of the head. Chemicals emitted by the water flea's insect predators provoke the change.

Figure 13.15 Environmental effect on phenotype of the water flea (*Daphnia pulex*).

the basis for any phenotypic differences among them. Genetically identical yarrow plants grow to different heights at different altitudes (**Figure 13.14**). More challenging temperature, soil, and water conditions are typically encountered at higher altitudes. Differences in altitude are also correlated with changes in the reproductive mode of yarrow: Plants at higher altitude tend to reproduce asexually, and plants at lower altitude reproduce sexually. Plasticity of phenotype gives these plants an ability to thrive in diverse habitats.

Alternative Phenotypes in Water Fleas The water flea is a microscopic freshwater relative of shrimp (**Figure 13.15A**). Water fleas have different phenotypes depending on whether the aquatic insects that prey on them are present (**Figure 13.15B**). Individual water fleas can also switch between asexual and sexual reproduction depending on environmental conditions. During the early spring, competition is scarce in their freshwater pond habitats. At that time, the fleas reproduce rapidly by asexual means, giving birth to large numbers of female offpsring that quickly fill the ponds. Later in the season, pond water becomes warmer, saltier, and more crowded; competition for resources is also more intense. Under these conditions, some of the water fleas start giving birth to males, and then reproducing sexually. The increased genetic diversity of sexually produced offspring may offer the population an advantage in an environment that presents a greater challenge to survival.

Mood Disorders in Humans We have known for a long time that environment is a factor in schizophrenia,

bipolar disorder, depression, and probably other mood disorders as well. Certain mutations are associated with these disorders, but not all people with the mutations end up with a mental illness; environment also plays a part. Recent discoveries in animal models are beginning to reveal mechanisms by which the environment influences our mental state, and the extent to which it does.

For example, learning and memory are correlated with dynamic and rapid chromatin modifications in brain cells. Mood is, too. Stress-induced depression causes methylation-based silencing of a particular nerve growth factor. Some antidepressants work by reversing this methylation. As another example, rats whose mothers are not very nurturing end up anxious and having a reduced resilience for stress as adults. The difference between these rats and ones who had nurturing maternal care is traceable to chromatin modifications that result in a lower than normal level of another nerve growth factor. Drugs can reverse these chromatin modifications—and their effects. We do not know which human genes are correlated with mental state, but the implication of such research is that future treatments for many disorders will involve deliberate modification of epigenetic marks in one's DNA.

Take-Home Message

Is genotype the only factor that gives rise to phenotype?

» The environment influences gene expression, and therefore can alter phenotype.

» Cell-signaling pathways link environmental cues with epigenetic marks such as methylation and other chromatin modifications.

13.7 Complex Variation in Traits

■ Individuals of most species vary in some of their shared traits. Many traits show a continuous range of variation.
■ Link to Genetically identical organisms 8.7

You know by now that individuals of a species typically vary in many of their shared traits. The pea plant phenotypes that Mendel studied appeared in two or three forms, which made them easy to track through generations. All are single-gene traits (one gene determines the trait). However, many other traits do not appear in distinct forms with predictable inheritance patterns. Such traits are often the result of complex interactions between several genes, with environmental influences on top of those interactions (we return to this topic in Chapter 17, as we consider some evolutionary consequences of variation in phenotype). Tracking traits with complex variation presents a special challenge, which is why the genetic basis of many of them has not yet been completely unraveled.

Continuous Variation

Some traits occur in a range of small differences that is called **continuous variation**. Continuous variation can be an outcome of polygenic inheritance (epistasis), in which multiple genes affect a single trait. The more genes and environmental factors that influence a trait, the more continuous is its variation.

How do we determine whether a trait varies continuously? First, we divide the total range of phenotypes into measurable categories, such as inches of height (**Figure 13.16A,B**). Next, we count how many individuals of a group fall into each category; this count gives the relative frequencies of phenotypes across our range of measurable values. Finally, we plot the data as a bar chart (**Figure 13.16C**). A graph line around the top of the bars shows the distribution of values for the trait. If the line is a bell-shaped curve, or **bell curve**, the trait varies continuously.

B Height (in inches) of male biology students

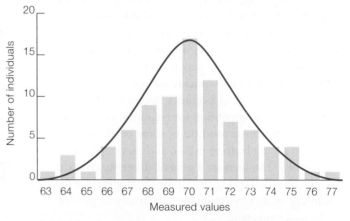

C Graphing the resulting data produces a bell-shaped curve, an indication that height varies continuously. This graph represents the data collected from male biology students, shown in (**B**).

A Height (in inches) of female biology students

Figure 13.16 Example of continuous variation. Biology students at the University of Florida were divided into categories of one-inch increments in height and counted.

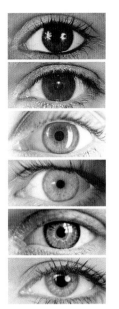

Human skin color varies continuously (a topic we return to in Chapter 14), as does human eye color. The colored part of the eye is the iris, a doughnut-shaped, pigmented structure. Iris color, like skin color, is the result of epistasis among gene products that make and distribute melanins. The more melanin deposited in the iris, the less light is reflected from it. Dark irises have dense melanin deposits that absorb almost all light, and reflect almost none. Melanin deposits are not as extensive in brown eyes, which reflect some light. Green and blue eyes have the least amount of melanin, so they reflect the most light.

Regarding the Unexpected Phenotype

Most organic molecules are made in metabolic pathways involving many enzymes. Genes encoding those enzymes can mutate in any number of ways, so their products may function within a spectrum of activity that ranges from excessive to not at all. Genes encoding products that regulate expression mutate as well, so the same gene product can be expressed at different levels in different individuals. Thus, the end product of a pathway can be produced within a range of concentration and activity. Such variations are associated

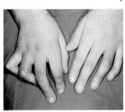

with unpredictable traits such as camptodactyly, in which finger shape and movement are abnormal. Any or all fingers on either or both hands may be affected, to varying degrees (*left*). Camptodactyly is heritable, but its manifestation can differ even within the same family tree.

bell curve Bell-shaped curve; typically results from graphing frequency versus distribution for a trait that varies continuously.
continuous variation Range of small differences in a shared trait.

Take-Home Message

Do all traits occur in distinct forms?

» The more genes and other factors that influence a trait, the more continuous is its range of variation.

» Unpredictable phenotypes arise from genes with a range of expression among individuals.

Menacing Mucus (revisited)

The allele most commonly associated with cystic fibrosis, $\Delta F508$, is eventually lethal in homozygous individuals, but not in those who are heterozygous. This allele is codominant with the normal one, so both copies of the gene are expressed in heterozygous individuals. Such individuals make enough of the normal CFTR protein to have normal chloride ion transport.

The $\Delta F508$ allele is at least 50,000 years old and very common: 1 in 25 people carry it in some populations. Why has this allele persisted for so long and at such high frequency if it is dangerous? The $\Delta F508$ allele may be the lesser of two evils because it offers heterozygous individuals a survival advantage against certain deadly infectious diseases. The unmutated CFTR protein triggers endocytosis when it binds to bacteria. This process is an essential part of the body's immune response to bacteria in the respiratory tract.

However, the same function of CFTR allows bacteria to enter cells of the gastrointestinal tract, where they can be deadly. For example, endocytosis of *Salmonella typhi* (shown at *left*) into epithelial cells lining the gut results in a dangerous infection called typhoid fever. The $\Delta F508$ mutation alters the CFTR protein so that bacteria can no longer be taken up by intestinal cells. People that carry it may have a decreased susceptibility to typhoid fever and other bacterial diseases that begin in the intestinal tract.

How would you vote? Tests for predisposition to genetic disorders are now available. Do you support legislation preventing discrimination based on the results of such tests?

Summary

Section 13.1 Cystic fibrosis is usually caused by a particular mutation. The allele persists at high frequency despite its devastating effects. Only those homozygous for the allele have the disorder.

Section 13.2 Each gene occurs at a particular **locus**, or location, on a chromosome. Individuals with identical alleles are **homozygous** for the allele. **Heterozygous** individuals, or **hybrids**, have two non-identical alleles. A **dominant** allele masks the effect of a **recessive** allele on the homologous chromosome. **Genotype** (an individual's particular set of alleles) gives rise to **phenotype**.

	P	p
P	PP	Pp
p	Pp	pp

Section 13.3 Crossing individuals that breed true for two forms of a trait yields identically heterozygous F_1 offspring. A cross between such offspring is a **monohybrid cross**. The frequency at which the traits appear in the offspring of **testcrosses** can reveal dominance relationships among the alleles associated

with those traits. **Punnett squares** are useful in determining the probability of the genotype and phenotype of the offspring of crosses. Mendel's monohybrid cross results led to his **law of segregation** (stated here in modern terms): During meiosis, the genes of each pair on homologous chromosomes separate, so each gamete gets one or the other gene.

Section 13.4 Crossing individuals that breed true for two forms of two traits yields F$_1$ offspring identically heterozygous for alleles governing those traits. A cross between such offspring is a **dihybrid cross**. Mendel's dihybrid cross results led to his **law of independent assortment** (stated here in modern terms): During meiosis, gene pairs on homologous chromosomes tend to sort into gametes independently of other gene pairs. Crossovers can break up **linkage groups**.

Section 13.5 With **incomplete dominance**, an allele is not fully dominant over its partner on a homologous chromosome, so an intermediate phenotype results. **Codominant** alleles are both fully expressed in heterozygotes; neither is dominant. Codominanace may occur in **multiple allele systems** such as the one underlying ABO blood typing. In **epistasis**, two or more genes often affect the same trait. A **pleiotropic** gene affects two or more traits.

Section 13.6 An individual's phenotype—including appearance as well as behavior—is influenced by environmental factors. Environmental cues alter gene expression by way of cell signaling pathways that ultimately affect gene controls such as chromatin remodeling or RNA interference.

Section 13.7 A trait that is influenced by the products of multiple genes often occurs in a range of small increments of phenotype called **continuous variation**. Continuous variation is indicated by a **bell curve** in the range of values.

Self-Quiz

Answers in Appendix III

1. A heterozygous individual has a _____ for a trait being studied.
   a. pair of identical alleles
   b. pair of nonidentical alleles
   c. haploid condition, in genetic terms

2. An organism's observable traits constitute its _____ .
   a. phenotype   c. genotype
   b. variation   d. pedigree

3. In genetics, F stands for filial, which means _____ .
   a. friendly   c. final
   b. offspring   d. hairlike

4. F$_1$ offspring of the cross $AA \times aa$ are _____ .
   a. all AA c. all Aa
   b. all aa d. 1/2 AA and 1/2 aa

5. The second-generation offspring of a cross between individuals who are homozygous for different alleles of a gene are called the _____ .
   a. F$_1$ generation   c. hybrid generation
   b. F$_2$ generation   d. none of the above

6. Refer to question 4. Assuming complete dominance, the F$_2$ generation will show a phenotypic ratio of _____ .
   a. 3:1   b. 9:1   c. 1:2:1   d. 9:3:3:1

7. A testcross is a way to determine _____ .
   a. phenotype   b. genotype   c. both a and b

8. Assuming complete dominance, crosses between two dihybrid F$_1$ pea plants, which are offspring from a cross $AABB \times aabb$, result in F$_2$ phenotype ratios of _____ .
   a. 1:2:1   b. 3:1   c. 1:1:1:1   d. 9:3:3:1

9. The probability of a crossover occurring between two genes on the same chromosome _____ .
   a. is unrelated to the distance between them
   b. decreases with the distance between them
   c. increases with the distance between them

10. A gene that affects three traits is _____ .
    a. epistatic   c. pleiotropic
    b. a multiple allele system   d. dominant

11. _____ alleles are both expressed.
    a. Dominant   c. Pleiotropic
    b. Codominant   d. Hybrid

12. A bell curve indicates _____ in a trait.
    a. epigenetic effects   c. incomplete dominance
    b. pleiotropy   d. continuous variation

13. True or false? All traits are inherited in a predictable way.

14. Match the terms with the best description.
    ___ dihybrid cross   a. bb
    ___ monohybrid cross   b. $AaBb \times AaBb$
    ___ homozygous condition   c. Aa
    ___ heterozygous condition d. $Aa \times Aa$

Genetics Problems

Answers in Appendix III

1. Assuming that independent assortment occurs during meiosis, what type(s) of gametes will form in individuals with the following genotypes?
   a. $AABB$ b. $AaBB$ c. $Aabb$ d. $AaBb$

2. Refer to problem 1. Determine the frequencies of each genotype among offspring from the following matings:
   a. $AABB \times aaBB$ c. $AaBb \times aabb$
   b. $AaBB \times AABb$ d. $AaBb \times AaBb$

3. Refer to problem 2. Assume a third gene has alleles C and c. For each genotype listed, what allele combinations will occur in gametes, assuming independent assortment?
   a. $AABBCC$ c. $AaBBCc$
   b. $AaBBcc$ d. $AaBbCc$

4. Some genes are so vital for development that mutations in them are lethal in homozygous recessives. Even so, heterozygotes can perpetuate recessive, lethal alleles. The allele *Manx* (M^L) in cats is an example. Homozygous cats (M^LM^L) die before birth. In heterozygotes (M^LM), the spine develops abnormally, and the cats end up with no tail (*above*). Two M^LM cats mate. What is the probability that any one of their surviving kittens will be heterozygous?

The Cystic Fibrosis Mutation and Typhoid Fever The $\Delta F508$ mutation disables the receptor function of the CFTR protein, so it inhibits endocytosis of bacteria into epithelial cells. Endocytosis is an important part of the respiratory tract's immune defenses against common *Pseudomonas* bacteria, which is why *Pseudomonas* infections are a chronic problem in cystic fibrosis patients.

The $\Delta F508$ mutation also inhibits endocytosis of *Salmonella typhi* into cells of the gastrointestinal tract, where internalization of this bacteria can cause typhoid fever. Typhoid fever is a common worldwide disease. Its symptoms include extreme fever and diarrhea, and the resulting dehydration causes delirium that may last several weeks. If untreated, it kills up to 30 percent of those infected. Around 600,000 people die annually from typhoid fever. Most of them are children.

In 1998, Gerald Pier and his colleagues compared the uptake of *S. typhi* by different types of epithelial cells: those homozygous for the normal allele, and those heterozygous for the $\Delta F508$ mutation (**Figure 13.17**). (Cells homozygous for the mutation do not take up *S. typhi* bacteria.)

1. Regarding the Ty2 strain of *S. typhi*, about how many more bacteria were able to enter normal cells (those expressing unmutated *CFTR*) than cells expressing the gene with the $\Delta F508$ deletion?

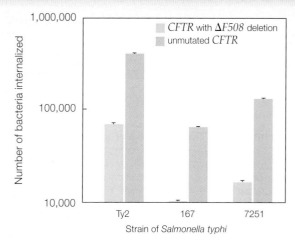

Figure 13.17 In epithelial cells, effect of the *CF* mutation on the uptake of three different strains of *Salmonella typhi* bacteria.

2. Which strain of bacteria entered normal epithelial cells most easily?

3. The $\Delta F508$ deletion inhibited the entry of all three *S. typhi* strains into epithelial cells. Can you tell which strain was most inhibited?

5. Sometimes the gene for tyrosinase mutates so its product is not functional. An individual who is homozygous recessive for such a mutation cannot make melanin. Albinism, the absence of melanin, results. Humans and many other organisms can have this phenotype (*left*). In the following situations, what are the probable genotypes of the father, the mother, and their children?

a. Both parents have normal phenotypes; some of their children are albino and others are unaffected.
b. Both parents are albino and have albino children.
c. The woman is unaffected, the man is albino, and they have one albino child and three unaffected children.

6. Several alleles affect traits of roses, such as plant form and bud shape. Alleles of one gene govern whether a plant will be a climber (dominant) or shrubby (recessive). All F_1 offspring from a cross between a true-breeding climber and a shrubby plant are climbers. If an F_1 plant is crossed with a shrubby plant, about 50 percent of the offspring will be shrubby; 50 percent will be climbers. Using symbols A and a for the dominant and recessive alleles, make a Punnett-square diagram of the expected genotypes and phenotypes in F_1 offspring and in offspring of a cross between an F_1 plant and a shrubby plant.

7. Mendel crossed a true-breeding pea plant with green pods and a true-breeding pea plant with yellow pods. All the F_1 plants had green pods. Which color is recessive?

8. Suppose you identify a new gene in mice. One of its alleles specifies white fur, another specifies brown. You want to see if the two interact in simple or incomplete dominance. What sorts of genetic crosses would give you the answer?

9. In sweet pea plants, an allele for purple flowers (P) is dominant to an allele for red flowers (p). An allele for long pollen grains (L) is dominant to an allele for round pollen grains (l). Bateson and Punnett crossed a plant having purple flowers/long pollen grains with one having white flowers/round pollen grains. All F_1 offspring had purple flowers and long pollen grains. Among the F_2 generation, the researchers observed the following phenotypes:

296 purple flowers/long pollen grains
19 purple flowers/round pollen grains
27 red flowers/long pollen grains
85 red flowers/round pollen grains

What is the best explanation for these results?

10. Red-flowering snapdragons are homozygous for allele R^1. White-flowering snapdragons are homozygous for a different allele (R^2). Heterozygous plants (R^1R^2) bear pink flowers. What phenotypes should appear among first-generation offspring of the crosses listed? What are the expected proportions for each phenotype?
a. $R^1R^1 \times R^1R^2$ c. $R^1R^2 \times R^1R^2$
b. $R^1R^1 \times R^2R^2$ d. $R^1R^2 \times R^2R^2$

(Incompletely dominant alleles are usually designated by superscript numerals, as shown, not by uppercase letters for dominance and lowercase letters for recessiveness.)

11. A single allele gives rise to the Hb^S form of hemoglobin. Homozygotes (Hb^SHb^S) develop sickle-cell anemia (Section 9.6). Heterozygotes (Hb^AHb^S) have few symptoms. A couple who are both heterozygous for the Hb^S allele plan to have children. For each of the pregnancies, state the probability that they will have a child who is:
a. homozygous for the Hb^S allele
b. homozygous for the normal allele (Hb^A)
c. heterozygous: Hb^AHb^S

LEARNING ROADMAP

Where you have been This chapter revisits dominance (Sections 13.2, 13.5, and 13.7), gene expression (9.2, 9.3), mutations (9.6), meiosis (12.3), gametes (12.5), and chromosome replication and repair (8.2, 8.5, 8.6). Sampling error (1.8), proteins (3.6), cell components (4.6, 4.8, 4.10, 4.11), metabolism (5.5), membrane receptors (5.7), pigments (6.2), gene control (10.4, 10.6), telomeres (11.5), and oncogenes (11.6) also turn up again.

Where you are now

Tracking Traits in Humans
Inheritance patterns in humans are revealed by traits that crop up in family trees. The traits are typically genetic abnormalities or syndromes associated with a genetic disorder.

Autosomal Inheritance
Traits associated with dominant alleles on human autosomes appear in every generation. Traits associated with recessive alleles on human autosomes can skip generations.

Sex-Linked Inheritance
Traits associated with alleles on the X chromosome tend to affect more men than women. Men cannot pass such alleles to a son; carrier mothers bridge affected generations.

Changes in Chromosome Structure and Number
In humans, large-scale changes in the structure or number of autosomes or sex chromosomes usually result in a genetic disorder.

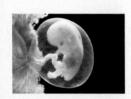

Genetic Testing
Genetic testing provides information about the risk of passing a harmful allele to offspring. Prenatal testing can reveal a genetic abnormality or disorder in a developing fetus.

Where you are going Individual genetic disorders such as muscular dystrophy are discussed in later chapters, in the context of the systems that they affect. Chapter 15 returns to the topic of human chromosomes as part of genomics and genetic engineering. Chapter 17 explores evolutionary adaptations and factors that influence the frequency of alleles in a population. The cells and other structural components of human skin are covered in detail in Section 31.8. Chapters 41 and 42 return to processes of human reproduction and development.

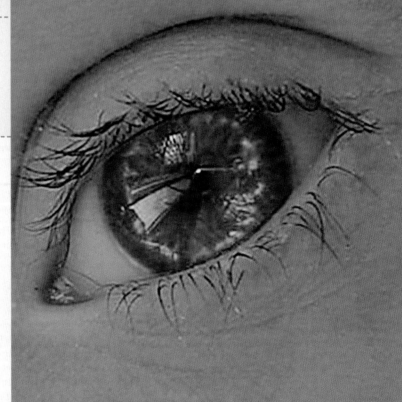

The color of human skin begins with skin cell organelles called melanosomes. Melanosomes make two types of melanin: one brownish-black; the other, reddish. Most people have about the same number of melanosomes in their skin cells. Variations in skin color occur because the kinds and amounts of melanins vary among people, as does the formation, transport, and distribution of the melanosomes in the skin.

Variations in skin color may have evolved as a balance between vitamin production and protection against harmful UV radiation. Dark skin would have been beneficial under the intense sunlight of the African savannas where humans first evolved. Melanin acts as a natural sunscreen because it prevents UV radiation in sunlight from breaking down folate, a vitamin essential for normal sperm formation and embryonic development. Children born to light-skinned women exposed to high levels of sunlight have a heightened risk of birth defects.

Early human groups that migrated to regions with colder climates were exposed to less sunlight. In these regions, lighter skin color would have been beneficial. Why? UV radiation stimulates skin cells to make a molecule the body converts to vitamin D. Where sunlight exposure is minimal, UV radiation damage is less of a risk than vitamin D deficiency, which has serious health consequences for developing fetuses and children. People with dark, UV-shielding skin have a high risk of this deficiency in regions where little sunlight reaches Earth's surface.

Skin color, like most other human traits, has a genetic basis. More than 100 gene products are involved in the synthesis of melanin, and the formation and deposition of melanosomes. Mutations in at least some of these genes may have contributed to regional variations in human skin color. Consider a gene on chromosome 15, *SLC24A5*, that encodes a transport protein in melanosome membranes. Nearly all people of African, Native American, or east Asian descent carry the same allele of this gene. Between 6,000 and 10,000 years ago, a mutation gave rise to a different allele. The mutation, a single base-pair substitution, changed the 111th amino acid of the transport protein from alanine to threonine. The change results in less melanin—and lighter skin color—than the original African allele does. Today, nearly all people of European descent carry this mutated allele.

A person of mixed ethnicity may make gametes that contain different combinations of alleles for dark and light skin. It is fairly rare that one of those gametes contains all of the alleles for dark skin, or all of the alleles for light skin, but it happens (**Figure 14.1**).

Skin color is only one of many human traits that vary as a result of single nucleotide mutations. The small scale of such changes offers a reminder that all of us share the genetic legacy of common ancestry.

Figure 14.1 Variation in human skin color (*left*) begins with differences in alleles inherited from parents. *Above*, fraternal twin girls Kian and Remee, born in 2006. Both of the children's grandmothers are of European descent, and have pale skin. Both of their grandfathers are of African descent, and have dark skin. The twins inherited different alleles of some of the genes that affect skin color from their mixed-race parents, who, given the appearance of their children, must be heterozygous for those alleles.

14.2 Human Chromosomes

- Geneticists study inheritance patterns in humans by tracking genetic disorders and abnormalities through families.
- Charting genetic connections with pedigrees reveals inheritance patterns of certain traits.
- Links to Sampling error 1.8, Chromosomes 8.2, Dominance 13.2, Complex inheritance patterns 13.7

Some organisms, including pea plants and fruit flies, are ideal for genetic analysis. They have relatively few chromosomes, they reproduce quickly under controlled conditions, and breeding them poses few ethical problems. It does not take long to track a trait through many generations. Humans, however, are a different story. Unlike flies grown in laboratories, we humans live under variable conditions, in different places, and we live as long as the geneticists who study us. Most of us select our own mates and reproduce if and when we want to. Our families tend to be on the small side, so sampling error (Section 1.8) is a major factor in human genetics studies. Thus, geneticists often use historical records to track genetic disorders and abnormalities that run in families. These researchers make and study standardized charts of genetic connections called **pedigrees** (Figure 14.2). Pedigree analyses can reveal whether a trait is associated with a dominant or recessive allele, and whether the allele is on an autosome or a sex chromosome. Pedigree analysis also allows geneticists to determine the probability that a trait will recur in future generations of a family or a population.

Types of Genetic Variation

Some easily observed human traits follow Mendelian inheritance patterns. Like the flower color of pea plants, these traits are controlled by a single gene with two alleles, one dominant and the other recessive. For example, some people have earlobes that attach at their base, and others have earlobes that dangle free. The allele for unattached earlobes is dominant; the allele for attached earlobes is recessive. Similarly, an allele that specifies a cleft chin is dominant over the allele for a smooth chin, and the allele for dimples is dominant over that for no dimples. Someone who is homozygous for two recessive alleles of the *MC1R* gene makes the reddish kind of melanin but not the brownish-black kind, so this person has red hair.

Single genes on autosomes or sex chromosomes also govern more than 6,000 genetic abnormalities

pedigree Chart of family connections that shows the appearance of a trait through generations.

□ male

○ female

□—○ marriage/mating

offspring

■ ● individual showing trait being studied

◇ sex not specified

I, II, III, IV... generation

A Standard symbols used in pedigrees.

B A pedigree for polydactyly, which is characterized by extra fingers, toes, or both. The *black* numbers signify the number of fingers on each hand; the *red* numbers signify the number of toes on each foot. Though it occurs on its own, polydactyly is also one of several symptoms of Ellis–van Creveld syndrome.

* Gene not expressed in this carrier.

C *Right*, pedigree for Huntington's disease, a progressive degeneration of the nervous system. Researcher Nancy Wexler and her team constructed this extended family tree for nearly 10,000 Venezuelans. Their analysis of unaffected and affected individuals revealed that a dominant allele on human chromosome 4 is the culprit. Wexler has a special interest in the disorder: It runs in her family.

Figure 14.2 Animated Pedigrees.

and disorders. **Table 14.1** lists a few examples. A genetic abnormality is a rare or uncommon version of a trait, such as having six fingers on a hand or having a web between two toes. Genetic abnormalities are not inherently life-threatening, and how you view them is a matter of opinion. By contrast, a genetic disorder sooner or later causes medical problems that may be severe. A genetic disorder is often characterized by a specific set of symptoms (a syndrome). In general, much more research focuses on genetic disorders than on other human traits, because what we learn helps us develop treatments for affected people.

The next two sections of this chapter focus on inheritance patterns of human single-gene disorders, which affect about 1 in 200 people. Keep in mind that these inheritance patterns are the least common kind. Most human traits, including skin color, are polygenic (influenced by multiple genes) and some have epigenetic contributions or causes. Many genetic disorders are like this, including diabetes, asthma, obesity, cancers, heart disease, and multiple sclerosis. The inheritance patterns of these disorders are complex, and despite intense research our understanding of the genetics behind them remains incomplete. For example, mutations on almost every chromosome have been found in people with autism, a developmental disorder, but not all people who carry these mutations have autism. Appendix VI shows a map of human chromosomes with the locations of some alleles known to play a role in genetic disorders and other human traits.

Alleles that give rise to severe genetic disorders are generally rare in populations, because they compromise the health and reproductive ability of their bearers. Why do they persist? Mutations periodically reintroduce them. In some cases, a codominant allele offers a survival advantage in a particular environment. You learned about one example, the Δ*F508* allele that causes cystic fibrosis, in Chapter 13: People heterozygous for this codominant allele are protected from infection by bacteria that cause typhoid fever. You will see additional examples in later chapters.

Take-Home Message

How do we study inheritance patterns in humans?

» Human inheritance patterns are often studied by tracking genetic abnormalities or disorders through family trees.

» A genetic abnormality is a rare version of an inherited trait. A genetic disorder is an inherited condition that causes medical problems.

» Some human genetic traits are governed by single genes and are inherited in a Mendelian fashion. Many others are influenced by multiple genes and epigenetics.

Table 14.1 Patterns of Inheritance for Some Genetic Abnormalities and Disorders

Disorder or Abnormality	Main Symptoms
Autosomal dominant inheritance pattern	
Achondroplasia	One form of dwarfism
Aniridia	Defects of the eyes
Camptodactyly	Rigid, bent fingers
Familial hypercholesterolemia	High cholesterol level; clogged arteries
Huntington's disease	Degeneration of the nervous system
Marfan syndrome	Abnormal or missing connective tissue
Polydactyly	Extra fingers, toes, or both
Progeria	Drastic premature aging
Neurofibromatosis	Tumors of nervous system, skin
Autosomal recessive inheritance pattern	
Albinism	Absence of pigmentation
Hereditary methemoglobinemia	Blue skin coloration
Cystic fibrosis	Abnormal glandular secretions leading to tissue and organ damage
Ellis–van Creveld syndrome	Dwarfism, heart defects, polydactyly
Fanconi anemia	Physical abnormalities, marrow failure
Friedreich's ataxia	Progressive loss of motor and sensory function
Galactosemia	Brain, liver, eye damage
Hereditary hemochromatosis	Iron overload damages joints, organs
Phenylketonuria (PKU)	Mental impairment
Sickle-cell anemia	Anemia, adverse pleiotropic effects
Tay–Sachs disease	Deterioration of mental and physical abilities; early death
X-linked recessive inheritance pattern	
Androgen insensitivity syndrome	XY individual but having some female traits; sterility
Red–green color blindness	Inability to distinguish red from green
Hemophilia	Impaired blood clotting ability
Muscular dystrophies	Progressive loss of muscle function
X-linked anhidrotic dysplasia	Mosaic skin (patches with or without sweat glands); other ill effects
X-linked dominant inheritance pattern	
Fragile X syndrome	Intellectual, emotional disability
Incontinentia pigmenti	Abnormalities of skin, hair, teeth, nails, eyes; neurological problems
Changes in chromosome number	
Down syndrome	Mental impairment; heart defects
Turner syndrome (XO)	Sterility; abnormal ovaries, sexual traits
Klinefelter syndrome	Sterility; mild mental impairment
XXX syndrome	Minimal abnormalities
XYY condition	Mild mental impairment or no effect
Changes in chromosome structure	
Chronic myelogenous leukemia (CML)	Overproduction of white blood cells; organ malfunctions
Cri-du-chat syndrome	Mental impairment; abnormal larynx

- An allele is inherited in an autosomal dominant pattern if the trait it specifies appears in heterozygous people.
- An allele is inherited in an autosomal recessive pattern if the trait it specifies appears only in homozygous people.
- Links to Protein structure 3.6, Nuclear envelope 4.6, Lysosomes in Tay–Sachs disease 4.8, Cytoskeletal elements 4.10, Cell membrane receptors 5.7, Autosomes 8.2, DNA replication 8.5, DNA repair 8.6, Gene expression 9.2, RNA processing 9.3, Mutations 9.6, Growth factors 11.6, Inheritance 13.2, Codominance and pleiotropy 13.5

The Autosomal Dominant Pattern

A dominant allele on an autosome is expressed in people who are heterozygous for it as well as those who are homozygous. Traits governed by such alleles tend to appear in every generation, and they affect both sexes equally. When one parent is heterozygous, and the other is homozygous for the recessive allele, each of their children has a 50 percent chance of inheriting the dominant allele and displaying the trait associated with it (Figure 14.3A).

Achondroplasia A form of hereditary dwarfism, achondroplasia, offers an example of an autosomal dominant disorder (a disorder caused by a dominant allele on an autosome). Mutations associated with achondroplasia occur in a gene for a growth hormone receptor. By an unknown mechanism, the mutations cause the receptor, a negative regulator of bone development, to be overly active. About 1 out of 10,000 people are heterozygous for one of these mutations. As adults, these people are, on average, about four feet, four inches (1.3 meters) tall, and have abnormally short arms and legs relative to other body parts (Figure 14.3B).

An allele that causes achondroplasia can be passed to children because its expression does not interfere with reproduction, at least in heterozygous people. The homozygous condition results in severe skeletal malformations that cause early death.

Huntington's Disease Another autosomal dominant allele causes Huntington's disease, in which involuntary muscle movements increase as the nervous system slowly deteriorates. Mutations associated with this disorder alter a gene for a cytoplasmic protein whose function is still unknown. The mutations are insertions in which the same three nucleotides become repeated many, many times in the gene's sequence. The mutated gene encodes an oversized protein product that is chopped into pieces inside nerve cells of the brain. The pieces clump together, and large aggregates that accumulate in the cytoplasm eventually prevent the cells from functioning properly. Brain

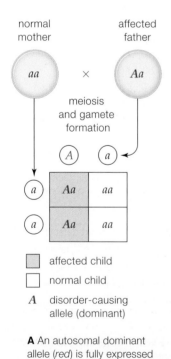

normal mother / affected father

aa × Aa

meiosis and gamete formation

A   a

	A	a
a	Aa	aa
a	Aa	aa

☐ affected child
☐ normal child

A disorder-causing allele (dominant)

A An autosomal dominant allele (*red*) is fully expressed in heterozygous people.

B Achondroplasia affects Ivy Broadhead (*left*), as well as her brother, father, and grandfather.

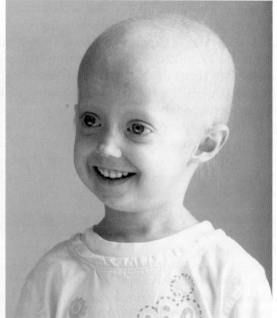

C Symptoms of Hutchinson–Gilford progeria are already evident in Megan Nighbor at age 5.

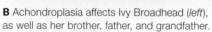

Figure 14.3 Animated Autosomal dominant inheritance.

cells involved in movement, thinking, and emotion are particularly affected. In the most common form of Huntington's, symptoms do not start until after the age of thirty, and affected people die during their forties or fifties. With this and other late-onset disorders, people tend to reproduce before symptoms appear, so the allele is often passed unknowingly to children.

Hutchinson–Gilford Progeria Hutchinson–Gilford progeria is an autosomal dominant disorder characterized by drastically accelerated aging. It is usually caused by a mutation in the gene for lamin A, a protein subunit of intermediate filaments that support the nuclear membrane. Lamins also have roles inside the nucleus, for example in mitosis, DNA synthesis and repair, and transcriptional regulation. The mutation is a base-pair substitution that adds a signal for a splice site. The resulting lamin A protein is too short and cannot be processed correctly after translation. Pleiotropic effects of the allele include a grossly abnormal architecture of the nuclear envelope. Nuclear pore complexes do not assemble properly, and membrane proteins localize to the wrong side of the envelope. The function of the nucleus as protector of chromosomes and gateway for transcription is severely impaired.

DNA damage accumulates quickly in the cells of affected people, and outward symptoms of the disorder begin to appear before age two. Skin that should be plump and resilient starts to thin, muscles weaken, and bones that should lengthen and grow stronger soften. Premature baldness is inevitable (**Figure 14.3C**). Most people with the disorder die in their early teens as a result of a stroke or heart attack brought on by hardened arteries, a condition typical of advanced age. Progeria does not run in families because affected people do not usually live long enough to reproduce.

The Autosomal Recessive Pattern

A recessive allele on an autosome is expressed only in homozygous people, so traits associated with the allele tend to skip generations. Both males and females are equally affected. People heterozygous for the allele are carriers, which means that they have the allele but not the trait. Any child of two carriers has a 25 percent chance of inheriting the allele from both parents (**Figure 14.4A**). Being homozygous for the allele, such children would have the trait.

Albinism Albinism, a genetic abnormality characterized by an abnormally low level of melanin, is inherited in an autosomal recessive pattern. Mutations

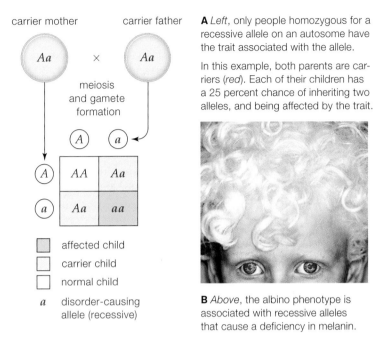

A *Left*, only people homozygous for a recessive allele on an autosome have the trait associated with the allele.

In this example, both parents are carriers (*red*). Each of their children has a 25 percent chance of inheriting two alleles, and being affected by the trait.

B *Above*, the albino phenotype is associated with recessive alleles that cause a deficiency in melanin.

Figure 14.4 Animated Autosomal recessive inheritance.

associated with the albino phenotype occur in genes involved in the production of melanin. Depending on which gene is mutated, skin, hair, or eye pigmentation may be reduced or missing. The most dramatic form of the phenotype is caused by mutations that disable the enzyme tyrosinase. The skin is typically very white and does not tan, and the hair is white. The irises lack pigment, so they appear red due to the visibility of the underlying blood vessels (**Figure 14.4B**). Melanin in the retina plays a role in vision, so people with this phenotype tend to have reduced visual acuity and other problems with vision.

Tay–Sachs Disease Tay–Sachs disease is another example of an autosomal recessive disorder. Mutations in the gene for an enzyme of lysosomes are responsible for the disease (Section 4.8). In the general population, about 1 in 300 people is a carrier for a Tay–Sachs allele, but the incidence is ten times higher in some groups, such as Jews of eastern European descent.

Take-Home Message

How do we know when a trait is associated with an allele on an autosome?

» Persons heterozygous for an allele inherited in an autosomal dominant pattern have the associated trait. The trait tends to appear in every generation.

» With an autosomal recessive inheritance pattern, only persons who are homozygous for an allele have the associated trait. The trait tends to skip generations.

■ Traits associated with recessive alleles on the X chromosome appear more frequently in men than in women.

■ A man cannot pass an X chromosome allele to a son.

■ Links to Cytoskeleton 4.10, Extracellular matrix 4.11, Pigments 6.2, X chromosome inactivation 10.4, Homozygous and heterozygous 13.2

Many genetic disorders are associated with alleles on the X chromosome (**Figure 14.5**). Most are inherited in a recessive pattern, probably because those caused by dominant X chromosome alleles tend to be lethal in male embryos.

The X-Linked Recessive Pattern

A recessive allele on the X chromosome (an X-linked recessive allele) leaves two clues when it causes a genetic disorder. First, the disorder appears in males more often than in females. This is because all males who carry the allele have the disorder, but heterozygous females do not (**Figure 14.6A**). Due to random X chromosome inactivation, only about half of a heterozygous female's cells express the recessive allele. The other half express the dominant, normal allele that she carries on her other X chromosome, and this expression can mask the effects of the recessive allele. Males,

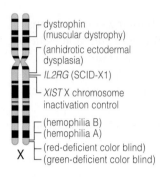

dystrophin (muscular dystrophy)
(anhidrotic ectodermal dysplasia)
IL2RG (SCID-X1)
XIST X chromosome inactivation control
(hemophilia B)
(hemophilia A)
(red-deficient color blind)
X (green-deficient color blind)

Figure 14.5 The human X chromosome. This chromosome carries about 2,000 genes—almost 10 percent of the total. Most X chromosome alleles that cause genetic disorders are inherited in a recessive pattern. A few disorders are listed (in parentheses).

having only one X chromosome, do not carry a normal allele that can mask the effects of a recessive allele. Second, an affected father does not pass an X-linked recessive allele to a son, because all children who inherit their father's X chromosome are female. Thus, a heterozygous female is always the bridge between an affected male and his affected grandson.

Red–Green Color Blindness The pattern of X-linked recessive inheritance shows up among individuals who have some degree of color blindness (**Figure 14.6B,C**). The term refers to a range of conditions in which an individual cannot distinguish among some or all colors in the spectrum of visible light. Color vision depends on the proper function of pigment-containing receptors in the eyes. Most of the genes

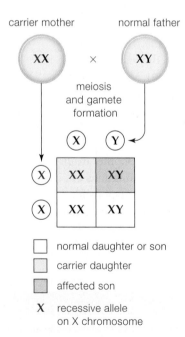

carrier mother   normal father

XX   ×   XY

meiosis and gamete formation

X   Y

	X	Y
X	XX	XY
X	XX	XY

☐ normal daughter or son

☐ carrier daughter

☐ affected son

X   recessive allele on X chromosome

A In this example of X-linked inheritance, the mother carries a recessive allele on one of her two X chromosomes (*red*).

B A view of color blindness. The image on the *left* shows how a person with red–green color blindness sees the image on the *right*. The perception of blues and yellows is normal; red and green appear similar.

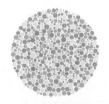

You may have one form of red–green color blindness if you see a 7 in this circle instead of a 29.

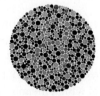

You may have another form of red–green color blindness if you see a 3 instead of an 8 in this circle.

C Part of a standardized test for color blindness. A set of 38 of these circles is commonly used to diagnose deficiencies in color perception.

Figure 14.6 Animated X-linked recessive inheritance.

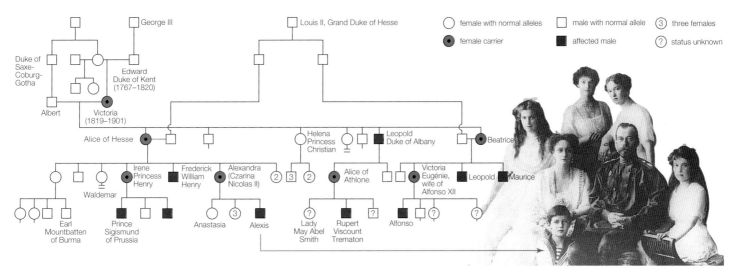

Figure 14.7 A classic case of X-linked recessive inheritance: a partial pedigree of the descendants of Queen Victoria of England. At one time, the recessive X-linked allele that resulted in hemophilia was present in eighteen of Victoria's sixty-nine descendants, who sometimes intermarried. Of the Russian royal family members shown, the mother (Alexandra Czarina Nicolas II) was a carrier.

Figure It Out: How many of Alexis's siblings were affected by hemophilia A?   Answer: None

involved in color vision are on the X chromosome, and mutations in those genes often result in altered or missing receptors. Normally, humans can sense the differences among 150 colors. People who have red–green color blindness see fewer than 25 colors because receptors that respond to red and green wavelengths are weakened or absent. Some confuse red and green; others see green as gray.

Duchenne Muscular Dystrophy Duchenne muscular dystrophy (DMD) is an X-linked recessive disorder characterized by muscle degeneration. It is caused by mutations in the X chromosome gene for dystrophin, a cystoskeletal protein that links actin microfilaments in cytoplasm to a complex of proteins in the plasma membrane. This complex structurally and functionally links the cell to extracellular matrix. When dystrophin is absent, the entire complex is unstable. Muscle cells, which are subject to stretching, are particularly affected. Their plasma membrane is easily damaged, and they become flooded with calcium ions. Eventually, the muscle cells die and become replaced by fat cells and connective tissue.

DMD affects about 1 in 3,500 people, almost all of them boys. Symptoms begin between ages three and seven. Anti-inflammatory drugs can slow the progression of DMD, but there is no cure. When an affected boy is about twelve, he will begin to use a wheelchair and his heart muscle will start to fail. Even with the best care, he will probably die before age thirty, from a heart disorder or respiratory failure (suffocation).

Hemophilia A Hemophilia A is an X-linked recessive disorder that interferes with blood clotting. Most of us have a blood clotting mechanism that quickly stops bleeding from minor injuries. That mechanism involves factor VIII, a protein product of a gene on the X chromosome. Bleeding can be prolonged in males who carry a mutation in this gene, or in females who are homozygous for one (heterozygous females make enough factor VIII to have a clotting time that is close to normal). Affected people tend to bruise very easily, but internal bleeding is their most serious problem. Repeated bleeding inside the joints disfigures them and causes chronic arthritis.

In the nineteenth century, the incidence of hemophilia A was relatively high in royal families of Europe and Russia, probably because the common practice of inbreeding kept the allele in their family trees (**Figure 14.7**). Today, about 1 in 7,500 people is affected, but that number may be rising because the disorder is now a treatable one. More affected people are living long enough to transmit the mutated allele to children.

Take-Home Message

How do we know when a trait is associated with an allele on an X chromosome?

» Men who have an X-linked recessive allele have the trait associated with the allele. Heterozygous women do not, because they have a dominant allele on their second X chromosome that can mask the affects of the recessive allele. Thus, the trait appears more often in men.

» Men transmit an X-linked allele to their daughters, but not to their sons.

14.5 Heritable Changes in Chromosome Structure

- Chromosome structure rarely changes, but when it does, the outcome can be severe or lethal.
- Links to Protein structure 3.6, Karyotyping 8.2, Deletions 9.6, *SRY* gene 10.4, Telomeres 11.5, Oncogenes 11.6, Meiosis 12.3

Large-scale changes in chromosome structure usually have drastic effects on health; about half of all miscarriages are due to chromosome abnormalities of the developing embryo. These changes occur spontaneously in nature, but can also be induced by exposure to chemicals or radiation. The scale of such changes often allows them to be detected by karyotyping.

Types of Chromosome Changes

Duplication Even normal chromosomes have DNA sequences that are repeated two or more times. These repetitions are called **duplications** (**Figure 14.8A**). Duplications are an outcome of unequal crossing over between homologous chromosomes during prophase I of meiosis. When homologous chromosomes align side by side, their DNA sequences may misalign at some point along their length. In this case, the crossover

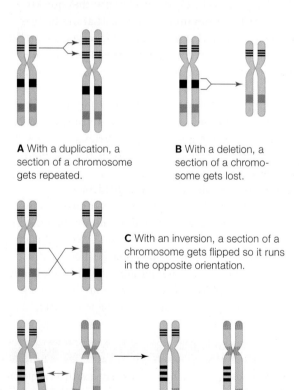

A With a duplication, a section of a chromosome gets repeated.

B With a deletion, a section of a chromosome gets lost.

C With an inversion, a section of a chromosome gets flipped so it runs in the opposite orientation.

D With a translocation, a broken piece of a chromosome gets reattached in the wrong place. This example shows a reciprocal translocation, in which two chromosomes exchange chunks.

Figure 14.8 Large-scale changes in chromosome structure.

deletes a stretch of DNA from one chromosome and splices it into the homologous partner. The probability of misalignment is greater in regions where the same sequence of nucleotides is repeated. Some duplications, such as the mutations that cause Huntington's, cause genetic abnormalities or disorders. Others have been evolutionarily important.

Deletion Deletions (**Figure 14.8B**) tend to have severe consequences on health. Most mutations associated with Duchenne muscular dystrophy are deletions in the X chromosome. A small deletion in chromosome 5 shortens life span, impairs mental functioning, and results in an abnormally shaped larynx. This disorder, cri-du-chat (French for "cat's cry"), is named for the sound that affected infants make when they cry.

Inversion With an **inversion**, part of the sequence of DNA within the chromosome becomes oriented in the reverse direction, with no molecular loss (**Figure 14.8C**). An inversion may not affect a carrier's health if it does not interrupt a gene or gene control region, because the individual's cells still have their full complement of genetic material. However, fertility may be affected because inverted chromosomes mispair during meiosis. Crossovers between mispaired chromosomes can produce large deletions or duplications that reduce the viability of forthcoming embryos. People who carry an inversion may not know about it until they are diagnosed with infertility and their karyotype is checked.

Translocation If a chromosome breaks, the broken part may get attached to a different chromosome, or to a different part of the same one. This structural change is called a **translocation**. Most translocations are reciprocal, or balanced, which means that two nonhomologous chromosomes exchange broken parts (**Figure 14.8D**). A reciprocal translocation between chromosomes 8 and 14 is the usual cause of Burkitt's lymphoma, an aggressive cancer of the immune system. This translocation moves a proto-oncogene to a region that is vigorously transcribed in immune cells, with disastrous results. Many other reciprocal translocations have no adverse effects on health, but, like inversions, they can affect fertility. During meiosis, translocated chromosomes pair abnormally and segregate improperly; about half of the resulting gametes carry major duplications or deletions. If one of these gametes unites with a normal gamete at fertilization, the resulting embryo almost always dies. As with inversions, people who carry a translocation may not know about it until they have difficulty with fertility.

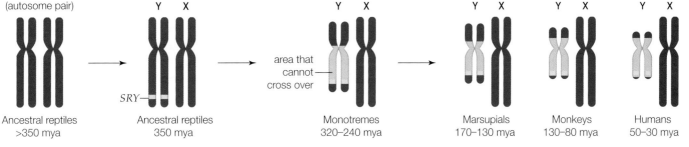

| (autosome pair) | | Y X | | Y X | | Y X | Y X | Y X |

Ancestral reptiles
>350 mya

Ancestral reptiles
350 mya

SRY

area that
cannot—
cross over

Monotremes
320–240 mya

Marsupials
170–130 mya

Monkeys
130–80 mya

Humans
50–30 mya

A Before 350 mya, sex was determined by temperature, not by chromosome differences.

B The *SRY* gene begins to evolve 350 mya. The DNA sequences of the chromosomes diverge as other mutations accumulate.

C By 320–240 mya, the DNA sequences of the chromosomes are so different that the pair can no longer cross over in one region. The Y chromosome begins to get shorter.

D Three more times, the pair stops crossing over in yet another region. Each time, the DNA sequences of the chromosomes diverge, and the Y chromosome shortens. Today, the pair crosses over only at a small region near the ends.

Figure 14.9 Evolution of the Y chromosome. Today, the *SRY* gene determines male sex. Homologous regions of the chromosomes are shown in *pink*; mya, million years ago. Monotremes are egg-laying mammals; marsupials are pouched mammals.

Chromosome Changes in Evolution

As you can see, large-scale alterations in chromosome structure may reduce an individual's fertility. Individuals who are heterozygous for such changes may not be able to produce offspring at all. However, individuals homozygous for an inversion sometimes become the founders of new species. It may seem as if this outcome would be exceedingly rare, but it is not. Speciation can and does occur by large-scale changes in chromosomes. Karyotyping and DNA sequence comparisons show that the chromosomes of all species contain evidence of major structural alterations. For example, duplications have often allowed a copy of a gene to mutate while the original carried out its unaltered function. The multiple and strikingly similar globin chain genes of humans and other primates apparently evolved by this process. Four globin chains associate in each hemoglobin molecule (Section 9.6). Different alleles specify different versions of the chains. Which versions of the chains get assembled into a hemoglobin molecule determines the oxygen-binding characteristics of the resulting protein.

As another example, X and Y chromosomes were once homologous autosomes in reptilelike ancestors of mammals (**Figure 14.9**). Ambient temperature probably determined the gender of those organisms, as it still does in turtles and some other modern reptiles. About 350 million years ago, a gene on one of the two homologous chromosomes mutated. The change, which was the beginning of the male sex determination gene *SRY*, interfered with crossing over during meiosis. A reduced frequency of crossing over allowed the chromosomes to diverge around the changed region. Mutations began to accumulate separately in

the two chromosomes. Over evolutionary time, the chromosomes became so different that they no longer crossed over at all in the changed region, so they diverged even more. Today, the Y chromosome is much smaller than the X, and only retains about 5 percent homology with it. The Y crosses over mainly with itself—by translocating duplicated regions of its own DNA.

Some chromosome structure changes contributed to differences among closely related organisms, such as apes and humans. Human somatic cells have twenty-three pairs of chromosomes, but those of chimpanzees, gorillas, and orangutans have twenty-four. Thirteen human chromosomes are almost identical with chimpanzee chromosomes. Nine more are similar, except for some inversions. One human chromosome matches up with two in chimpanzees and the other great apes (**Figure 14.10**). During human evolution, two chromosomes evidently fused end to end and formed our chromosome 2. How do we know? The region where the fusion occurred contains the remnants of a telomere.

telomere
sequence

human   chimpanzee

Figure 14.10
Human chromosome 2 compared with chimpanzee chromosomes 2A and 2B.

duplication Repeated section of a chromosome.
inversion Structural rearrangement of a chromosome in which part of it becomes oriented in the reverse direction.
translocation Structural change of a chromosome in which a broken piece gets reattached in the wrong location.

Take-Home Message

Does chromosome structure change?

» A segment of a chromosome may be duplicated, deleted, inverted, or translocated. Such a change is usually harmful or lethal, but may be conserved in the rare circumstance that it has a neutral or beneficial effect.

■ Occasionally, abnormal events occur before or during meiosis, and new individuals end up with the wrong chromosome number. Consequences range from minor to lethal changes in form and function.

■ Links to Bias 1.8, Chromosomes 8.2, X chromosome inactivation 10.4, Meiosis 12.3, Gamete formation 12.5

About 70 percent of flowering plant species, and some insects, fishes, and other animals, are **polyploid**, which means that they have three or more complete sets of chromosomes. Cells in some adult human tissues are normally polyploid, but inheriting more than two full sets of chromosomes is invariably fatal in humans.

Less than 1 percent of children are born with a diploid chromosome number that differs from the normal 46. Chromosome number changes often arise through **nondisjunction**, in which chromosomes do not separate properly during nuclear division. Nondisjunction during meiosis (**Figure 14.11**) can affect chromosome number at fertilization. For example, if a normal gamete (n) fuses with one that has an extra chromosome ($n+1$), the resulting zygote will have three copies of one type of chromosome and two of every other type ($2n+1$), a condition called trisomy. If an $n-1$ gamete fuses with a normal n gamete, the zygote will be $2n-1$, a condition called monosomy. Trisomy and monosomy are examples of **aneuploidy**, in which an individual's cells have too many or too few copies of a chromosome.

Autosomal Change and Down Syndrome

Autosomal aneuploidy is usually fatal in humans, but there are exceptions. A few trisomic humans are born alive, but only those with trisomy 21 have a high probability of surviving infancy. A newborn with three chromosomes 21 will develop Down syndrome (**Figure 14.12**). This disorder occurs once in 800 to 1,000 births, and it affects more than 350,000 people in the United States alone. The risk increases with maternal age.

Individuals with Down syndrome have upward-slanting eyes, a fold of skin that starts at the inner corner of each eye, a deep crease across the sole of each palm and foot, one (instead of two) horizontal furrows on their fifth fingers, and other outward symptoms. Not all of these outward symptoms develop in every individual. That said, people with the disorder tend to have moderate to severe mental impairment and heart problems. Their skeleton grows and develops abnormally, so older children have short body parts, loose joints, and misaligned bones of the fingers, toes, and hips. The muscles and reflexes are weak, and motor skills such as speech develop slowly. With medical care, affected individuals live about fifty-five years. Early training can help them learn to care for themselves and to take part in normal activities.

Change in the Sex Chromosome Number

Nondisjunction also causes alterations in the number of X and Y chromosomes, with a frequency of about 1 in 400 live births. Most often, such alterations lead to mild difficulties in learning and impaired motor skills such as a speech delay. These problems may be so subtle that the condition is never diagnosed.

Female Sex Chromosome Abnormalities Individuals with Turner syndrome have an X chromosome and no corresponding X or Y chromosome (XO). The

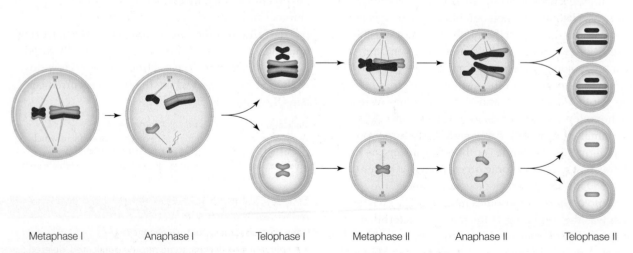

| Metaphase I | Anaphase I | Telophase I | Metaphase II | Anaphase II | Telophase II |

Figure 14.11 Nondisjunction, which can occur during anaphase I or II of meiosis. Of the two pairs of homologous chromosomes shown here, one fails to separate during anaphase I. A zygote that forms from one of the resulting gametes will have an abnormal chromosome number.

15.1 Personal DNA Testing

About 99 percent of your DNA is exactly the same as everyone else's. If you compared your DNA with your neighbor's, about 2.97 billion nucleotides of the two sequences would be identical; the remaining 30 million or so are sprinkled throughout your chromosomes, mainly as single nucleotide differences. The sprinkling is not entirely random because some regions of DNA vary less than others. Such conserved regions are of particular interest to researchers because they are the ones most likely to have an essential function. When a conserved sequence does vary among people, the variation tends to be in particular nucleotides. A nucleotide difference carried by a measurable percentage of a population, usually above 1 percent, is called a **single-nucleotide polymorphism**, or SNP (pronounced "snip").

Alleles of most genes differ by single nucleotides, and differences in alleles are the basis of the variation in human traits that makes each individual unique (Section 12.2). Thus, SNPs account for many of the differences in the way humans look, and they also have a lot to do with differences in the way our bodies work—how we age, respond to drugs, weather assaults by pathogens and toxins, and so on.

Consider a gene, *APOE*, that specifies apolipoprotein E, a protein component of lipoprotein particles (Section 3.5). One allele of this gene, *ε4*, is carried by about 25 percent of people. Nucleotide 4,874 of this

allele is a cytosine instead of the normal thymine, a SNP that results in a single amino acid change in the protein product of the gene. How this change affects the function of apolipoprotein E is unclear, but we do know that having the *ε4* allele increases one's risk of developing Alzheimer's disease later in life, particularly in people homozygous for it.

About 4.5 million SNPs in human DNA have been identified, and that number is growing every day. A few companies are now offering to determine some of the SNPs you carry (**Figure 15.1**). The companies extract your DNA from the cells in a few drops of spit, then analyze it for SNPs.

Personal genetic testing may soon revolutionize medicine by allowing physicians to customize treatments on the basis of an individual's genetic makeup. For example, an allele associated with a heightened risk of a particular medical condition could be identified long before symptoms actually appear. People with that allele could then be encouraged to make lifestyle changes known to delay the onset of the condition. For some conditions, treatment that begins early enough may prevent symptoms from developing at all. Physicians could design treatments to fit the way a condition is likely to progress in the individual, and also to prescribe only those drugs that will work in the person's body.

single-nucleotide polymorphism (**SNP**) One-base DNA sequence variation carried by a measurable percentage of a population.

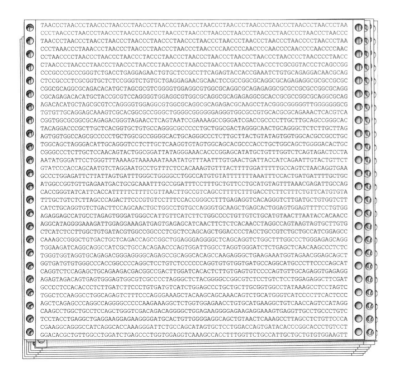

Figure 15.1 Personal genetic testing. *Above*, a SNP-chip. Personal DNA testing companies use chips like this one to analyze their customers' chromosomes for SNPs. This chip, shown actual size, reveals which versions of 1,140,419 SNPs occur in the DNA of four individuals at a time. Only about 1 percent of the 3 billion bases in a person's DNA (*left*) are unique to the individual.

15.2 Cloning DNA

- Researchers cut up DNA from different sources, then paste the resulting fragments together.
- Cloning vectors can carry foreign DNA into host cells.
- Links to Plasmids 4.4, Clones 8.1, Discovery of DNA structure 8.3, Base pairing and directionality of DNA strands 8.4, DNA ligase 8.5, mRNA 9.2, Introns 9.3, The lac operon 10.5

Cutting and Pasting DNA

In the 1950s, excitement over the discovery of DNA's structure gave way to frustration: No one could determine the order of nucleotides in a molecule of DNA. Identifying a single base among thousands or millions of others turned out to be a huge technical challenge. A seemingly unrelated discovery offered a solution. Werner Arber, Hamilton Smith, and their coworkers discovered why some bacteria are resistant to infection by bacteriophage (Section 8.3). Special enzymes inside these bacteria chop up any injected viral DNA before it has a chance to integrate into the bacterial chromosome. The enzymes restrict viral growth; hence their name, **restriction enzymes**. A restriction enzyme cuts DNA wherever a specific nucleotide sequence occurs (**Figure 15.2**). For example, the enzyme *Eco*RI (named after *E. coli*, the bacteria from which it was isolated) cuts DNA at the sequence GAATTC ❶. Other restriction enzymes cut different sequences.

The discovery of restriction enzymes allowed researchers to cut chromosomal DNA into manageable chunks. It also allowed them to combine DNA fragments from different organisms. How? Many restriction enzymes, including *Eco*RI, leave single-stranded tails on DNA fragments ❷. Researchers realized that complementary tails will base-pair, regardless of the source of the DNA ❸. The tails are called "sticky

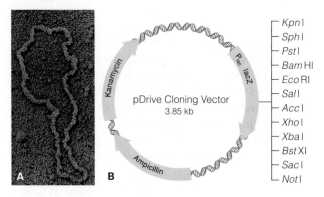

Figure 15.3 Plasmid cloning vectors. (**A**) Micrograph of a plasmid. (**B**) A commercial plasmid cloning vector. Restriction enzyme recognition sequences are indicated on the *right* by the name of the enzyme that cuts them. Foreign DNA can be inserted into the vector at these sites. Bacterial genes (*gold*) help researchers identify host cells that take up a vector with inserted DNA. This vector carries two antibiotic resistance genes and the lac operon.

ends," because two DNA fragments stick together when their matching tails base-pair. The enzyme DNA ligase (Section 8.5) can be used to seal the gaps between base-paired sticky ends, so continuous DNA strands form ❹.

Thus, using appropriate restriction enzymes and DNA ligase, researchers can cut and paste DNA from different sources. The result, a hybrid molecule of DNA from two or more organisms, is called **recombinant DNA**. Making recombinant DNA is the first step in **DNA cloning**, a set of laboratory methods that uses living cells to mass-produce specific DNA fragments. For example, researchers often insert fragments of eukaryotic DNA into plasmids (Section 4.4). Before a bacterium divides, it copies any plasmids it carries along with its chromosome, so both descendant cells get one of each. If a plasmid carries a fragment of for-

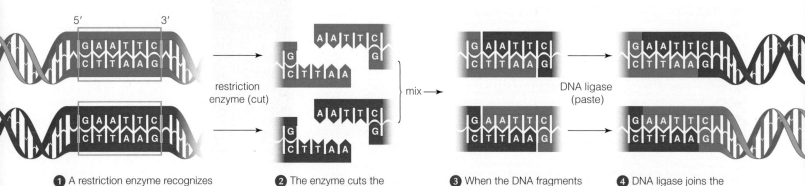

❶ A restriction enzyme recognizes a specific base sequence (*orange* boxes) in DNA from two sources.

❷ The enzyme cuts the DNA into fragments. This enzyme leaves sticky ends.

❸ When the DNA fragments from the two sources are mixed together, matching sticky ends base-pair with each other.

❹ DNA ligase joins the base-paired DNA fragments. Molecules of recombinant DNA are the result.

Figure 15.2 Animated Using restriction enzymes to make recombinant DNA.
Figure It Out: Why did the enzyme cut both strands of DNA?
Answer: Because the recognition sequence occurs on both strands.

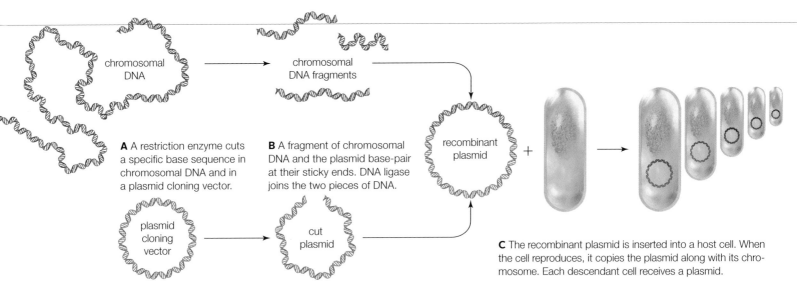

A A restriction enzyme cuts a specific base sequence in chromosomal DNA and in a plasmid cloning vector.

B A fragment of chromosomal DNA and the plasmid base-pair at their sticky ends. DNA ligase joins the two pieces of DNA.

C The recombinant plasmid is inserted into a host cell. When the cell reproduces, it copies the plasmid along with its chromosome. Each descendant cell receives a plasmid.

chromosomal DNA

chromosomal DNA fragments

plasmid cloning vector

cut plasmid

recombinant plasmid

Figure 15.4 Animated An example of cloning. Here, a fragment of chromosomal DNA is inserted into a plasmid.

eign DNA, that fragment gets copied and distributed to descendant cells along with the plasmid. Thus, plasmids can be used as **cloning vectors**, which are molecules that carry foreign DNA into host cells (**Figure 15.3**). A host cell into which a cloning vector has been inserted can be grown in the laboratory (cultured) to yield a huge population of genetically identical cells, or clones (Section 8.1). Each clone contains a copy of the vector (**Figure 15.4**). A DNA fragment carried by the vector can be collected in large quantities by harvesting it from the many clones.

cDNA Cloning

Remember from Section 9.3 that eukaryotic DNA contains introns. Unless you are a eukaryotic cell, it is not very easy to determine which parts of eukaryotic DNA encode gene products. Researchers who study gene expression in eukaryotes often start with mature mRNA, because intron sequences are excised during post-transcriptional processing.

An mRNA cannot be cut with restriction enzymes or pasted with DNA ligase, because these enzymes

cDNA DNA synthesized from an RNA template by the enzyme reverse transcriptase.
cloning vector A DNA molecule that can accept foreign DNA and get replicated inside a host cell.
DNA cloning Set of procedures that uses living cells to make many identical copies of a DNA fragment.
recombinant DNA A DNA molecule that contains genetic material from more than one organism.
restriction enzyme Type of enzyme that cuts specific nucleotide sequences in DNA.
reverse transcriptase An enzyme that uses mRNA as a template to make a strand of cDNA.

work only on double-stranded DNA. Thus, research with mRNA often involves transcribing the mRNA back into DNA. This process requires **reverse transcriptase**, a replication enzyme that uses an RNA template to assemble a strand of complementary DNA, or **cDNA**:

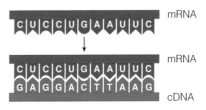

DNA polymerase added to the mixture strips the RNA from the hybrid molecule as it copies the cDNA into a second strand of DNA. The outcome is a double-stranded DNA version of the original mRNA:

Like any other double-stranded DNA, this one may be cut with restriction enzymes and pasted into a cloning vector using DNA ligase.

Take-Home Message

What is DNA cloning?

» DNA cloning uses living cells to mass-produce particular DNA fragments. Restriction enzymes cut DNA into fragments, then DNA ligase seals the fragments into cloning vectors. Recombinant DNA molecules result.

» A cloning vector that holds foreign DNA can be introduced into a living cell. When the host cell divides, it gives rise to huge populations of genetically identical cells (clones), each of which contains a copy of the foreign DNA.

15.3 Isolating Genes

- DNA libraries and the polymerase chain reaction (PCR) help researchers isolate particular DNA fragments.
- Links to Tracers 2.2, Denaturation 3.7, Base pairing 8.4, DNA replication 8.5

A Individual bacterial cells from a DNA library are spread over the surface of a solid growth medium. The cells divide repeatedly and form colonies—clusters of millions of genetically identical descendant cells.

B A piece of special paper pressed onto the surface of the growth medium will bind some cells from each colony.

C The paper is soaked in a solution that ruptures the cells and releases their DNA. The DNA clings to the paper in spots mirroring the distribution of colonies.

D A radioactive probe is added to the liquid bathing the paper. The probe hybridizes (base-pairs) with any spot of DNA that contains a complementary sequence.

E The paper is pressed against x-ray film. The radioactive probe darkens the film in a spot where it has hybridized. The spot's position is compared to the positions of the original bacterial colonies. Cells from the colony that corresponds to the spot are cultured, and their DNA is harvested.

Figure 15.5 Animated Nucleic acid hybridization. In this example, a radioactive probe helps identify a colony of bacteria that host a targeted sequence of DNA.

DNA Libraries

The entire set of genetic material—the **genome**—of most organisms consists of thousands of genes. To study or manipulate a single gene, researchers must first separate the gene from all of the others. They often begin by cutting an organism's DNA into pieces, and then cloning all the pieces. The result is a genomic library, a set of clones that collectively contain all of the DNA in a genome. Researchers also harvest mRNA, make cDNA copies of it, and then clone the cDNA. The resulting cDNA library represents only those genes being expressed at the time the mRNA was harvested.

Genomic and cDNA libraries are **DNA libraries**, sets of cells that host various cloned DNA fragments. In such libraries, a cell that contains a particular DNA fragment of interest is mixed up with thousands or millions of others that do not. One way to find that one clone among the others—a needle in a haystack—involves **probes**, which are fragments of DNA or RNA labeled with a tracer (Section 2.2). For example, researchers may synthesize a short chain of nucleotides based on the known DNA sequence of a gene, then attach a radioactive phosphate group to it. The nucleotide sequences of the probe and the gene are complementary, so the two can base-pair. Base pairing between DNA (or DNA and RNA) from more than one source is called **nucleic acid hybridization**. When the probe is mixed with DNA from a library, it will base-pair with (hybridize to) the gene, but not to other DNA (**Figure 15.5**). Researchers pinpoint a clone that hosts the gene by detecting the label on the probe. That clone is isolated and cultured, and its DNA can be extracted in bulk.

PCR

The **polymerase chain reaction (PCR)** is a technique used to mass-produce copies of a particular section of DNA without having to clone it in living cells (**Figure 15.6**). The reaction can transform a needle in a haystack—that one-in-a-million fragment of DNA—into a huge

DNA library Collection of cells that host different fragments of foreign DNA, often representing an organism's entire genome.
genome An organism's complete set of genetic material.
nucleic acid hybridization Base-pairing between DNA or RNA from different sources.
polymerase chain reaction (PCR) Method that rapidly generates many copies of a specific section of DNA.
primer Short, single strand of DNA designed to hybridize with a DNA fragment.
probe Short fragment of DNA labeled with a tracer; designed to hybridize with a nucleotide sequence of interest.

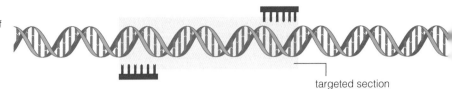

Figure 15.6 Animated Two rounds of PCR. Each cycle of this reaction can double the number of copies of a targeted sequence of DNA. Thirty cycles can make a billion copies.

targeted section

stack of needles with a little hay in it. The starting material for PCR is any sample of DNA with at least one molecule of a targeted sequence. It might be DNA from, say, a mixture of 10 million different clones, a sperm, a hair left at a crime scene, or a mummy—essentially any sample that has DNA in it.

The PCR reaction is based on DNA replication (Section 8.5). First, the starting material is mixed with DNA polymerase, nucleotides, and primers. **Primers** are short single strands of DNA that base-pair with a certain DNA sequence. In PCR, two primers are made. Each base-pairs with one end of the section of DNA to be amplified, or mass-produced ❶. Researchers expose the reaction mixture to repeated cycles of high and low temperatures. High temperatures disrupt the hydrogen bonds that hold the two strands of a DNA double helix together (Section 8.4), so every molecule of DNA unwinds and becomes single-stranded ❷. As the temperature of the reaction mixture is lowered, the single DNA strands hybridize with complementary partner strands, and double-stranded DNA forms again.

The DNA polymerases of most organisms denature at the high temperatures required to separate DNA strands. The kind that is used in PCR reactions, *Taq* polymerase, is from *Thermus aquaticus*. This bacterial species lives in hot springs and hydrothermal vents, so its DNA polymerase is necessarily heat-tolerant. *Taq* polymerase recognizes hybridized primers as places to start DNA synthesis ❸. Synthesis proceeds along the template strand until the temperature rises and the DNA separates into single strands ❹. The newly synthesized DNA is a copy of the targeted section.

When the mixture cools, the primers rehybridize, and DNA synthesis begins again. The number of copies of the targeted section of DNA can double with each cycle of heating and cooling ❺. Thirty PCR cycles may amplify that number a billionfold.

Take-Home Message

How do researchers study one gene in the context of many?

» Researchers isolate one gene from the many other genes in a genome by making DNA libraries or with PCR.

» Probes are used to identify one clone that hosts a DNA fragment of interest among many other clones in a DNA library.

» PCR quickly mass-produces copies of a particular section of DNA.

❶ DNA (*blue*) with a targeted sequence is mixed with primers (*pink*), nucleotides, and heat-tolerant *Taq* DNA polymerase.

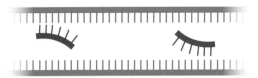

❷ When the mixture is heated, the double-stranded DNA separates into single strands. When the mixture is cooled, some of the primers base-pair with the DNA at opposite ends of the targeted sequence.

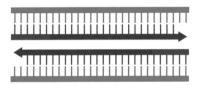

❸ *Taq* polymerase begins DNA synthesis at the primers, so it produces complementary strands of the targeted DNA sequence.

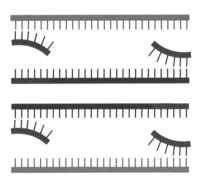

❹ The mixture is heated again, so all double-stranded DNA separates into single strands. When it is cooled, primers base-pair with the targeted sequence in the original template DNA and in the new DNA strands.

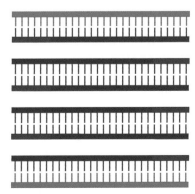

❺ Each round of PCR reactions can double the number of copies of the targeted DNA section.

15.4 DNA Sequencing

- DNA sequencing reveals the order of nucleotide bases in a section of DNA.
- Links to Tracers 2.2, Nucleotides 8.4, DNA replication 8.5

Technique

Researchers determine the order of bases in DNA with **DNA sequencing** (Figure 15.7). The most common method is similar to DNA replication. The DNA to be sequenced (the template) is mixed with nucleotides, a primer, and DNA polymerase. Starting at the primer, the polymerase joins free nucleotides into a new complementary strand of DNA, in the order dictated by the sequence of the template. Remember that DNA polymerase can add a nucleotide only to the hydroxyl group on the 3′ carbon of a DNA strand. The sequencing reaction mixture includes four kinds of dideoxynucleotides, which have no hydroxyl group on their 3′ carbon ❶. Each kind (A, C, G, or T) is labeled with a different colored pigment. During the reaction, the polymerase randomly adds either a regular nucleotide or a dideoxynucleotide to the end of a growing DNA strand. If it adds a dideoxynucleotide, the 3′ carbon of the strand will not have a hydroxyl group, so synthesis of the strand ends there ❷.

The reaction produces millions of DNA fragments of different lengths—incomplete copies of the starting DNA ❸. Each fragment ends with a dideoxynucleotide that is complementary to the template sequence. For example, if the tenth base in the template DNA was thymine, then any newly synthesized fragment that is 10 bases long ends with a dideoxyadenine.

The fragments are then separated by **electrophoresis**. With this technique, an electric field pulls the DNA fragments through a semisolid gel. DNA fragments of different sizes move through the gel at different rates. The shorter the fragment, the faster it moves, because shorter fragments slip through the tangled molecules of the gel faster than longer fragments do. All fragments of the same length move through the gel at the same speed, so they gather into bands. All fragments in a given band have the same dideoxynucleotide at their ends, and the pigment labels now impart distinct colors to the bands ❹. Each color designates one of the four dideoxynucleotides, so the order of colored bands in the gel represents the DNA sequence ❺.

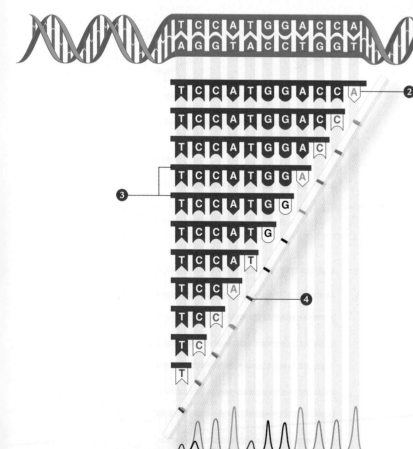

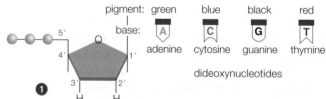

Figure 15.7 **Animated** DNA sequencing method in which DNA polymerase is used to incompletely replicate a section of DNA.

❶ Sequencing depends on dideoxynucleotides (compare **Figure 9.2**) to terminate DNA replication. Each of the four dideoxynucleotides (A, C, G, or T) is labeled with a different colored pigment.

❷ DNA polymerase uses a section of DNA as a template to synthesize new strands of DNA. Synthesis of each new strand stops when a dideoxynucleotide is added.

❸ Millions of DNA fragments of different lengths form. All are copies of the template DNA, but most are incomplete.

❹ Electrophoresis separates the fragments according to length. Fragments of identical length gather into bands. All of the DNA strands in each band end with the same dideoxynucleotide; thus, each band is the color of that dideoxynucleotide's tracer pigment.

❺ A computer detects and records the color of successive bands on the gel (see **Figure 15.8** for an example). The order of colors of the bands represents the sequence of the template DNA.

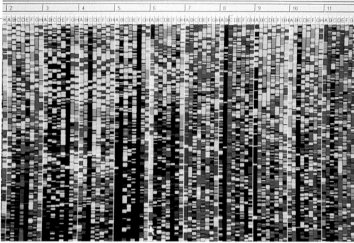

Figure 15.8 Human genome sequencing. *Left*, some of the supercomputers used to assemble the sequence of the human genome at Venter's Celera Genomics in Maryland. Information in Celera's SNP database is the basis of many new genetic tests. *Right*, a human DNA sequence, raw data.

The Human Genome Project

The sequencing method described above was invented in 1975. Ten years later, it had become so routine that people were debating about sequencing the entire human genome. Knowing the sequence would have huge potential payoffs for medicine and research, but sequencing 3 billion bases was a daunting proposition. It would require at least 6 million sequencing reactions, a task that, given techniques available at the time, would have taken 50 years to complete. Opponents said the effort would divert too much attention and funding from more urgent research. However, sequencing techniques kept getting better, and with each improvement more bases could be sequenced in less time. Automated (robotic) DNA sequencing and PCR had just been invented. Both were still too cumbersome and expensive to be useful in routine applications, but they would not be so for long. Waiting for faster technologies seemed the most efficient way to sequence the genome, but just how fast did they need to be before the project should begin?

A few privately owned companies decided not to wait, and started sequencing. One of them intended to patent the sequence after it was determined. This development provoked widespread outrage, but it also spurred commitments in the public sector. In 1988, the National Institutes of Health (NIH) effectively annexed the project by hiring James Watson (of DNA structure fame) to head an official Human Genome Project, and providing $200 million per year to fund it. A consortium formed between the NIH and international institutions that were sequencing different parts of the genome. Watson set aside 3 percent of the funding for studies of ethical and social issues arising from the

research. He later resigned over a patent disagreement, and geneticist Francis Collins took his place.

Amid ongoing squabbles over patent issues, Celera Genomics formed in 1998. With biologist Craig Venter at its helm, the company intended to commercialize genetic information. Celera invented faster techniques for sequencing genomic DNA (**Figure 15.8**), because the first to have the complete sequence had a legal basis for patenting it. The competition motivated the public consortium to move its efforts into high gear.

Then, in 2000, U.S. President Bill Clinton and British Prime Minister Tony Blair jointly declared that the sequence of the human genome could not be patented. Celera kept sequencing anyway. Celera and the public consortium separately published about 90 percent of the sequence in 2001. By 2003, fifty years after the discovery of the structure of DNA, the sequence of the human genome was officially completed. Researchers have not discovered what all of the genes encode, only where they are in the genome. What do we do with this vast amount of data? The next step is to find out what the sequence means.

DNA sequencing Method of determining the order of nucleotides in DNA.
electrophoresis Technique that separates DNA fragments by size.

Take-Home Message

How is the order of nucleotides in DNA determined?

» With DNA sequencing, a strand of DNA is partially replicated. Electrophoresis is used to separate the resulting fragments by length.

» Improved sequencing techniques and worldwide efforts allowed the human genome sequence to be determined.

15.5 Genomics

- Comparing the human genome sequence with that of other species is helping us understand how the human body works.
- Unique sequences of genomic DNA can be used to distinguish an individual from all others.
- Links to Lipoproteins 3.6, DNA replication 8.5, Knockouts 10.3, Locus 13.2, Complex variation in traits 13.7, Chromosome structural changes 14.5

It took 15 years to sequence the human genome for the first time, but the techniques have improved so much that sequencing an entire genome now takes a few weeks. Anyone can now pay to have their genome sequenced. However, even though we are able to determine the sequence of an individual's genome, it will be a long time before we understand all the information coded within that sequence.

The human genome contains a massive amount of seemingly cryptic data. One way to decipher it is by comparing it to genomes of other organisms, the premise being that all organisms are descended from shared ancestors, so all genomes are related to some extent. We see evidence of such genetic relationships simply by comparing the raw sequence data, which, in some regions, is extremely similar across many species (Figure 15.9).

The study of genomes is called **genomics**, a broad field that encompasses whole-genome comparisons, structural analysis of gene products, and surveys of small-scale variations in sequence. Genomics is providing powerful insights into evolution. For example, comparing primate genomes revealed how speciation can occur by structural changes in chromosomes (Section 14.5). Comparing genomes also showed that changes in chromosome structure do not occur randomly. Rather, if a chromosome breaks, it tends to do so in a particular spot. Human, mouse, rat, cow, pig, dog, cat, and horse chromosomes have undergone several translocations at these breakage hot spots during evolution. In humans, chromosome abnormalities that contribute to the progression of cancer also occur at the very same hot spots.

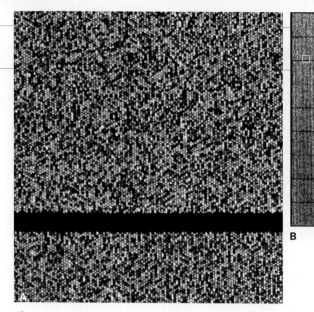

Figure 15.10 SNP-chip analysis. (**A**) Each spot is a region where the individual's genomic DNA has hybridized with one SNP. A *red* or *green* dot means that the individual is homozygous for a SNP; a combined signal (*yellow* dot) indicates heterozygosity. (**B**) The entire chip tests for 550,000 SNPs. The small *white* box indicates the magnified portion shown in **A**.

Comparing the coding regions of genomes offers medical benefits. We have learned the function of many human genes by studying their counterpart genes in other species. For instance, researchers comparing human and mouse genomes discovered a human version of a mouse gene, *APOA5*, that encodes a lipoprotein (Section 3.6). Mice with an *APOA5* knockout have four times the normal level of triglycerides in their blood. The researchers then looked for—and found—a correlation between *APOA5* mutations and high triglyceride levels in humans. High triglycerides are a risk factor for coronary artery disease.

DNA Profiling

As you learned in Section 15.1, only about 1 percent of your DNA is unique. The shared part is what makes you human; the differences make you a unique member of the species. In fact, those differences are so unique that they can be used to identify you. Identifying an individual by his or her DNA is called **DNA profiling**.

```
758 GATAATCCTGTTTTGAACAAAAGGTCAAATTGCTGAATAGAAA-GTCTTGATTAACTAAAAGATGTACAAAGTGGAATTA 836   Human
752 GATAATCCTGTTTTGAACAAAAGGTCAAATTGCTGAATAGAAA-GTCTTGATTAACTAAAAGATGTACAAAGTGGAATTA 830   Mouse
751 GATAATCCTGTTTTGAACAAAAGGTCAAATTGCTGAATAGAAA-GTCTTGATTAACTAAAAGATGTACAAAGTGGAATTA 829   Rat
754 GATAATCCTGTTTTGAACAAAAGGTCAAATTGCTGAATAGAAA-GTCTTGATTAACTAAAAGATGTACAAAGTGGAATTA 832   Dog
782 GATAATCCTGTTTTGAACAAAAGGTCAAATTGCTGAATAGAAA-GTCTTGATTAACTAAAAGATGTACAAAGTGGAATTA 860   Chicken
758 GATAATCCTGTTTTGAACAAAAGGTCAAATTGCTGAATAGAAA-GTCTTGATTAAGTAAAAGATGTACAAAGTGGAATTA 836   Frog
823 GATAATCCTGTTTTGAACAAAAGGTCAGATTGCTGAATAGAAAAGGCTTGATTAAAGCAGAGATGTACAAAGTGGACGCA 902   Zebrafish
763 GATAATCCTGTTTTGAACAAAAGGTCAAATTGTTGAATAGAGACGCTTTGATAAAGCGGAGGAGGTACAAAGTGGGACC- 841   Pufferfish
```

Figure 15.9 Genomic DNA alignment. This is a region of the gene for a DNA polymerase. Differences are highlighted. The chance that any two of these sequences would randomly match is about 1 in 10^{46}.

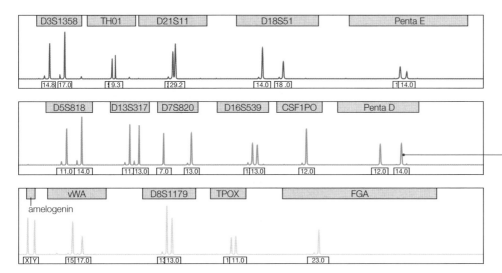

Figure 15.11 An individual's short tandem repeat profile. Tested regions of the individual's DNA are indicated in gray boxes (D3S1358, TH01, and so on).

A peak's location on the x-axis corresponds to the length of the DNA fragment amplified (a measure of the number of repeats). Peak height reflects the amount of DNA.

Remember, human body cells are diploid. Two peaks appear on a profile when the two members of a chromosome pair carry a different number of repeats. For example, this individual has 12 repeats at the Penta D region on one chromosome, and 14 repeats on the other.

One type of DNA profiling involves SNP-chips (one is shown in **Figure 15.1**). A SNP-chip contains a tiny glass plate with microscopic spots of DNA stamped on it. The DNA sample in each spot is a short, synthetic single strand with a unique SNP sequence. When an individual's genomic DNA is washed over a SNP-chip, it hybridizes only with DNA spots that have a matching SNP sequence. Probes reveal where the genomic DNA has hybridized—and which SNPs are carried by the individual (**Figure 15.10**).

Another method of DNA profiling involves analysis of **short tandem repeats**, sections of DNA in which a series of 4 or 5 nucleotides is repeated several times in a row. Short tandem repeats tend to occur in predictable spots, but the number of repeats in each spot differs among individuals. For example, one person's DNA may have fifteen repeats of the bases TTTTC at a certain locus. Another person's DNA may have this sequence repeated only twice in the same locus. Such repeats slip spontaneously into DNA during replication, and their numbers grow or shrink over generations. Unless two people are identical twins, the chance that they have identical short tandem repeats in even three regions of DNA is 1 in a quintillion (10^{18}), which is far more than the number of people on Earth. Thus, an individual's array of short tandem repeats is, for all practical purposes, unique.

Analyzing a person's short tandem repeats begins with PCR, which is used to copy ten to thirteen particular regions of chromosomal DNA known to have repeats. The lengths of the copied DNA fragments differ among most individuals, because the number of tandem repeats in those regions also differs. Thus, electrophoresis can be used to reveal an individual's unique array of short tandem repeats (**Figure 15.11**).

Short tandem repeat analysis will soon be replaced by full genome sequencing, but for now it continues to be a common DNA profiling method. Geneticists compare short tandem repeats on Y chromosomes to determine relationships among male relatives, and to trace an individual's ethnic heritage. They also track mutations that accumulate in populations over time by comparing DNA profiles of living humans with those of ancient ones. Such studies are allowing us to reconstruct population dispersals that happened long ago.

Short tandem repeat profiles are routinely used to resolve kinship disputes, and as evidence in criminal cases. Within the context of a criminal or forensic investigation, DNA profiling is called DNA fingerprinting. As of May 2011, the database of DNA fingerprints maintained by the Federal Bureau of Investigation contained the short tandem repeat profiles of 9.5 million offenders, and had been used in over 100,000 criminal investigations. DNA fingerprints have also been used to identify the remains of almost 300,000 people, including the individuals who died in the World Trade Center on September 11, 2001.

DNA profiling Identifying an individual by analyzing the unique parts of his or her DNA.
genomics The study of genomes.
short tandem repeats In chromosomal DNA, sequences of 4 or 5 bases repeated multiple times in a row.

Take-Home Message

How do we use what we know about human gene sequences?

» Analysis of the human genome sequence is yielding new information about human genes and how they work.

» DNA profiling identifies individuals by the unique parts of their DNA.

15.6 Genetic Engineering

■ Bacteria and yeast are the most common genetically engineered organisms.
■ Link to Gene expression 9.2

Traditional cross-breeding methods can alter genomes, but only if individuals with the desired traits will interbreed. Genetic engineering takes gene-swapping to an entirely different level. **Genetic engineering** is a laboratory process by which an individual's genome is deliberately modified. A gene may be altered and reinserted into an individual of the same species, or a gene from one species may be transferred to another to produce an organism that is **transgenic**. Both methods result in a **genetically modified organism**, or **GMO**.

The most common GMOs are bacteria and yeast. These cells have the metabolic machinery to make complex organic molecules, and they are easily modified. For example, the *E. coli* on the *left* have been modified to produce a fluorescent protein from jellyfish. The cells are genetically identical, so the visible variation in fluorescence among them reveals differences in gene expression. Such differences may help us discover why some bacteria of a population become dangerously resistant to antibiotics, and others do not.

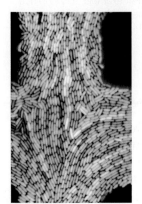

Genetically modified bacteria expressing a jellyfish gene emit green light.

Bacteria and yeast have been modified to produce medically important proteins. People with diabetes were among the first beneficiaries of such organisms. Insulin for their injections was once extracted from animals, but it provoked an allergic reaction in some people. Human insulin, which does not provoke allergic reactions, has been produced by transgenic *E. coli* since 1982. Slight modifications of the gene have also yielded fast-acting and slow-release forms of human insulin.

Engineered microorganisms also produce proteins used in food manufacturing. For example, cheese is traditionally made with an extract of calf stomachs, which contain the enzyme chymotrypsin. Most cheese manufacturers now use chymotrypsin produced by genetically engineered bacteria. Other GMO-produced enzymes improve the taste and clarity of beer and fruit juice, slow bread staling, or modify fats.

Take-Home Message

What is genetic engineering?

» Genetic engineering is the deliberate alteration of an individual's genome, and it results in a genetically modified organism (GMO).

» A transgenic organism carries a gene from a different species. Transgenic bacteria and yeast are used in research, medicine, and industry.

15.7 Designer Plants

■ Genetically engineered crop plants are widespread in the United States.
■ Links to β-carotene 6.2, Promoters 9.3

As crop production expands to keep pace with human population growth, it places unavoidable pressure on ecosystems everywhere. Irrigation leaves mineral and salt residues in soils. Tilled soil erodes, taking topsoil with it. Runoff clogs rivers, and fertilizer in it causes algae to grow so fast that fish suffocate. Pesticides can be harmful to humans, other animals, and beneficial insects such as bees.

Pressured to produce more food at lower cost and with less damage to the environment, many farmers have begun to rely on genetically modified crop plants. Genes can be introduced into plant cells by way of electric or chemical shocks, or by blasting them with DNA-coated micropellets. Another way involves *Agrobacterium tumefaciens* bacteria. *A. tumefaciens* carries a plasmid with genes that cause tumors to form on infected plants; hence the name Ti plasmid (for Tumor-inducing). Researchers replace the tumor-inducing genes with foreign or modified genes, then use the plasmid as a vector to deliver the desired genes into plant cells. Whole plants can be grown from plant cells that integrate a recombinant plasmid into their chromosomes (**Figure 15.12**).

Some genetically modified crop plants carry genes that impart resistance to devastating plant diseases. Others offer improved yields, such as a strain of transgenic wheat that has twice the yield of unmodified wheat. GMO crops such as Bt corn and soy help farmers use smaller amounts of toxic pesticides. Organic farmers often spray their crops with spores of Bt (*Bacillus thuringiensis*), a bacterial species that makes a protein toxic only to some insect larvae. Researchers transferred the gene encoding the Bt protein into plants. The engineered plants produce the Bt protein, but otherwise they are essentially identical to unmodified plants. Larvae die shortly after eating their first and only GMO meal. Farmers can use much less pesticide on crops that make their own (**Figure 15.13**).

Transgenic crop plants are also being developed for Africa and other impoverished regions of the world. Genes that confer drought tolerance and insect resistance are being introduced into plants such as corn,

genetic engineering Process by which deliberate changes are introduced into an individual's genome.
genetically modified organism (GMO) Organism whose genome has been modified by genetic engineering.
transgenic Refers to a genetically modified organism that carries a gene from a different species.

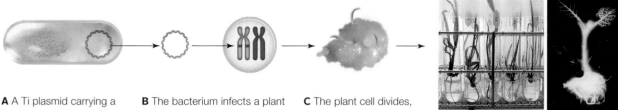

A A Ti plasmid carrying a foreign gene is inserted into an *Agrobacterium tumefaciens* bacterium.

B The bacterium infects a plant cell and transfers the Ti plasmid into it. The plasmid DNA becomes integrated into one of the cell's chromosomes.

C The plant cell divides, and its descendants form an embryo. Several embryos are sprouting from this mass of cells.

D Each embryo develops into a transgenic plant that expresses the foreign gene. The glowing tobacco plant is expressing a gene from fireflies.

Figure 15.12 **Animated** Using the Ti plasmid to make a transgenic plant.

beans, sugarcane, cassava, cowpeas, banana, and wheat. The resulting GMO crops may help people who rely on agriculture for food and income.

Genetic modifications can make food plants more nutritious. For example, rice plants have been engineered to make β-carotene, an orange photosynthetic pigment that is remodeled by cells of the small intestine into vitamin A. These rice plants carry two genes in the β-carotene synthesis pathway: one from corn, the other from bacteria. One cup of the engineered rice seeds—grains of Golden Rice—has enough β-carotene to satisfy a child's daily need for vitamin A.

The USDA Animal and Plant Health Inspection Service (APHIS) regulates the introduction of GMOs into the environment. At this writing, APHIS has deregulated seventy-eight crop plants, which means the plants are approved for unregulated use in the United States. Worldwide, more than 330 million acres are currently planted in GMO crops, the majority of which are corn, sorghum, cotton, soy, canola, and alfalfa engineered for resistance to the herbicide glyphosate. Rather than tilling the soil to control weeds, farmers can spray their fields with glyphosate, which kills the weeds but not the engineered crops.

After long-term, widespread use of glyphosate, weeds resistant to the herbicide are becoming more common. The engineered gene is also appearing in wild plants and in nonengineered crops, which means that recombinant genes can (and do) escape into the environment. The genes are probably being transferred from transgenic plants to nontransgenic ones via pollen carried by wind or insects.

Many people are opposed to any GMO. Some worry that our ability to tinker with genetics has surpassed our ability to understand the impact of the tinkering. Controversy raised by GMO use invites you to read the research and form your own opinions. The alternative is to be swayed by media hype (the term "Frankenfood," for instance), or by reports from possibly biased sources (such as herbicide manufacturers).

Figure 15.13 Genetically modified crops can help farmers use less pesticide. *Top*, the Bt gene conferred insect resistance to the genetically modified plants that produced this corn. *Bottom*, corn produced by unmodified plants is more vulnerable to insect pests.

Take-Home Message

Are genetically modified plants used as commercial crops?

» Genetically modified crop plants can help farmers be more productive while reducing overall costs.

» The widespread use of GMO crops has had unintended environmental effects. Herbicide resistant weeds are now common, and recombinant genes have spread to wild plants and non-GMO crops.

15.8 Biotech Barnyards

■ Genetically engineered animals are invaluable in medical research and in other applications.

■ Links to Knockout experiments 10.3, Human genetic disorders Chapter 14

Traditional cross-breeding has produced animals so unusual that transgenic animals may seem a bit mundane by comparison (**Figure 15.14A**). Cross-breeding is also a form of genetic manipulation, but many transgenic animals would probably never have occurred without laboratory intervention (**Figure 15.14B,C**).

The first genetically modified animals were mice. Today, such mice are commonplace, and they are invaluable in research (**Figure 15.15**). For example, we have discovered the function of human genes (including the *APOA5* gene discussed in Section 15.5) by inactivating their counterparts in mice. Genetically modified mice are also used as models of human diseases. For example, researchers inactivated the molecules involved in the control of glucose metabolism, one by one, in mice. Studying the effects of the knockouts has resulted in much of our current understanding of how diabetes works in humans.

Genetically modified animals also make proteins that have medical and industrial applications. Various transgenic goats produce proteins used to treat cystic fibrosis, heart attacks, blood clotting disorders, and even nerve gas exposure. Milk from goats transgenic for lysozyme, an antibacterial protein in human milk, may protect infants and children in developing countries from acute diarrheal disease. Goats transgenic for a spider silk gene produce the silk protein in their milk; researchers can spin this protein

into nanofibers that are useful in medical and electronics applications. Rabbits make human interleukin-2, a protein that triggers divisions of immune cells. Genetic engineering has also given us dairy goats with heart-healthy milk, pigs with heart-healthy fat and environmentally friendly low-phosphate feces, extra-large sheep, and cows that are resistant to mad cow disease.

Many people think that genetically engineering livestock is unconscionable. Others see it as an extension of thousands of years of acceptable animal husbandry practices. The techniques have changed, but not the intent: We humans continue to have a vested interest in improving our livestock.

Knockouts and Organ Factories

Millions of people suffer with organs or tissues that are damaged beyond repair. In any given year, more than 80,000 of them are on waiting lists for an organ transplant in the United States alone. Human donors are in such short supply that illegal organ trafficking is now a common problem.

Pigs are a potential source of organs for transplantation, because pig and human organs are about the same in both size and function. However, the human immune system battles anything it recognizes as nonself. It rejects a pig organ at once, because it recognizes proteins and carbohydrates on the plasma membrane of pig cells. Within a few hours, blood coagulates inside the organ's vessels and dooms the transplant. Drugs can suppress the immune response, but they also render organ recipients particularly vulnerable to infection. Researchers have produced genetically

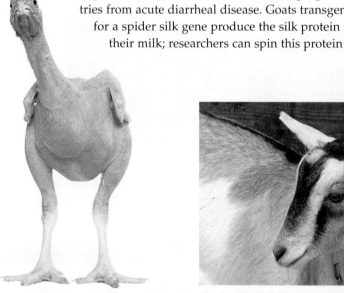

A The genes of this chicken were modified using a traditional method: cross-breeding. Featherless chickens survive in deserts where cooling systems are not an option.

B This genetically engineered goat is transgenic for a human gene. Its milk contains human antithrombin III (a protein that inhibits blood clotting).

C The genetically engineered pig on the *left* is transgenic for a bacterial gene. The gene's product, a yellow fluorescent protein, is visible in all of its tissues. A nontransgenic littermate is on the *right*.

Figure 15.14 Examples of genetically modified animals.

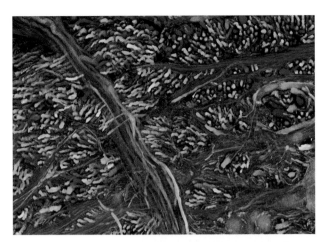

Figure 15.15 Example of how genetically engineered animals are useful in research. Mice transgenic for multiple pigments ("brainbow mice") are allowing researchers to map the complex neural circuitry of the brain. Individual nerve cells in the brain stem of a brainbow mouse are visible in this fluorescence micrograph.

modified pigs that lack the offending molecules on their cells. The human immune system may not reject tissues or organs transplanted from these pigs.

Transferring an organ from one species into another is called **xenotransplantation**. Critics of xenotransplantation are concerned that, among other things, pig-to-human transplants would invite pig viruses to cross the species barrier and infect humans, perhaps with catastrophic results. Their concerns are not unfounded. Evidence suggests that some of the worst pandemics arose when animal viruses adapted to replicate in human hosts.

Tinkering with the genes of animals raises a host of ethical dilemmas. For example, mice, monkeys, and other animals that have been genetically modified to carry mutations associated with certain human diseases often suffer the same terrible symptoms of these conditions as humans do. However, these animals are allowing researchers to study—and test treatments for—conditions such as multiple sclerosis, cystic fibrosis, diabetes, cancer, and Huntington's disease without experimenting on humans.

xenotransplantation Transplantation of an organ from one species into another.

15.9 Safety Issues

■ The first transfer of foreign DNA into bacteria ignited an ongoing debate about potential dangers of transgenic organisms that enter the environment.

When James Watson and Francis Crick presented their model of DNA in 1953, they ignited a global blaze of optimism. The very book of life seemed to be open for scrutiny. In reality, no one could read it. New techniques would have to be invented before that book would become readable.

Twenty years later, Paul Berg and his coworkers discovered how to make recombinant organisms by fusing DNA from two species of bacteria. Researchers now had the tools to be able to study its sequence in detail. They began to clone DNA from many different organisms. The technique of genetic engineering was born, and suddenly everyone was worried about it. Researchers knew that DNA itself was not toxic, but they could not predict with certainty what would happen each time they fused genetic material from different organisms. Would they accidentally make a superpathogen? Could they make a new, dangerous form of life by fusing DNA of two normally harmless organisms? What if an engineered organism escaped from the laboratory and transformed other organisms?

In a remarkably quick and responsible display of self-regulation, scientists reached a consensus on new safety guidelines for DNA research. Adopted at once by the NIH, these guidelines included precautions for laboratory procedures. They covered the design and use of host organisms that could survive only under the narrow range of conditions inside the laboratory. Researchers stopped using DNA from pathogenic or toxic organisms for recombinant DNA experiments until proper containment facilities were developed.

Now, all genetic engineering should be done under these laboratory guidelines, but the rules are not a guarantee of safety. We are still learning about escaped GMOs and their effects, and enforcement is a problem. For example, the expense of deregulating a GMO is prohibitive for endeavors in the public sector. Thus, most commercial GMOs were produced by large, private companies—the same ones that typically wield tremendous political influence over the very government agencies charged with regulating them.

15.10 Genetically Modified Humans

■ We as a society continue to work our way through the ethical implications of applying new DNA technologies.

■ The manipulation of individual genomes continues even as we are weighing the risks and benefits of this research.

■ Links to Proto-oncogenes and cancer 11.6, Locus 13.2, Human genetic disorders 14.2

Getting Better

We know of more than 15,000 serious genetic disorders. Collectively, they cause 20 to 30 percent of infant deaths each year, and account for half of all mentally impaired patients and a fourth of all hospital admissions. They also contribute to many age-related disorders, including cancer, Parkinson's disease, and diabetes. Drugs and other treatments can minimize the symptoms of some genetic disorders, but gene therapy is the only cure. **Gene therapy** is the transfer of recombinant DNA into an individual's body cells, with the intent to correct a genetic defect or treat a disease. The transfer, which occurs by way of lipid clusters or genetically engineered viruses, inserts an unmutated gene into an individual's chromosomes.

Human gene therapy is a compelling reason to embrace genetic engineering research. It is now being tested as a treatment for heart attack, sickle-cell anemia, cystic fibrosis, hemophilia A, Parkinson's disease, Alzheimer's disease, several types of cancer, and inherited diseases of the eye, the ear, and the immune system. The results are encouraging. For example, little Rhys Evans (**Figure 15.16**) was born with SCID-X1, a severe X-linked genetic disorder that stems from a mutated allele of the *IL2RG* gene. The gene encodes a receptor for an immune signaling molecule. Children affected by this disorder can survive only in germ-free isolation tents, because they cannot fight infections. In the late 1990s, researchers used a genetically engineered virus to insert unmutated copies of *IL2RG* into cells taken from the bone marrow of twenty boys with SCID-X1. Each child's modified cells were infused back into his bone marrow. Within months of their treatment, eighteen of the boys left their isolation tents for good. Rhys was one of them. Gene therapy had permanently repaired their immune systems.

Getting Worse

Manipulating a gene within the context of a living individual is unpredictable even when we know its sequence and locus. No one, for example, can predict where a virus-injected gene will become integrated into a chromosome. Its insertion might disrupt other

Figure 15.16
Rhys Evans, who was born with SCID-X1. His immune system has been permanently repaired by gene therapy.

genes. If it interrupts a gene that is part of the controls over cell division, then cancer might be the outcome. Five of the twenty boys treated with gene therapy for SCID-X1 have since developed a type of bone marrow cancer called leukemia, and one of them has died. The researchers had wrongly predicted that cancer related to the gene therapy would be rare. Research now implicates the very gene targeted for repair, especially when combined with the virus that delivered it. Apparently, integration of the modified viral DNA activated nearby proto-oncogenes (Section 11.6) in the children's chromosomes.

Getting Perfect

The idea of selecting the most desirable human traits, **eugenics**, is an old one. It has been used as a justification for some of the most horrific episodes in human history, including the genocide of 6 million Jews during World War II. Thus, it continues to be a hotly debated social issue. For example, using gene therapy to cure human genetic disorders seems like a socially acceptable goal to most people. However, imagine taking this idea a bit further. Would it also be acceptable to engineer the genome of an individual who is within a normal range of phenotype in order to modify a particular trait? Researchers have already produced mice that have improved memory, enhanced learning ability, bigger muscles, and longer lives. Why not people?

Given the pace of genetics research, the eugenics debate is no longer about how we would engineer desirable traits, but how we would choose the traits that are desirable. Realistically, cures for many severe but rare genetic disorders will not be found, because the financial return will not even cover the cost of the research. Eugenics, however, might just turn a profit.

Personal DNA Testing (revisited)

The results of SNP analysis by a personal DNA testing company also include estimated risks of developing conditions associated with your particular set of SNPs. For example, the test will probably determine whether you are homozygous for one allele of the *MC1R* gene. If you are, the company's report will tell you that you have red hair. Very few SNPs have such a clear cause-and-effect relationship as the *MC1R* allele for red hair, however. Most human traits are polygenic, and many are also influenced by environmental factors such as lifestyle (Section 13.6). Thus, although a DNA test can reliably determine the SNPs in an individual's genome, it cannot reliably predict the effect of those SNPs on the individual.

For example, if you carry one ε4 allele of the *APOE* gene, a DNA testing company cannot tell you whether you will develop Alzheimer's disease later in life. Instead, the company will report your lifetime risk of developing the disease, which is about 29 percent, as compared with about 9 percent for someone who has no ε4 allele.

What does a 29 percent lifetime risk of developing Alzheimer's disease mean? The number is a probability statistic; it means that, on aver-

age, 29 of every 100 people who have the ε4 allele eventually get the disease. Having a high risk does not mean you are certain to end up with Alzheimer's, however. Not everyone who develops the disease has the ε4 allele, and not everyone with the ε4 allele develops Alzheimer's disease. Other factors, including epigenetic methylation of DNA, contribute to the disease.

How would you vote? The plunging cost of genetic testing has spurred an explosion of companies offering personal DNA sequencing and SNP profiling. The results of such testing may in some cases be of clinical use, for example in diagnosis of early-onset genetic disorders, or in predicting how an individual will respond to certain medications. However, we are still at an extremely early stage in our understanding of how genes contribute to most conditions, particularly age-related disorders such as Alzheimer's disease. Geneticists believe that it will be five to ten more years before we can use genotype to accurately predict an individual's risk of these conditions. Until then, should genetic testing companies be prohibited from informing clients of their estimated risk of developing such disorders based on SNPs?

How much would potential parents pay to be sure that their child will be tall or blue-eyed? Would it be okay to engineer "superhumans" with breathtaking strength or intelligence? How about a treatment that can help you lose that extra weight, and keep it off permanently? The gray area between interesting and abhorrent can be very different depending on who is asked. In a survey conducted in the United States, more than 40 percent of those interviewed said it would be fine to use gene therapy to make smarter and cuter babies. In one poll of British parents, 18 percent would be willing to use it to keep a child from being aggressive, and 10 percent would use it to keep a child from growing up to be homosexual.

Getting There

Some people are adamant that we must never alter the

DNA of anything. The concern is that gene therapy puts us on a slippery slope that may result in irreversible damage to ourselves and to the biosphere. We as a society may not have the wisdom to know how to stop once we set foot on that slope. One is reminded of our peculiar human tendency to leap before we look. And yet, something about the human experience allows us to dream of such things as wings of our own making, a capacity that carried us into space.

In this brave new world, the questions before you are these: What do we stand to lose if serious risks are not taken? And, do we have the right to impose the consequences of taking such risks on those who would choose not to take them?

eugenics Idea of deliberately improving the genetic qualities of the human race.
gene therapy Treating a genetic defect or disorder by transferring a normal or modified gene into the affected individual.

Take-Home Message

Can people be genetically modified?

» Genes can be transferred into a person's cells to correct a genetic defect or treat a disease. However, the outcome of altering a person's genome remains unpredictable given our current understanding of how the genome works.

Summary

Section 15.1 Personal DNA testing companies identify a person's unique array of **single-nucleotide polymorphisms**. Personal genetic testing may soon revolutionize the way medicine is practiced.

Section 15.2 In **DNA cloning**, **restriction enzymes** cut DNA into pieces, then DNA ligase splices the pieces into plasmids or other **cloning vectors**. The resulting **recombinant DNA** molecules are inserted into host cells such as bacteria. When a host cell divides, it forms huge populations of genetically identical descendant cells (clones), each with a copy of the foreign DNA. The enzyme **reverse transcriptase** is used to transcribe RNA into **cDNA** for cloning.

Section 15.3 A **DNA library** is a collection of cells that host different fragments of DNA, often representing an organism's entire **genome**. Researchers can use **probes** to identify cells that host a specific fragment of DNA. Base pairing between nucleic acids from different sources is called **nucleic acid hybridization**. The **polymerase chain reaction** (**PCR**) uses **primers** and a heat-resistant DNA polymerase to rapidly increase the number of copies of a targeted section of DNA.

Section 15.4 DNA sequencing reveals the order of bases in a section of DNA. DNA polymerase is used to partially replicate a DNA template. The reaction produces a mixture of DNA fragments of all different lengths. **Electrophoresis** separates the fragments by length into bands. The entire genomes of several organisms have now been sequenced.

Section 15.5 Genomics is providing insights into the function of the human genome. Similarities between genomes of different organisms are evidence of evolutionary relationships, and can be used as a predictive tool in research. **DNA profiling** identifies a person by the unique parts of his or her DNA. An example is the determination of an individual's array of **short tandem repeats** (SNPs). Within the context of a criminal investigation, a DNA profile is called a DNA fingerprint.

Sections 15.6–15.9 Recombinant DNA technology is the basis of **genetic engineering**, the directed modification of an organism's genetic makeup with the intent to modify its phenotype. A gene is modified and reinserted into an individual of the same species, or a gene from one species is inserted into an individual of a different species to make a **transgenic** organism. The result of either process is a **genetically modified organism** (**GMO**). Transgenic bacteria and yeast produce medically valuable proteins. Transgenic crop plants are helping farmers produce food more efficiently.

Genetically modified animals produce human proteins, and may one day provide a source of organs and tissues for **xenotransplantation** into humans.

Section 15.10 With **gene therapy**, a gene is transferred into body cells to correct a genetic defect or treat a disease. Potential benefits of genetically modifying humans must be weighed against potential risks. The practice raises ethical issues such as whether **eugenics** is desirable in some circumstances.

Self-Quiz
Answers in Appendix III

1. _____ cut(s) DNA molecules at specific sites.
   a. DNA polymerase   c. Restriction enzymes
   b. DNA probes   d. Reverse transcriptase

2. A _____ is a small circle of bacterial DNA that contains a few genes and is separate from the chromosome.
   a. plasmid   c. nucleus
   b. chromosome   d. double helix

3. Reverse transcriptase assembles a(n) _____ on a(n) _____ template.
   a. mRNA; DNA   c. DNA; ribosome
   b. cDNA; mRNA   d. protein; mRNA

4. For each species, all _____ in the complete set of chromosomes is the _____ .
   a. genomes; phenotype   c. mRNA; start of cDNA
   b. DNA; genome   d. cDNA; start of mRNA

5. A set of cells that host various DNA fragments collectively representing an organism's entire set of genetic information is a _____ .
   a. genome   c. genomic library
   b. clone   d. GMO

6. _____ is a technique to determine the order of nucleotide bases in a fragment of DNA.

7. Fragments of DNA can be separated by electrophoresis according to _____ .
   a. sequence   c. species
   b. length   d. composition

8. PCR can be used _____ .
   a. to increase the number of specific DNA fragments
   b. in DNA fingerprinting
   c. to modify a human genome
   d. a and b are correct

9. An individual's set of unique _____ can be used as a DNA profile.
   a. DNA sequences   c. SNPs
   b. short tandem repeats d. all of the above

10. A transgenic organism _____ .
    a. carries a gene from another species
    b. has been genetically modified
    c. both a and b

11. Which of the following can be used to carry foreign DNA into host cells? Choose all correct answers.
    a. RNA   e. lipid clusters
    b. viruses   f. blasts of pellets
    c. PCR   g. xenotransplantation
    d. plasmids   h. sequencing

Enhanced Spatial Learning in Mice with Autism Mutation

Autism is a neurobiological disorder with a range of symptoms that include impaired social interactions and stereotyped patterns of behavior such as hand-flapping or rocking. A relatively high proportion of autistic people—around 10 percent—have an extraordinary skill or talent such as greatly enhanced memory.

Mutations in neuroligin 3, a cell adhesion protein (Section 5.7) that connects brain cells to one another, have been associated with autism. One mutation changes amino acid 451 from arginine to cysteine. Mouse and human neuroligin 3 are very similar. In 2007, Katsuhiko Tabuchi and his colleagues genetically modified mice to carry the same arginine-to-cysteine substitution in their neuroligin 3. The mutation caused an increase in transmission of some types of signals between brain cells. Mice with the mutation had impaired social behavior, and, unexpectedly, enhanced spatial learning ability (**Figure 15.17**).

1. In the first test, how many days did unmodified mice need to learn to find the location of a hidden platform within 10 seconds?

2. Did the modified or the unmodified mice learn the location of the platform faster in the first test?

3. Which mice learned faster the second time around?

4. Which mice had the greatest improvement in memory?

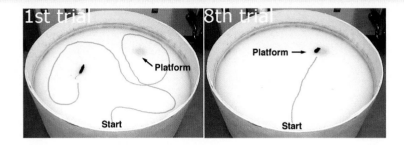

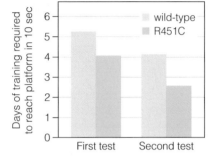

A Mice were tested in a water maze, in which a platform is submerged slightly below the surface of a deep pool of warm water. The platform is not visible to swimming mice. Mice do not particularly enjoy swimming, so they locate a hidden platform as fast as they can. When tested again in the same pool, they use visual cues around the pool's edge to remember the platform's location.

Figure 15.17 Enhanced spatial learning ability in mice with a mutation in neuroligin 3 (R451C), compared with unmodified (wild-type) mice.

B How quickly mice remember the location of a hidden platform in a water maze is a measure of spatial learning ability.

---

12. Transgenic _____ can pass a foreign gene to offspring.
    a. plants   c. a and b
    b. animals   d. none are correct

13. _____ can be used to correct a genetic defect.
    a. Cloning vectors   d. Xenotransplantation
    b. Gene therapy   e. a and b
    c. Cloning   f. all of the above

14. Match the method with the appropriate enzyme.
    ___ PCR   a. *Taq* polymerase
    ___ cutting DNA   b. DNA ligase
    ___ cDNA synthesis   c. reverse transcriptase
    ___ DNA sequencing   d. restriction enzyme
    ___ pasting DNA   e. DNA polymerase (not *Taq*)

15. Match the terms with the most suitable description.
    ___ DNA profile   a. GMO with a foreign gene
    ___ Ti plasmid   b. an allele contains one
    ___ nucleic acid   c. a person's unique collection
        hybridization   of short tandem repeats
    ___ eugenics   d. base pairing of DNA or
    ___ SNP   DNA and RNA from
    ___ transgenic   different sources
    ___ GMO   e. selecting "desirable" traits
                           f. genetically modified
                           g. used in some gene transfers

Critical Thinking

1. Restriction enzymes in bacterial cytoplasm cut injected bacteriophage DNA wherever certain sequences occur. Why do you think these enzymes do not chop up the bacterial chromosome, which is exposed to the enzymes in cytoplasm?

2. The *FOXP2* gene encodes a transcription factor associated with vocal learning in mice, bats, birds, and humans. The chimpanzee, gorilla, and rhesus FOXP2 proteins are identical; the human version differs in only 2 of 715 amino acids, a change thought to have contributed to the development of spoken language. In humans, loss-of-function mutations in *FOXP2* result in severe speech and language disorders. In mice, they hamper brain function and impair vocalizations. Mice genetically engineered to carry the human version of *FOXP2* show changes in vocal patterns, and more growth and greater adaptability of neurons involved in memory and learning. Biologists do not anticipate that a similar experiment in chimpanzees would confer the ability to speak, because spoken language is an extremely complex trait. If genetic engineering could produce a talking chimp, how do you think the debate about using animals for research would change?

3. In 1918, an influenza pandemic that originated with avian flu killed 50 million people. Researchers isolated samples of that virus from bodies of infected people preserved in Alaskan permafrost since 1918. From the samples, they sequenced the viral genome, then reconstructed the virus. The reconstructed virus is 39,000 times more infectious than modern influenza strains, and 100 percent lethal in mice.

Understanding how this virus works can help us defend ourselves against other strains that may arise. For example, discovering what makes it is so infectious and deadly would help us design more effective vaccines. Critics of the research are concerned: If the virus escapes the containment facilities (even though it has not done so yet), it might cause another pandemic. Worse, terrorists could use the published DNA sequence and methods to make the virus for horrific purposes. Do you think this research makes us more or less safe?

Appendix I. The Amino Acids

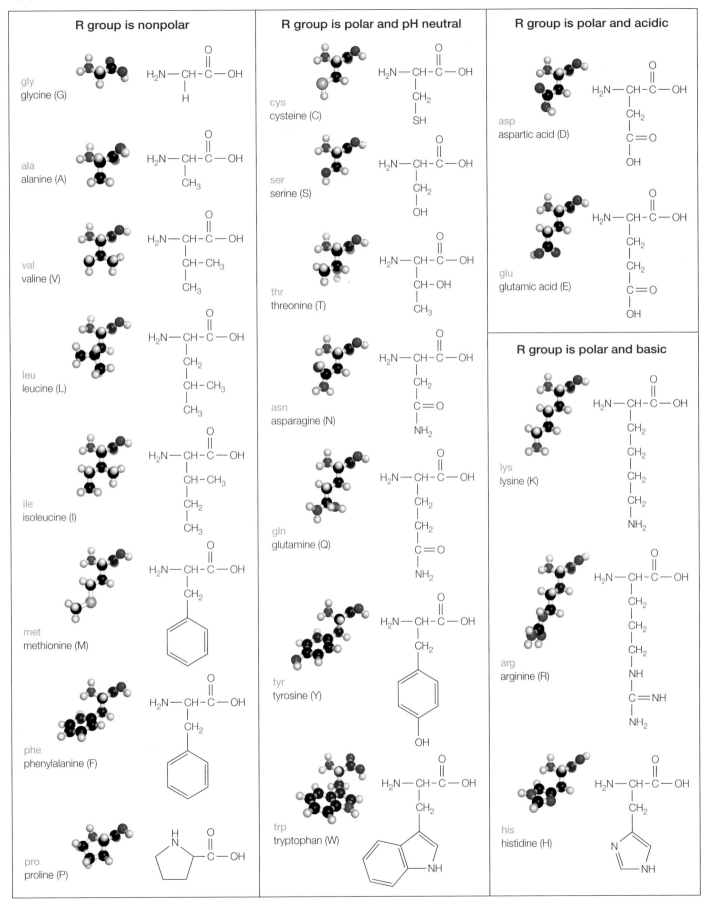

R group is nonpolar

gly
glycine (G)

$$H_2N-CH-C-OH$$ with O double bond and H

ala
alanine (A)

$$H_2N-CH-C-OH$$ with O double bond and CH_3

val
valine (V)

$$H_2N-CH-C-OH$$ with O double bond, $CH-CH_3$, CH_3

leu
leucine (L)

$$H_2N-CH-C-OH$$ with O double bond, CH_2, $CH-CH_3$, CH_3

ile
isoleucine (I)

$$H_2N-CH-C-OH$$ with O double bond, $CH-CH_3$, CH_2, CH_3

met
methionine (M)

$$H_2N-CH-C-OH$$ with O double bond, CH_2, benzene ring

phe
phenylalanine (F)

$$H_2N-CH-C-OH$$ with O double bond, CH_2, benzene ring

pro
proline (P)

$$C-OH$$ with O double bond, ring with NH

R group is polar and pH neutral

cys
cysteine (C)

$$H_2N-CH-C-OH$$ with O double bond, CH_2, SH

ser
serine (S)

$$H_2N-CH-C-OH$$ with O double bond, CH_2, OH

thr
threonine (T)

$$H_2N-CH-C-OH$$ with O double bond, $CH-OH$, CH_3

asn
asparagine (N)

$$H_2N-CH-C-OH$$ with O double bond, CH_2, $C=O$, NH_2

gln
glutamine (Q)

$$H_2N-CH-C-OH$$ with O double bond, CH_2, CH_2, $C=O$, NH_2

tyr
tyrosine (Y)

$$H_2N-CH-C-OH$$ with O double bond, CH_2, benzene ring, OH

trp
tryptophan (W)

$$H_2N-CH-C-OH$$ with O double bond, CH_2, indole ring with NH

R group is polar and acidic

asp
aspartic acid (D)

$$H_2N-CH-C-OH$$ with O double bond, CH_2, $C=O$, OH

glu
glutamic acid (E)

$$H_2N-CH-C-OH$$ with O double bond, CH_2, CH_2, $C=O$, OH

R group is polar and basic

lys
lysine (K)

$$H_2N-CH-C-OH$$ with O double bond, CH_2, CH_2, CH_2, CH_2, NH_2

arg
arginine (R)

$$H_2N-CH-C-OH$$ with O double bond, CH_2, CH_2, CH_2, NH, $C=NH$, NH_2

his
histidine (H)

$$H_2N-CH-C-OH$$ with O double bond, CH_2, imidazole ring with N and NH

Appendix II. Annotations to A Journal Article

This journal article reports on the movements of a female wolf during the summer of 2002 in northwestern Canada. It also reports on a scientific process of inquiry, observation and interpretation to learn where, how and why the wolf traveled as she did. In some ways, this article reflects the story of "how to do science" told in section 1.5 of this textbook. These notes are intended to help you read and understand how scientists work and how they report on their work.

---

① ARCTIC

② VOL. 57, NO. 2 (JUNE 2004) P. 196–203

③ Long Foraging Movement of a Denning Tundra Wolf

④ Paul F. Frame,[1,2] David S. Hik,[1] H. Dean Cluff,[3] and Paul C. Paquet[4]

⑤ (Received 3 September 2003; accepted in revised form 16 January 2004)

---

⑥ **ABSTRACT.** Wolves (*Canis lupus*) on the Canadian barrens are intimately linked to migrating herds of barren-ground caribou (*Rangifer tarandus*). We deployed a Global Positioning System (GPS) radio collar on an adult female wolf to record her movements in response to changing caribou densities near her den during summer. This wolf and two other females were observed nursing a group of 11 pups. She traveled a minimum of 341 km during a 14-day excursion. The straight-line distance from the den to the farthest location was 103 km, and the overall minimum rate of travel was 3.1 km/h. The distance between the wolf and the radio-collared caribou decreased from 242 km one week before the excursion to 8 km four days into the excursion. We discuss several possible explanations for the long foraging bout.

⑦ *Key words:* wolf, GPS tracking, movements, *Canis lupus*, foraging, caribou, Northwest Territories

⑧ **RÉSUMÉ.** Les loups (*Canis lupus*) dans la toundra canadienne sont étroitement liés aux hardes de caribous des toundras (*Rangifer tarandus*). On a équipé une louve adulte d'un collier émetteur muni d'un système de positionnement mondial (GPS) afin d'enregistrer ses déplacements en réponse au changement de densité du caribou près de sa tanière durant l'été. On a observé cette louve ainsi que deux autres en train d'allaiter un groupe de 11 louveteaux. Elle a parcouru un minimum de 341 km durant une sortie de 14 jours. La distance en ligne droite de la tanière à l'endroit le plus éloigné était de 103 km, et la vitesse minimum durant tout le voyage était de 3,1 km/h. La distance entre la louve et le caribou muni du collier émetteur a diminué de 242 km une semaine avant la sortie à 8 km quatre jours après la sortie. On commente diverses explications possibles pour ce long épisode de recherche de nourriture.

Mots clés: loup, repérage GPS, déplacements, *Canis lupus*, recherche de nourriture, caribou, Territoires du Nord-Ouest

Traduit pour la revue *Arctic* par Nésida Loyer.

---

⑨ Introduction

Wolves (*Canis lupus*) that den on the central barrens of mainland Canada follow the seasonal movements of their main prey, migratory barren-ground caribou (*Rangifer tarandus*) (Kuyt, 1962; Kelsall, 1968; Walton et al., 2001). However, most wolves do not den near caribou calving grounds, but select sites farther south, closer to the tree line (Heard and Williams, 1992). Most caribou migrate beyond primary wolf denning areas by mid-June and do not return until mid-to-late July (Heard et al., 1996; Gunn et al., 2001). Conse- quently, caribou density near dens is low for part of the summer.

During this period of spatial separation from the main caribou herds, wolves must either search near the homesite for scarce caribou or alternative prey (or both), travel to where prey are abundant, or use a combination of these strategies.

Walton et al. (2001) postulated that the travel of tundra wolves outside their normal summer ranges is a response to low caribou availability rather than a pre-dispersal exploration like that observed in terri- torial wolves (Fritts and Mech, 1981; Messier, 1985). The authors postulated this because most such travel was directed toward caribou calving grounds. We report details of such a long-distance excursion by a breeding female tundra wolf wearing a GPS radio collar. We discuss the relationship of the excursion to movements of satellite-collared caribou (Gunn et al., 2001), supporting the hypothesis that tundra wolves make directional, rapid, long-distance movements in response to seasonal prey availability.

[1] Department of Biological Sciences, University of Alberta, Edmonton, Alberta T6G 2E9, Canada
[2] Corresponding author: pframe@ualberta.ca
[3] Department of Resources, Wildlife, and Economic Development, North Slave Region, Government of the Northwest Territories, P.O. Box 2668, 3803 Bretzlaff Dr., Yellowknife, Northwest Territories X1A 2P9, Canada; Dean_Cluff@gov.nt.ca
[4] Faculty of Environmental Design, University of Calgary, Calgary, Alberta T2N 1N4, Canada; current address: P.O. Box 150, Meacham, Saskatchewan S0K 2V0, Canada

196

Annotations (right column):

1 Title of the journal, which reports on science taking place in Arctic regions.

2 Volume number, issue number and date of the journal, and page numbers of the article.

3 Title of the article: a concise but specific description of the subject of study—one episode of long-range travel by a wolf hunting for food on the Arctic tundra.

4 Authors of the article: scientists working at the institutions listed in the footnotes below. Note #2 indicates that P. F. Frame is the corresponding author—the person to contact with questions or comments. His email address is provided.

5 Date on which a draft of the article was received by the journal editor, followed by date one which a revised draft was accepted for publication. Between these dates, the article was reviewed and critiqued by other scientists, a process called peer review. The authors revised the article to make it clearer, according to those reviews.

6 ABSTRACT: A brief description of the study containing all basic elements of this report. First sentence summarizes the *background* material. Second sentence encapsulates the *methods* used. The rest of the paragraph sums up the *results*. Authors introduce the main *subject* of the study—a female wolf (#388) with pups in a den—and refer to later *discussion* of possible explanations for her behavior.

7 Key words are listed to help researchers using computer data- bases. Searching the databases using these key words will yield a list of studies related to this one.

8 RÉSUMÉ: The French translation of the abstract and key words. Many researchers in this field are French Canadian. Some journals provide such translations in French or in other languages.

9 INTRODUCTION: Gives the back- ground for this wolf study. This para- graph tells of known or suspected wolf behavior that is important for this study. Note that (a) major species mentioned are always accompanied by scientific names, and (b) statements of fact or *postulations* (claims or assumptions about what is likely to be true) are followed by references to studies that established those facts or supported the postulations.

10 This paragraph focuses directly on the wolf behaviors that were studied here.

11 This paragraph starts with a state- ment of the *hypothesis* being tested, one that originated in other studies and is supported by this one. The hypothesis is restated more succinctly in the last sentence of this paragraph. This is the *inquiry* part of the scientific process— asking questions and suggesting possible answers.

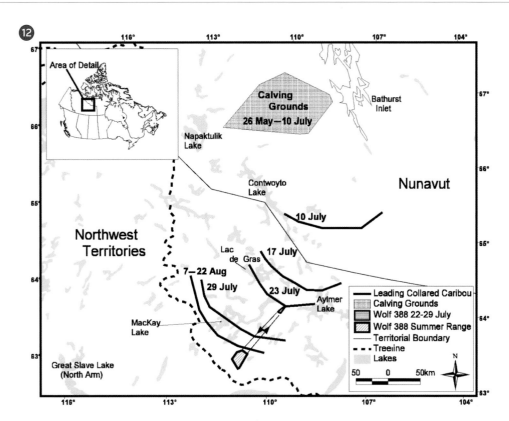

Figure 1. Map showing the movements of satellite radio-collared caribou with respect to female wolf 388's summer range and long foraging movement, in summer 2002.

12 This map shows the study area and depicts wolf and caribou locations and movements during one summer. Some of this information is explained below.

13 STUDY AREA: This section sets the stage for the study, locating it precisely with latitude and longitude coordinates and describing the area (illustrated by the map in Figure 1).

14 Here begins the story of how prey (caribou) and predators (wolves) interact on the tundra. Authors describe movements of these nomadic animals throughout the year.

15 We focus on the denning season (summer) and learn how wolves locate their dens and travel according to the movements of caribou herds.

⑬ Study Area

Our study took place in the northern boreal forest–low Arctic tundra transition zone (63° 30′ N, 110° 00′ W; Figure 1; Timoney et al., 1992). Permafrost in the area changes from discontinuous to continuous (Harris, 1986). Patches of spruce (*Picea mariana, P. glauca*) occur in the southern portion and give way to open tundra to the northeast. Eskers, kames, and other glacial deposits are scattered throughout the study area. Standing water and exposed bedrock are characteristic of the area.

⑭ *Details of the Caribou-Wolf System*

The Bathurst caribou herd uses this study area. Most caribou cows have begun migrating by late April, reaching calving grounds by June (Gunn et al., 2001;

Figure 1). Calving peaks by 15 June (Gunn et al., 2001), and calves begin to travel with the herd by one week of age (Kelsall, 1968). The movement patterns of bulls are less known, but bulls frequent areas near calving grounds by mid-June (Heard et al., 1996; Gunn et al., 2001). In summer, Bathurst caribou cows generally travel south from their calving grounds and then, parallel to the tree line, to the northwest. The rut usually takes place at the tree line in October (Gunn et al., 2001). The winter range of the Bathurst herd varies among years, ranging through the taiga and along the tree line from south of Great Bear Lake to southeast of Great Slave Lake. Some caribou spend the winter on the tundra (Gunn et al., 2001; Thorpe et al., 2001).

In winter, wolves that prey on Bathurst caribou do ⑮ not behave territorially. Instead, they follow the herd throughout its winter range (Walton et al., 2001; Musiani, 2003). However, during denning (May–

16 Other variables are considered—prey other than caribou and their relative abundance in 2002.

17 METHODS: There is no one scientific method. Procedures for each and every study must be explained carefully.

18 Authors explain when and how they tracked caribou and wolves, including tools used and the exact procedures followed.

19 This important subsection explains what data were calculated (average distance ...) and how, including the software used and where it came from. (The calculations are listed in Table 1.) Note that the behavior measured (traveling) is carefully defined.

20 RESULTS: The heart of the report and the *observation* part of the scientific process. This section is organized parallel to the Methods section.

21 This subsection is broken down by periods of observation. Pre-excursion period covers the time between 388's capture and the start of her long-distance travel. The investigators used visual observations as well as telemetry (measurements taken using the global positioning system (GPS)) to gather data. They looked at how 388 cared for her pups, interacted with other adults, and moved about the den area.

Table 1. Daily distances from wolf 388 and the den to the nearest radio-collared caribou during a long excursion in summer 2002.

Date (2002)	Mean distance from caribou to wolf (km)	Daily distance from closest caribou to den
12 July	242	241
13 July	210	209
14 July	200	199
15 July	186	180
16 July	163	162
17 July	151	148
18 July	144	137
19 July[1]	126	124
20 July	103	130
21 July	73	130
22 July	40	110
23 July[2]	9	104
29 July[3]	16	43
30 July	32	43
31 July	28	44
1 August	29	46
2 August[4]	54	52
3 August	53	53
4 August	74	74
5 August	75	75
6 August	74	75
7 August	72	75
8 August	76	75
9 August	79	79

[1] Excursion starts.
[2] Wolf closest to collared caribou.
[3] Previous five days' caribou locations not available.
[4] Excursion ends.

August, parturition late May to mid-June), wolf movements are limited by the need to return food to the den. To maximize access to migrating caribou, many wolves select den sites closer to the tree line than to caribou calving grounds (Heard and Williams, 1992). Because of caribou movement patterns, tundra denning wolves are separated from the main caribou herds by several hundred kilometers at some time during summer (Williams, 1990:19; Figure 1; Table 1).

 Muskoxen do not occur in the study area (Fournier and Gunn, 1998), and there are few moose there (H.D. Cluff, pers. obs.). Therefore, alternative prey for wolves includes waterfowl, other ground-nesting birds, their eggs, rodents, and hares (Kuyt, 1972; Williams, 1990:16; H.D. Cluff and P.F. Frame, unpubl. data). During 56 hours of den observations, we saw no ground squirrels or hares, only birds. It appears that the abundance of alternative prey was relatively low in 2002.

Methods

Wolf Monitoring

We captured female wolf 388 near her den on 22 June 2002, using a helicopter net-gun (Walton et al., 2001). She was fitted with a releasable GPS radio collar (Merrill et al., 1998) programmed to acquire locations at 30-minute intervals. The collar was electronically released (e.g., Mech and Gese, 1992) on 20 August 2002. From 27 June to 3 July 2002, we observed 388's den with a 78 mm spotting scope at a distance of 390 m.

Caribou Monitoring

In spring of 2002, ten female caribou were captured by helicopter net-gun and fitted with satellite radio collars, bringing the total number of collared Bathurst cows to 19. Eight of these spent the summer of 2002 south of Queen Maud Gulf, well east of normal Bathurst caribou range. Therefore, we used 11 caribou for this analysis. The collars provided one location per day during our study, except for five days from 24 to 28 July. Locations of satellite collars were obtained from Service Argos, Inc. (Landover, Maryland).

Data Analysis

Location data were analyzed by ArcView GIS software (Environmental Systems Research Institute Inc., Redlands, California). We calculated the average distance from the nearest collared caribou to the wolf and the den for each day of the study.

Wolf foraging bouts were calculated from the time 388 exited a buffer zone (500 m radius around the den) until she re-entered it. We considered her to be traveling when two consecutive locations were spatially separated by more than 100 m. Minimum distance traveled was the sum of distances between each location and the next during the excursion.

We compared pre- and post-excursion data using Analysis of Variance (ANOVA; Zar, 1999). We first tested for homogeneity of variances with Levene's test (Brown and Forsythe, 1974). No transformations of these data were required.

Results

Wolf Monitoring

Pre-Excursion Period: Wolf 388 was lactating when captured on 22 June. We observed her and two other females nursing a group of 11 pups between 27 June and 3 July. During our observations, the pack consisted of at least four adults (3 females and 1 male) and 11 pups. On 30 June, three pups were moved to a location 310 m from the other eight and cared for by an uncollared female. The male was not seen at the den after the evening of 30 June.

Before the excursion, telemetry indicated 18 foraging bouts. The mean distance traveled during these bouts was 25.29 km (± 4.5 SE, range 3.1–82.5 km). Mean greatest distance from the den on foraging

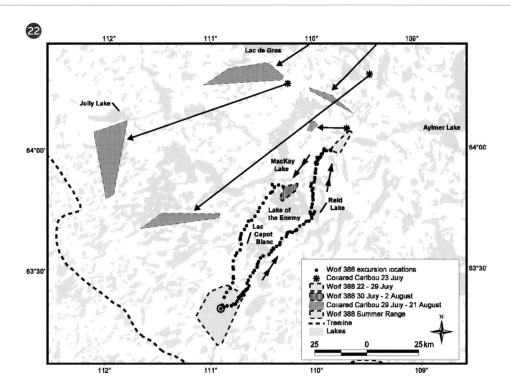

22 The key in the lower right-hand corner of the map shows areas (shaded) within which the wolves and caribou moved, and the dotted trail of 388 during her excursion. From the results depicted on this map, the investigators tried to determine when and where 388 might have encountered caribou and how their locations affected her traveling behavior.

23 The wolf's excursion (her long trip away from the den area) is the focus of this study. These paragraphs present detailed measurements of daily movements during her two-week trip—how far she traveled, how far she was from collared caribou, her time spent traveling and resting, and her rate of speed. Authors use the phrase "minimum distance traveled" to acknowledge they couldn't track every step but were measuring samples of her movements. They knew that she went at least as far as they measured. This shows how scientists try to be exact when reporting results. Results of this study are depicted graphically in the map in Figure 2.

Figure 2. Details of a long foraging movement by female wolf 388 between 19 July and 2 August 2002. Also shown are locations and movements of three satellite radio-collared caribou from 23 July to 21 August 2002. On 23 July, the wolf was 8 km from a collared caribou. The farthest point from the den (103 km distant) was recorded on 27 July. Arrows indicate direction of travel.

bouts was 7.1 km (± 0.9 SE, range 1.7–17.0 km). The average duration of foraging bouts for the period was 20.9 h (± 4.5 SE, range 1–71 h).

The average daily distance between the wolf and the nearest collared caribou decreased from 242 km on 12 July, one week before the excursion period, to 126 km on 19 July, the day the excursion began (Table 1).

Excursion Period: On 19 July at 2203, after spending 14 h at the den, 388 began moving to the northeast and did not return for 336 h (14 d; Figure 2). Whether she traveled alone or with other wolves is unknown. During the excursion, 476 (71%) of 672 possible locations were recorded. The wolf crossed the southeast end of Lac Capot Blanc on a small land bridge, where she paused for 4.5 h after traveling for 19.5 h (37.5

km). Following this rest, she traveled for 9 h (26.3 km) onto a peninsula in Reid Lake, where she spent 2 h before backtracking and stopping for 8 h just off the peninsula. Her next period of travel lasted 16.5 h (32.7 km), terminating in a pause of 9.5 h just 3.8 km from a concentration of locations at the far end of her excursion, where we presume she encountered caribou. The mean duration of these three movement periods was 15.7 h (± 2.5 SE), and that of the pauses, 7.3 h (± 1.5). The wolf required 72.5 h (3.0 d) to travel a minimum of 95 km from her den to this area near caribou (Figure 2). She remained there (35.5 km2) for 151.5 h (6.3 d) and then moved south to Lake of the Enemy, where she stayed (31.9 km^2) for 74 h (3.1 d) before returning to her den. Her greatest distance from the den, 103 km, was recorded 174.5 h (7.3 d) after the excursion

24 Post-excursion measurements of 388's movements were made to compare with those of the pre-excursion period. In order to compare, scientists often use *means*, or averages, of a series of measurements—mean distances, mean duration, etc.

25 In the comparison, authors used statistical calculations (F and df) to determine that the differences between pre- and post-excursion measurements were *statistically insignificant*, or close enough to be considered essentially the same or similar.

26 As with wolf 388, the investigators measured the movements of caribou during the study period. The areas within which the caribou moved are shown in Figure 2 by shaded polygons mentioned in the second paragraph of this subsection.

27 This subsection summarizes how distances separating predators and prey varied during the study period.

28 DISCUSSION: This section is the *interpretation* part of the scientific process.

29 This subsection reviews observations from other studies and suggests that this study fits with patterns of those observations.

30 Authors discuss a prevailing *theory* (CBFT) which might explain why a wolf would travel far to meet her own energy needs while taking food caught closer to the den back to her pups. The results of this study seem to fit that pattern.

---

began, at 0433 on 27 July. She was 8 km from a collared caribou on 23 July, four days after the excursion began (Table 1).

The return trip began at 0403 on 2 August, 318 h (13.2 d) after leaving the den. She followed a relatively direct path for 18 h back to the den, a distance of 75 km.

The minimum distance traveled during the excursion was 339 km. The estimated overall minimum travel rate was 3.1 km/h, 2.6 km/h away from the den and 4.2 km/h on the return trip.

(24) Post-Excursion Period: We saw three pups when recovering the collar on 20 August, but others may have been hiding in vegetation.

Telemetry recorded 13 foraging bouts in the post-excursion period. The mean distance traveled during these bouts was 18.3 km (+ 2.7 SE, range 1.2–47.7 km), and mean greatest distance from the den was 7.1 km (+ 0.7 SE, range 1.1–11.0 km). The mean duration of these post-excursion foraging bouts was 10.9 h (+ 2.4 SE, range 1–33 h).

When 388 reached her den on 2 August, the distance to the nearest collared caribou was 54 km. On 9 August, one week after she returned, the distance was 79 km (Table 1).

Pre- and Post-Excursion Comparison

(25) We found no differences in the mean distance of foraging bouts before and after the excursion period (F = 1.5, df = 1, 29, p = 0.24). Likewise, the mean greatest distance from the den was similar pre- and post-excursion (F = 0.004, df = 1, 29, p = 0.95). However, the mean duration of 388's foraging bouts decreased by 10.0 h after her long excursion (F = 3.1, df = 1, 29, p = 0.09).

(26) *Caribou Monitoring*

Summer Movements: On 10 July, 5 of 11 collared caribou were dispersed over a distance of 10 km, 140 km south of their calving grounds (Figure 1). On the same day, three caribou were still on the calving grounds, two were between the calving grounds and the leaders, and one was missing. One week later (17 July), the leading radio-collared cows were 100 km farther south (Figure 1). Two were within 5 km of each other in front of the rest, who were more dispersed. All radio-collared cows had left the calving grounds by this time. On 23 July, the leading radio-collared caribou had moved 35 km farther south, and all of them were more widely dispersed. The two cows closest to the leader were 26 km and 33 km away, with 37 km between them. On the next location (29 July), the most southerly caribou were 60 km

farther south. All of the caribou were now in the areas where they remained for the duration of the study (Figure 2).

A Minimum Convex Polygon (Mohr and Stumpf, 1966) around all caribou locations acquired during the study encompassed 85 119 km².

Relative to the Wolf Den: The distance from the **(27)** nearest collared caribou to the den decreased from 241 km one week before the excursion to 124 km the day it began. The nearest a collared caribou came to the den was 43 km away, on 29 and 30 July. During the study, four collared caribou were located within 100 km of the den. Each of these four was closest to the wolf on at least one day during the period reported.

(28) Discussion

Prey Abundance

Caribou are the single most important prey of tundra **(29)** wolves (Clark, 1971; Kuyt, 1972; Stephenson and James, 1982; Williams, 1990). Caribou range over vast areas, and for part of the summer, they are scarce or absent in wolf home ranges (Heard et al., 1996). Both the long distance between radio-collared caribou and the den the week before the excursion and the increased time spent foraging by wolf 388 indicate that caribou availability near the den was low. Observations of the pups' being left alone for up to 18 h, presumably while adults were searching for food, provide additional support for low caribou availability locally. Mean foraging bout duration decreased by 10.0 h after the excursion, when collared caribou were closer to the den, suggesting an increase in caribou availability nearby.

Foraging Excursion

One aspect of central place foraging theory (CPFT) **(30)** deals with the optimality of returning different-sized food loads from varying distances to dependents at a central place (i.e., the den) (Orians and Pearson, 1979). Carlson (1985) tested CPFT and found that the predator usually consumed prey captured far from the central place, while feeding prey captured nearby to dependants. Wolf 388 spent 7.2 days in one area near caribou before moving to a location 23 km back towards the den, where she spent an additional 3.1 days, likely hunting caribou. She began her return trip from this closer location, traveling directly to the den. While away, she may have made one or more successful kills and spent time meeting her own energetic needs before returning to the den. Alternatively, it may have taken several attempts to make a kill, which she then fed on before beginning her return trip. We do not know if she returned food to the pups, but such behavior would be supported by CPFT.

31 Other workers have reported wolves' making long round trips and referred to them as "extraterritorial" or "pre-dispersal" forays (Fritts and Mech, 1981; Messier, 1985; Ballard et al., 1997; Merrill and Mech, 2000). These movements are most often made by young wolves (1–3 years old), in areas where annual territories are maintained and prey are relatively sedentary (Fritts and Mech, 1981; Messier, 1985). The long excursion of 388 differs in that tundra wolves do not maintain annual territories (Walton et al., 2001), and the main prey migrate over vast areas (Gunn et al., 2001).

Another difference between 388's excursion and those reported earlier is that she is a mature, breeding female. No study of territorial wolves has reported reproductive adults making extraterritorial movements in summer (Fritts and Mech, 1981; Messier, 1985; Ballard et al., 1997; Merrill and Mech, 2001). However, Walton et al. (2001) also report that breeding female tundra wolves made excursions.

Direction of Movement

32 Possible explanations for the relatively direct route 388 took to the caribou include landscape influence and experience. Considering the timing of 388's trip and the locations of caribou, had the wolf moved northwest, she might have missed the caribou entirely, or the encounter might have been delayed.

A reasonable possibility is that the land directed 388's route. The barrens are crisscrossed with trails worn into the tundra over centuries by hundreds of thousands of caribou and other animals (Kelsall, 1968; Thorpe et al., 2001). At river crossings, lakes, or narrow peninsulas, trails converge and funnel towards and away from caribou calving grounds and summer range. Wolves use trails for travel (Paquet et al., 1996; Mech and Boitani, 2003; P. Frame, pers. observation). Thus, the landscape may direct an animal's movements and lead it to where cues, such as the odor of caribou on the wind or scent marks of other wolves, may lead it to caribou.

33 Another possibility is that 388 knew where to find caribou in summer. Sexually immature tundra wolves sometimes follow caribou to calving grounds (D. Heard, unpubl. data). Possibly, 388 had made such journeys in previous years and killed caribou. If this were the case, then in times of local prey scarcity she might travel to areas where she had hunted successfully before. Continued monitoring of tundra wolves may answer questions about how their food needs are met in times of low caribou abundance near dens.

Caribou often form large groups while moving **34** south to the tree line (Kelsall, 1968). After a large aggregation of caribou moves through an area, its scent can linger for weeks (Thorpe et al., 2001:104). It is conceivable that 388 detected caribou scent on the wind, which was blowing from the northeast on 19–21 July (Environment Canada, 2003), at the same time her excursion began. Many factors, such as odor strength and wind direction and strength, make systematic study of scent detection in wolves difficult under field conditions (Harrington and Asa, 2003). However, humans are able to smell odors such as forest fires or oil refineries more than 100 km away. The olfactory capabilities of dogs, which are similar to wolves, are thought to be 100 to 1 million times that of humans (Harrington and Asa, 2003). Therefore, it is reasonable to think that under the right wind conditions, the scent of many caribou traveling together could be detected by wolves from great distances, thus triggering a long foraging bout.

Rate of Travel

Mech (1994) reported the rate of travel of Arctic **35** wolves on barren ground was 8.7 km/h during regular travel and 10.0 km/h when returning to the den, a difference of 1.3 km/h. These rates are based on direct observation and exclude periods when wolves moved slowly or not at all. Our calculated travel rates are assumed to include periods of slow movement or no movement. However, the pattern we report is similar to that reported by Mech (1994), in that homeward travel was faster than regular travel by 1.6 km/h. The faster rate on return may be explained by the need to return food to the den. Pup survival can increase with the number of adults in a pack available to deliver food to pups (Harrington et al., 1983). Therefore, an increased rate of travel on homeward trips could improve a wolf's reproductive fitness by getting food to pups more quickly.

Fate of 388's Pups

Wolf 388 was caring for pups during den observa- **36** tions. The pups were estimated to be six weeks old, and were seen ranging as far as 800 m from the den. They received some regurgitated food from two of the females, but were unattended for long periods. The excursion started 16 days after our observations, and it is improbable that the pups could have traveled the distance that 388 moved. If the pups died, this would have removed parental responsibility, allowing the long movement.

Our observations and the locations of radio-collared caribou indicate that prey became scarce in

31 Here our authors note other possible explanations for wolves' excursions presented by other investigators, but this study does not seem to support those ideas.

32 Authors discuss possible reasons for why 388 traveled directly to where caribou were located. They take what they learned from earlier studies and apply it to this case, suggesting that the lay of the land played a role. Note that their description paints a clear picture of the landscape.

33 Authors suggest that 388 may have learned in traveling during previous summers where the caribou were. The last two sentences suggest ideas for future studies.

34 Or maybe 388 followed the scent of the caribou. Authors acknowledge difficulties of proving this, but they suggest another area where future studies might be done.

35 Authors suggest that results of this study support previous studies about how fast wolves travel to and from the den. In the last sentence, they speculate on how these observed patterns would fit into the theory of evolution.

36 Authors also speculate on the fate of 388's pups while she was traveling. This leads to . . .

37 Discussion of cooperative rearing of pups and, in turn, to speculation on how this study and what is known about cooperative rearing might fit into the animal's strategies for survival of the species. Again, the authors approach the broader theory of evolution and how it might explain some of their results.

38 And again, they suggest that this study points to several areas where further study will shed some light.

39 In conclusion, the authors suggest that their study supports the hypothesis being tested here. And they touch on the implications of increased human activity on the tundra predicted by their results.

40 ACKNOWLEDGEMENTS: Authors note the support of institutions, companies and individuals. They thank their reviewers ad list permits under which their research was carried on.

41 REFERENCES: List of all studies cited in the report. This may seem tedious, but is a vitally important part of scientific reporting. It is a record of the sources of information on which this study is based. It provides readers with a wealth of resources for further reading on this topic. Much of it will form the foundation of future scientific studies like this one.

---

the area of the den as summer progressed. Wolf 388 may have abandoned her pups to seek food for herself. However, she returned to the den after the excursion, where she was seen near pups. In fact, she foraged in a similar pattern before and after the excursion, suggesting that she again was providing for pups after her return to the den.

37 A more likely possibility is that one or both of the other lactating females cared for the pups during 388's absence. The three females at this den were not seen with the pups at the same time. However, two weeks earlier, at a different den, we observed three females cooperatively caring for a group of six pups. At that den, the three lactating females were observed providing food for each other and trading places while nursing pups. Such a situation at the den of 388 could have created conditions that allowed one or more of the lactating females to range far from the den for a period, returning to her parental duties afterwards. However, the pups would have been weaned by eight weeks of age (Packard et al., 1992), so nonlactating adults could also have cared for them, as often happens in wolf packs (Packard et al., 1992; Mech et al., 1999).

Cooperative rearing of multiple litters by a pack could create opportunities for long-distance foraging movements by some reproductive wolves during summer periods of local food scarcity. We have recorded multiple lactating females at one or more tundra wolf dens per year since 1997. This reproductive strategy may be an adaptation to temporally and spatially unpredictable food resources. All of these possibilities require further study, but emphasize both **38** the adaptability of wolves living on the barrens and their dependence on caribou.

Long-range wolf movement in response to caribou availability has been suggested by other researchers **39** (Kuyt, 1972; Walton et al., 2001) and traditional ecological knowledge (Thorpe et al., 2001). Our report demonstrates the rapid and extreme response of wolves to caribou distribution and movements in summer. Increased human activity on the tundra (mining, road building, pipelines, ecotourism) may influence caribou movement patterns and change the interactions between wolves and caribou in the region. Continued monitoring of both species will help us to assess whether the association is being affected adversely by anthropogenic change.

40 ## Acknowledgements

This research was supported by the Department of Resources, Wildlife, and Economic Development, Government of the Northwest Territories; the Department of Biological Sciences at the University of Alberta; the Natural Sciences and Engineering Research Council of Canada; the Department of Indian and Northern Affairs Canada; the Canadian Circumpolar Institute; and DeBeers Canada, Ltd. Lorna Ruechel assisted with den observations. A. Gunn provided caribou location data. We thank Dave Mech for the use of GPS collars. M. Nelson, A. Gunn, and three anonymous reviewers made helpful comments on earlier drafts of the manuscript. This work was done under Wildlife Research Permit – WL002948 issued by the Government of the Northwest Territories, Department of Resources, Wildlife, and Economic Development.

41 ## References

BALLARD, W.B., AYRES, L.A., KRAUSMAN, P.R., REED, D.J., and FANCY, S.G. 1997. Ecology of wolves in relation to a migratory caribou herd in northwest Alaska. Wildlife Monographs 135. 47 p.

BROWN, M.B., and FORSYTHE, A.B. 1974. Robust tests for the equality of variances. Journal of the American Statistical Association 69:364–367.

CARLSON, A. 1985. Central place foraging in the red-backed shrike (*Lanius collurio* L.): Allocation of prey between forager and sedentary consumer. Animal Behaviour 33:664–666.

CLARK, K.R.F. 1971. Food habits and behavior of the tundra wolf on central Baffin Island. Ph.D. Thesis, University of Toronto, Ontario, Canada.

ENVIRONMENT CANADA. 2003. National climate data information archive. Available online: http://www.climate.weatheroffice.ec.gc.ca/Welcome_e.html

FOURNIER, B., and GUNN, A. 1998. Musk ox numbers and distribution in the NWT, 1997. File Report No. 121. Yellowknife: Department of Resources, Wildlife, and Economic Development, Government of the Northwest Territories. 55 p.

FRITTS, S.H., and MECH, L.D. 1981. Dynamics, movements, and feeding ecology of a newly protected wolf population in northwestern Minnesota. Wildlife Monographs 80. 79 p.

GUNN, A., DRAGON, J., and BOULANGER, J. 2001. Seasonal movements of satellite-collared caribou from the Bathurst herd. Final Report to the West Kitikmeot Slave Study Society, Yellowknife, NWT. 80 p. Available online: http://www.wkss.nt.ca/HTML/08_ProjectsReports/PDF/Seasonal MovementsFinal.pdf

HARRINGTON, F.H., and ASA, C.S. 2003. Wolf communication. In: Mech, L.D., and Boitani, L., eds. Wolves: Behavior, ecology, and conservation. Chicago: University of Chicago Press. 66–103.

HARRINGTON, F.H., MECH, L.D., and FRITTS, S.H. 1983. Pack size and wolf pup survival: Their relationship under varying ecological conditions. Behavioral Ecology and Sociobiology 13:19–26.

HARRIS, S.A. 1986. Permafrost distribution, zonation and stability along the eastern ranges of the cordillera of North America. Arctic 39(1):29–38.

HEARD, D.C., and WILLIAMS, T.M. 1992. Distribution of wolf dens on migratory caribou ranges in the Northwest

Territories, Canada. Canadian Journal of Zoology 70:1504–1510.

HEARD, D.C., WILLIAMS, T.M., and MELTON, D.A. 1996. The relationship between food intake and predation risk in migratory caribou and implication to caribou and wolf population dynamics. Rangifer Special Issue No. 2:37–44.

KELSALL, J.P. 1968. The migratory barren-ground caribou of Canada. Canadian Wildlife Service Monograph Series 3. Ottawa: Queen's Printer. 340 p.

KUYT, E. 1962. Movements of young wolves in the Northwest Territories of Canada. Journal of Mammalogy 43:270–271.

———. 1972. Food habits and ecology of wolves on barren-ground caribou range in the Northwest Territories. Canadian Wildlife Service Report Series 21. Ottawa: Information Canada. 36 p.

MECH, L.D. 1994. Regular and homeward travel speeds of Arctic wolves. Journal of Mammalogy 75:741–742.

MECH, L.D., and BOITANI, L. 2003. Wolf social ecology. In: Mech, L.D., and Boitani, L., eds. Wolves: Behavior, ecology, and conservation. Chicago: University of Chicago Press. 1–34.

MECH, L.D., and GESE, E.M. 1992. Field testing the Wildlink capture collar on wolves. Wildlife Society Bulletin 20:249–256.

MECH, L.D., WOLFE, P., and PACKARD, J.M. 1999. Regurgitative food transfer among wild wolves. Canadian Journal of Zoology 77:1192–1195.

MERRILL, S.B., and MECH, L.D. 2000. Details of extensive movements by Minnesota wolves (Canis lupus). American Midland Naturalist 144:428–433.

MERRILL, S.B., ADAMS, L.G., NELSON, M.E., and MECH, L.D. 1998. Testing releasable GPS radiocollars on wolves and white-tailed deer. Wildlife Society Bulletin 26:830–835.

MESSIER, F. 1985. Solitary living and extraterritorial movements of wolves in relation to social status and prey abundance. Canadian Journal of Zoology 63:239–245.

MOHR, C.O., and STUMPF, W.A. 1966. Comparison of methods for calculating areas of animal activity. Journal of Wildlife Management 30:293–304.

MUSIANI, M. 2003. Conservation biology and management of wolves and wolf-human conflicts in western North America. Ph.D. Thesis, University of Calgary, Calgary, Alberta, Canada.

ORIANS, G.H., and PEARSON, N.E. 1979. On the theory of central place foraging. In: Mitchell, R.D., and Stairs, G.F., eds. Analysis of ecological systems. Columbus: Ohio State University Press. 154–177.

PACKARD, J.M., MECH, L.D., and REAM, R.R. 1992. Weaning in an arctic wolf pack: Behavioral mechanisms. Canadian Journal of Zoology 70:1269–1275.

PAQUET, P.C., WIERZCHOWSKI, J., and CALLAGHAN, C. 1996. Summary report on the effects of human activity on gray wolves in the Bow River Valley, Banff National Park, Alberta. In: Green, J., Pacas, C., Bayley, S., and Cornwell, L., eds. A cumulative effects assessment and futures outlook for the Banff Bow Valley. Prepared for the Banff Bow Valley Study. Ottawa: Department of Canadian Heritage.

STEPHENSON, R.O., and JAMES, D. 1982. Wolf movements and food habits in northwest Alaska. In: Harrington, F.H., and Paquet, P.C., eds. Wolves of the world. New Jersey: Noyes Publications. 223–237.

THORPE, N., EYEGETOK, S., HAKONGAK, N., and QITIRMIUT ELDERS. 2001. The Tuktu and Nogak Project: A caribou chronicle. Final Report to the West Kitikmeot/Slave Study Society, Ikaluktuuttiak, NWT. 160 p.

TIMONEY, K.P., LA ROI, G.H., ZOLTAI, S.C., and ROBINSON, A.L. 1992. The high subarctic forest-tundra of northwestern Canada: Position, width, and vegetation gradients in relation to climate. Arctic 45(1):1–9.

WALTON, L.R., CLUFF, H.D., PAQUET, P.C., and RAMSAY, M.A. 2001. Movement patterns of barren-ground wolves in the central Canadian Arctic. Journal of Mammalogy 82:867–876.

WILLIAMS, T.M. 1990. Summer diet and behavior of wolves denning on barren-ground caribou range in the Northwest Territories, Canada. M.Sc. Thesis, University of Alberta, Edmonton, Alberta, Canada.

ZAR, J.H. 1999. Biostatistical analysis. 4th ed. New Jersey: Prentice Hall. 663 p.

Appendix III. Answers to Self-Quizzes and Genetics Problems

Italicized numbers refer to relevant section numbers

CHAPTER 1

1.	a	1.2
2.	c	1.2
3.	c	1.3
4.	Homeostasis	1.3
5.	d	1.3
6.	reproduction	1.3
7.	Inheritance	1.3
8.	b	1.4
9.	a, c, d, e	1.2, 1.3, 1.4
10.	a, b	1.2, 1.4
11.	theory	1.9
12.	b	1.9
13.	b	1.6
14.	b	1.8
15.	c	1.2
	b	1.5
	d	1.9
	e	1.6
	a	1.6
	f	1.8

CHAPTER 2

1.	False (ions are atoms with different numbers of protons and electrons)	2.3
2.	d	2.2
3.	b	2.2
4.	a	2.4
5.	a	2.4
6.	polar covalent	2.4
7.	c, b, a	2.4
8.	c	2.5
9.	H+ or OH−	2.5
10.	b	2.5
11.	d	2.6
12.	a	2.6
13.	c	2.6
14.	a (hydrogen ion)	2.6
15.	c	2.5
	b	2.2
	d	2.5
	h	2.2
	g	2.5
	f	2.3
	e	2.3
	a	2.3

CHAPTER 3

1.	c	3.2
2.	4	3.2
3.	a	3.3
4.	e	3.4, 3.8
5.	a	3.4
6.	c	3.5
7.	False	3.1, 3.5
8.	b	3.5
9.	e	3.5
10.	d	3.6, 3.8
11.	d	3.7
12.	d	3.8
13.	a	3.8
14.	a (amino acids)	3.6
	b (carbohydrate)	3.4
	c (polypeptide)	3.6
	d (fatty acid)	3.5
15.	c	3.5
	e	3.4
	f	3.5
	d	3.8
	a	3.6
	b	3.8
16.	g	3.6
	a	3.5
	b	3.6
	c	3.5
	d	3.8
	e	3.4
	f	3.8
	i	3.6
	h	3.4
	j	3.5
	k	3.4

CHAPTER 4

1.	c	4.2
2.	c	4.2
3.	b	4.2
4.	False	4.4, 4.7
5.	c	4.4
6.	False (Protists are eukaryotes; by definition, all eukaryotes start life with a nucleus.)	4.5
7.	c	4.2
8.	a	4.7
9.	c, b, d, a	4.7
10.	lipids and proteins	4.7
11.	False (Many cells have walls that surround the plasma membrane.)	
12.	b	4.6, 4.7, 4.9
13.	d	4.10
14.	d	4.11
15.	a	4.11
16.	a	4.10
17.	c	4.10
	d	4.11
	e	4.11
	b	4.11
	a	4.3
18.	g	4.9
	f	4.9
	a	4.4
	e	4.7
	d	4.7
	b	4.7
	c	4.7
	i	4.9
	h	4.10

CHAPTER 5

1.	c	5.2
2.	b	5.2
3.	d	5.2
4.	a	5.3
5.	c	5.3
6.	d	5.4
7.	c	5.3
8.	a	5.4
9.	d	5.5
10.	c	5.5
11.	a	5.6
12.	more/less	5.8
13.	c	5.9
14.	b	5.9
15.	b	5.8
16.	e	5.10
17.	c	5.3
	e	5.10
	f	5.2
	b	5.3
	a	5.6
	g	5.8
	h	5.9
	d	5.9

CHAPTER 6

1.	weed (autotroph; all others heterotrophs)	6.1
2.	a	6.1
3.	b	6.7, 6.1 revisited
4.	a	6.2
5.	a	6.4
6.	d	6.5
7.	b	6.5
8.	b	6.5
9.	c	6.5
10.	c	6.7
11.	b	6.7
12.	e	6.7
13.	f	6.7
14.	f	6.7
	h	6.7
	g	6.5
	d	6.5, 6.6
	e	6.8
	b	6.1, 6.4
	a	6.2
	c	6.1

CHAPTER 7

1.	False	7.2
2.	d	7.2, 7.3
3.	a	7.2, 7.6
4.	c	7.3
5.	b	7.2, 7.4, 7.5
6.	e	7.4
7.	b	7.4
8.	c	7.5
9.	c	7.6
10.	b	7.6
11.	c	7.5
12.	d	7.7
13.	f	7.6
14.	b	7.3
	c	7.6
	a	7.4, 7.6
	d	7.5
15.	b	7.4
	d	7.3
	a	7.3, 7.6
	c	7.2, 7.5
	f	7.5
	e	7.2, 7.4
	g	7.1

CHAPTER 8

1.	b	8.2
2.	centromere	8.2
3.	b	8.2
4.	d	8.2
5.	c	8.4
6.	d	8.4
7.	c	8.5
8.	a	8.5
9.	f	8.5
10.	3'-CCAAAGAAGTTCTCT-5'	8.5
11.	d	8.7
12.	c	8.4
13.	d	8.6
14.	b	8.7
15.	d	8.7
16.	d	8.3
	b	8.1, 8.7
	a	8.4
	f	8.2
	e	8.5
	g	8.5
	c	8.2
	h	8.6

CHAPTER 9

1.	c	9.2
2.	b	9.3
3.	d	9.3
4.	a	9.2
5.	c	9.2
6.	b	9.4
7.	c	9.4
8.	a	9.4
9.	15	9.4
10.	a	9.3
11.	a	9.3
12.	c	9.5
13.	c	9.5
14.	a	9.4, 9.5
15.	e	9.6
16.	c	9.4
	g	9.3
	e	9.5
	a	9.3
	f	9.4
	d	9.3
	b	9.6

CHAPTER 10

1.	d	10.2
2.	d	10.2
3.	d	10.2
4.	b	10.2
5.	b	10.2
6.	g	10.2
7.	c	10.2
8.	d	10.3
9.	d	10.2
10.	c	10.3
11.	b	10.3
12.	b	10.4
13.	c	10.4
14.	b	10.5
15.	True (epigenetics)	10.6
16.	f	10.2
	a	10.5
	b	10.5
	e	10.4
	c	10.2
	d	10.6

CHAPTER 11

1.	d	11.2
2.	b	11.2
3.	d	11.2
4.	e	11.2
5.	f	11.4
6.	c	11.3
7.	d	11.2
8.	a	11.2
9.	c	11.5
10.	b	11.2, 11.3
11.	a	11.2
12.	d	11.6
13.	b	11.6
14.	c	11.4
	f	11.3
	a	11.6
	g	11.4
	b	11.4
	e	11.6
	h	11.5
	d	11.3
15.	d, b, c, a	11.3

CHAPTER 12

1.	b	12.1, 12.2
2.	c	12.1, 12.2
3.	d	12.2, 12.5
4.	b	12.2, 12.3
5.	b	12.4
6.	prophase I	12.3, 12.4
7.	c	12.3
8.	a	12.2
9.	Sister chromatids are still attached (chromosomes are duplicated)	12.3
10.	e	12.4
11.	b	12.3, 12.4, 12.6
12.	b	12.2, 12.3
	c	12.3
	a	12.2
	e	12.5
	d	12.2

CHAPTER 13

1.	b	13.2
2.	a	13.2
3.	b	13.3
4.	c	13.3
5.	b	13.3
6.	a	13.3
7.	b	13.3
8.	d	13.4
9.	c	13.4
10.	c	13.5
11.	b	13.5
12.	d	13.7
13.	False	13.7
14.	b	13.4
	d	13.3
	a	13.2
	c	13.2

CHAPTER 14

1.	b	14.2
2.	b	14.2
3.	a	14.2
4.	b	14.3
5.	False	14.4
6.	d	14.3, 14.4
7.	d	14.4
8.	Y-linked dominant	14.4, 14.5
9.	X mom, Y dad	14.4
10.	d	14.6
11.	b	14.6
12.	d	14.6
13.	True	14.6
14.	c	14.6
15.	c	14.6
	e	14.5
	f	14.6
	b	14.5
	a	14.2
	d	14.6

CHAPTER 15

1.	c	15.2
2.	a	15.2
3.	b	15.2
4.	b	15.3
5.	c	15.3
6.	DNA sequencing	15.4
7.	b	15.4
8.	d	15.4
9.	d	15.5
10.	c	15.6
11.	b	15.8, 15.10
	d	15.2, 15.7
	e	15.10
	f	15.7
12.	c	15.6
13.	b	15.10
14.	a	15.3
	d	15.2
	c	15.2
	e	15.4
	b	15.2
15.	c	15.5
	g	15.7
	d	15.3
	e	15.10
	b	15.1
	a	15.6
	f	15.6

1. a. AB

   b. AB, aB

   c. Ab, ab

   d. AB, Ab, aB, ab

2. a. All offspring will be $AaBB$.

   b. 1/4 $AABB$ (25% each genotype)
      1/4 $AABb$
      1/4 $AaBB$
      1/4 $AaBb$

   c. 1/4 $AaBb$ (25% each genotype)
      1/4 $Aabb$
      1/4 $aaBb$
      1/4 $aabb$

   d. 1/16 $AABB$ (6.25% of genotype)
      1/8 $AaBB$ (12.5%)
      1/16 $aaBB$ (6.25%)
      1/8 $AABb$ (12.5%)
      1/4 $AaBb$ (25%)
      1/8 $aaBb$ (12.5%)
      1/16 $AAbb$ (6.25%)
      1/8 $Aabb$ (12.5%)
      1/16 $aabb$ (6.25%)

3. a. ABC

   b. ABC, aBC

   c. ABC, aBC, ABc, aBc

   d. ABC
      aBC
      AbC
      abC
      ABc
      aBc
      Abc
      abc

4. A mating of two M^L cats yields 1/4 MM, 1/2 M^LM, and 1/4 M^LM^L. Because M^LM^L is lethal, the probability that any one kitten among the survivors will be heterozygous is 2/3.

5. a. Both parents are heterozygotes (Aa). Their children may be albino (aa) or unaffected (AA or Aa).

   b. All are homozygous recessive (aa).

   c. Homozygous recessive (aa) father, and heterozygous (Aa) mother. The albino child is aa, the unaffected children Aa.

6. Possible outcomes of an experimental cross between F_1 rose plants heterozygous for height (Aa):

3:1 possible ratio of genotypes and phenotypes in F_2 generation

Possible outcomes of a testcross between an F_1 rose plant heterozygous for height and a shrubby rose plant:

Gametes F_1 hybrid:

1:1 possible ratio of genotypes and phenotypes in F_2 generation

7. Yellow is recessive. Because F_1 plants have a green phenotype and must be heterozygous, green must be dominant over the recessive yellow.

8. A mating between a mouse from a true-breeding, white-furred strain and a mouse from a true-breeding, brown-furred strain would provide you with the most direct evidence. Because true-breeding strains of organisms typically are homozygous for a trait being studied, all F_1 offspring from this mating should be heterozygous. Record the phenotype of each F_1 mouse, then let them mate with one another. Assuming only one gene locus is involved, these are possible outcomes for the F_1 offspring:

   a. All F_1 mice are brown, and their F_2 offspring segregate: 3 brown : 1 white. *Conclusion*: Brown is dominant to white.

   b. All F_1 mice are white, and their F_2 offspring segregate: 3 white : 1 brown. *Conclusion*: White is dominant to brown.

   c. All F_1 mice are tan, and the F_2 offspring segregate: 1 brown : 2 tan : 1 white. *Conclusion*: The alleles at this locus show incomplete dominance.

9. The data reveal that these genes do not assort independently because the observed ratio is very far from the 9:3:3:1 ratio expected with independent assortment. Instead, the results can be explained if the genes are located close to each other on the same chromosome, which is called linkage.

10. a. 1/2 red   1/2 pink   white
    b.   red   All pink   white
    c. 1/4 red   1/2 pink   1/4 white
    d.   red   1/2 pink   1/2 white

11. Because both parents are heterozygotes (Hb^AHb^S), the following are the probabilities for each child:

   a. 1/4 Hb^SHb^S

   b. 1/4 Hb^AHb^A

   c. 1/2 Hb^AHb^S

1. The phenotype appeared in every generation shown in the diagram, so this must be a pattern of autosomal dominant inheritance.

2. a. Human males (XY) inherit their X chromosome from their mother.

b. A male can produce two kinds of gametes. Half carry an X chromosome and half carry a Y chromosome. All the gametes that carry the X chromosome carry the same X-linked allele.

c. A female homozygous for an X-linked allele produces only one kind of gamete.

d. Fifty percent of the gametes of a female who is heterozygous for an X-linked allele carry one of the two alleles at that locus; the other fifty percent carry its partner allele for that locus.

3. Because Marfan syndrome is a case of autosomal dominant inheritance and because one parent bears the allele, the probability that any child of theirs will inherit the mutant allele is 50 percent.

4. a. Nondisjunction might occur during anaphase I or anaphase II of meiosis.

b. As a result of translocation, chromosome 21 may get attached to the end of chromosome 14. The new individual's chromosome number would still be 46, but its somatic cells would have the translocated chromosome 21 in addition to two normal chromosomes 21.

5. A daughter could develop this muscular dystrophy only if she inherited two X-linked recessive alleles—one from each parent. Males who carry the allele are unlikely to father children because they develop the disorder and die early in life.

6. In the mother, a crossover between the two genes at meiosis generates an X chromosome that carries neither mutant allele.

Appendix IV. Periodic Table of the Elements

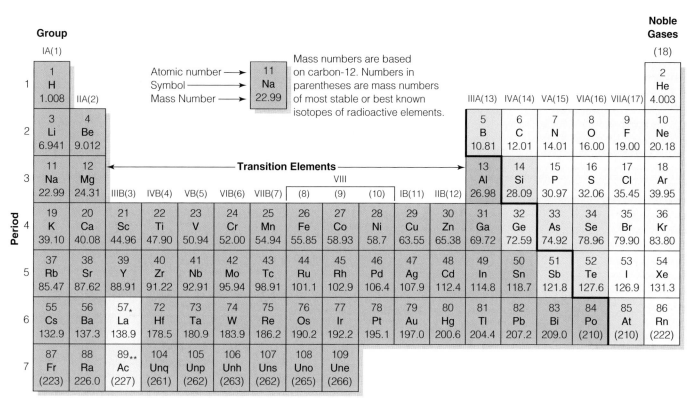

Group

Noble Gases

Period

IA(1)																	(18)
1 **H** 1.008	IIA(2)											IIIA(13)	IVA(14)	VA(15)	VIA(16)	VIIA(17)	2 **He** 4.003

Atomic number → 11
Symbol → Na
Mass Number → 22.99

Mass numbers are based on carbon-12. Numbers in parentheses are mass numbers of most stable or best known isotopes of radioactive elements.

Transition Elements

VIII

3 Li 6.941	4 Be 9.012											5 B 10.81	6 C 12.01	7 N 14.01	8 O 16.00	9 F 19.00	10 Ne 20.18
11 Na 22.99	12 Mg 24.31	IIIB(3)	IVB(4)	VB(5)	VIB(6)	VIIB(7)	(8)	(9)	(10)	IB(11)	IIB(12)	13 Al 26.98	14 Si 28.09	15 P 30.97	16 S 32.06	17 Cl 35.45	18 Ar 39.95
19 K 39.10	20 Ca 40.08	21 Sc 44.96	22 Ti 47.90	23 V 50.94	24 Cr 52.00	25 Mn 54.94	26 Fe 55.85	27 Co 58.93	28 Ni 58.7	29 Cu 63.55	30 Zn 65.38	31 Ga 69.72	32 Ge 72.59	33 As 74.92	34 Se 78.96	35 Br 79.90	36 Kr 83.80
37 Rb 85.47	38 Sr 87.62	39 Y 88.91	40 Zr 91.22	41 Nb 92.91	42 Mo 95.94	43 Tc 98.91	44 Ru 101.1	45 Rh 102.9	46 Pd 106.4	47 Ag 107.9	48 Cd 112.4	49 In 114.8	50 Sn 118.7	51 Sb 121.8	52 Te 127.6	53 I 126.9	54 Xe 131.3
55 Cs 132.9	56 Ba 137.3	57* La 138.9	72 Hf 178.5	73 Ta 180.9	74 W 183.9	75 Re 186.2	76 Os 190.2	77 Ir 192.2	78 Pt 195.1	79 Au 197.0	80 Hg 200.6	81 Tl 204.4	82 Pb 207.2	83 Bi 209.0	84 Po (210)	85 At (210)	86 Rn (222)
87 Fr (223)	88 Ra 226.0	89** Ac (227)	104 Unq (261)	105 Unp (262)	106 Unh (263)	107 Uns (262)	108 Uno (265)	109 Une (266)									

Inner Transition Elements

Lanthanide Series 6 *

58 Ce 140.1	59 Pr 140.9	60 Nd 144.2	61 Pm (145)	62 Sm 150.4	63 Eu 152.0	64 Gd 157.3	65 Tb 158.9	66 Dy 162.5	67 Ho 164.9	68 Er 167.3	69 Tm 168.9	70 Yb 173.0	71 Lu 175.0

Actinide Series 7 **

90 Th 232.0	91 Pa 231.0	92 U 238.0	93 Np 237.0	94 Pu (244)	95 Am (243)	96 Cm (247)	97 Bk (247)	98 Cf (251)	99 Es (252)	100 Fm (257)	101 Md (258)	102 No (259)	103 Lr (260)

Glycolysis

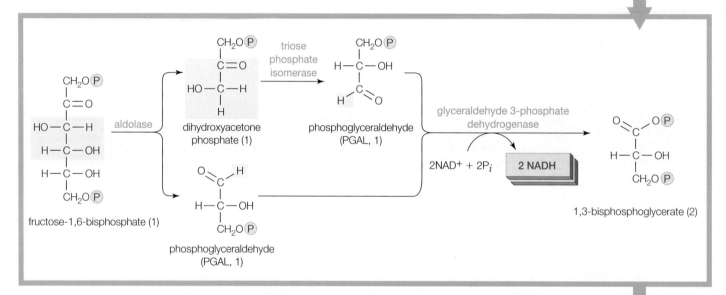

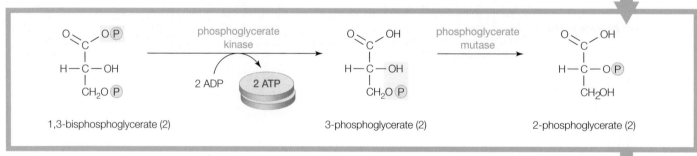

Figure A Glycolysis breaks down one glucose molecule into two 3-carbon pyruvate molecules for a net yield of two ATP. Enzyme names are indicated in *green*; parts of substrate molecules undergoing chemical change are highlighted *blue*.

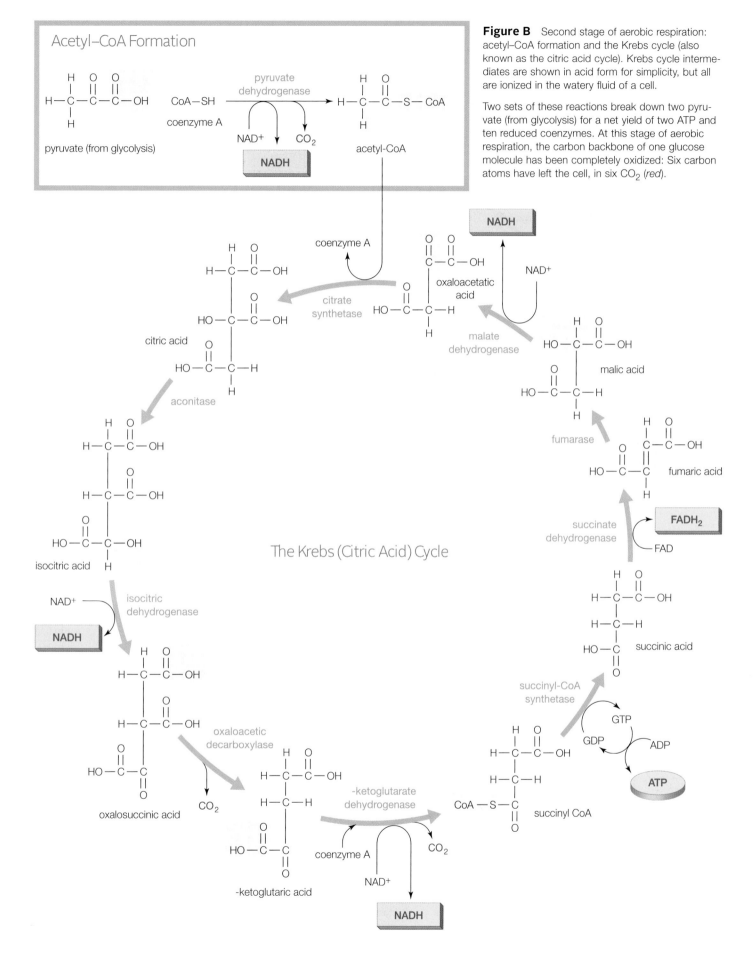

Acetyl–CoA Formation

pyruvate (from glycolysis)

coenzyme A

pyruvate dehydrogenase

NAD^+ CO_2

NADH

acetyl-CoA

Figure B Second stage of aerobic respiration: acetyl–CoA formation and the Krebs cycle (also known as the citric acid cycle). Krebs cycle intermediates are shown in acid form for simplicity, but all are ionized in the watery fluid of a cell.

Two sets of these reactions break down two pyruvate (from glycolysis) for a net yield of two ATP and ten reduced coenzymes. At this stage of aerobic respiration, the carbon backbone of one glucose molecule has been completely oxidized: Six carbon atoms have left the cell, in six CO_2 (*red*).

NADH

coenzyme A

citrate synthetase

citric acid

aconitase

isocitric acid

NAD^+ — isocitric dehydrogenase

NADH

oxaloacetic decarboxylase

oxalosuccinic acid

CO_2

-ketoglutaric acid

coenzyme A

NAD^+

NADH

-ketoglutarate dehydrogenase

CO_2

The Krebs (Citric Acid) Cycle

oxaloacetic acid

NAD^+

malate dehydrogenase

malic acid

fumarase

fumaric acid

succinate dehydrogenase

FADH$_2$

FAD

succinic acid

succinyl-CoA synthetase

GTP

GDP   ADP

ATP

succinyl CoA

CoA — S — C

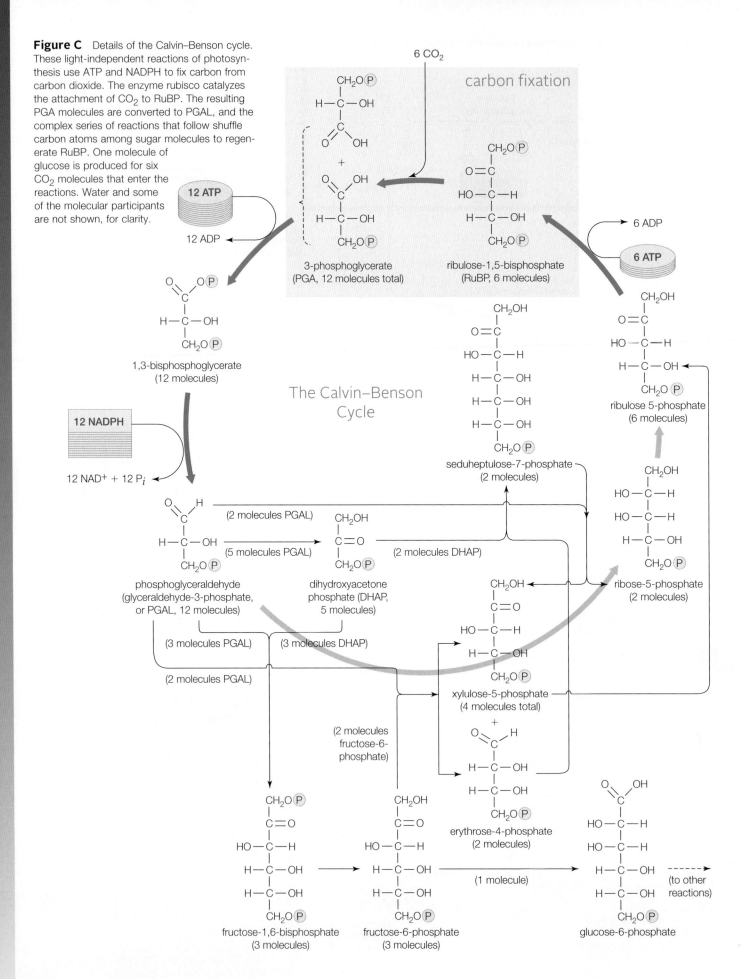

Figure C Details of the Calvin–Benson cycle. These light-independent reactions of photosynthesis use ATP and NADPH to fix carbon from carbon dioxide. The enzyme rubisco catalyzes the attachment of CO_2 to RuBP. The resulting PGA molecules are converted to PGAL, and the complex series of reactions that follow shuffle carbon atoms among sugar molecules to regenerate RuBP. One molecule of glucose is produced for six CO_2 molecules that enter the reactions. Water and some of the molecular participants are not shown, for clarity.

Appendix VI. A Plain English Map of the Human Chromosomes

Haploid set of human chromosomes. The banding patterns characteristic of each type of chromosome appear after staining with a reagent called Giemsa. The locations of some of the 20,065 known genes (as of November, 2005) are indicated. Also shown are locations that, when mutated, cause some of the genetic diseases discussed in the text.

Length
1 kilometer (km) = 0.62 miles (mi)
1 meter (m) = 39.37 inches (in)
1 centimeter (cm) = 0.39 inches

To convert	multiply by	to obtain
inches	2.25	centimeters
feet	30.48	centimeters
centimeters	0.39	inches
millimeters	0.039	inches

Area
1 square kilometer = 0.386 square miles
1 square meter = 1.196 square yards
1 square centimeter = 0.155 square inches

Volume
1 cubic meter = 35.31 cubic feet
1 liter = 1.06 quarts
1 milliliter = 0.034 fluid ounces = 1/5 teaspoon

To convert	multiply by	to obtain
quarts	0.95	liters
fluid ounces	28.41	milliliters
liters	1.06	quarts
milliliters	0.03	fluid ounces

Weight
1 metric ton (mt) = 2,205 pounds (lb) = 1.1 tons (t)
1 kilogram (kg) = 2.205 pounds (lb)
1 gram (g) = 0.035 ounces (oz)

To convert	multiply by	to obtain
pounds	0.454	kilograms
pounds	454	grams
ounces	28.35	grams
kilograms	2.205	pounds
grams	0.035	ounces

Temperature
Celcius (°C) to Fahrenheit (°F):
$$°F = 1.8 (°C) + 32$$

Fahrenheit (°F) to Celsius:
$$°C = \frac{(°F - 32)}{1.8}$$

	°C	°F
Water boils	100	212
Human body temperature	37	98.6
Water freezes	0	32

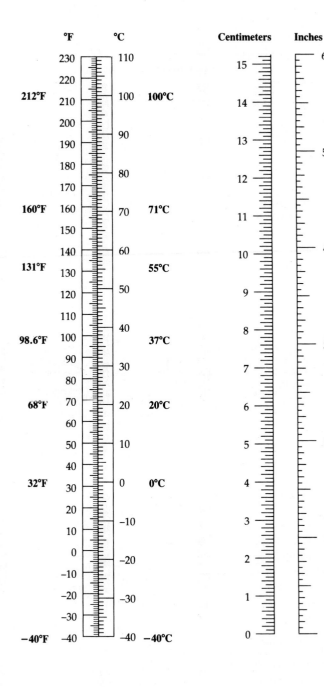

Glossary of Biological Terms

acid Substance that releases hydrogen ions in water. 32

activation energy Minimum amount of energy required to start a reaction. 80

activator Regulatory protein that increases the rate of transcription when it binds to a promoter or enhancer. 164

active site Of an enzyme, pocket in which substrates bind and a reaction occurs. 82

active transport Energy-requiring mechanism in which a transport protein pumps a solute across a cell membrane against its concentration gradient. 92

adenine *See* nucleotide.

adhering junction Cell junction composed of adhesion proteins; anchors cells to each other and extracellular matrix. 71

adhesion protein Membrane protein that helps cells stick together in animal tissues. 89

aerobic Involving or occurring in the presence of oxygen. 117

aerobic respiration Oxygen-requiring metabolic pathway that breaks down carbohydrates to produce ATP. Includes glycolysis, acetyl–CoA formation, the Krebs cycle, and electron transfer phosphorylation. 118

alcoholic fermentation Anaerobic carbohydrate breakdown pathway that produces ATP and ethanol. 126

alleles Forms of a gene with slightly different DNA sequences; may encode slightly different versions of the gene's product. Basis of variation in shared traits among sexual reproducers. 190

allosteric Describes a region of an enzyme that can bind a regulatory molecule and is not the active site. 85

alternative splicing RNA processing event in which some exons are removed or joined in alternate combinations. 153

amino acid Small organic compound consists of a carboxyl group, an amine group, and a characteristic side group (R), all typically bonded to the same carbon atom. Twenty kinds are common subunits of proteins. 46

anaerobic Occurring in the absence of oxygen; oxygen-free. 117

anaphase Stage of mitosis during which sister chromatids separate and move to opposite spindle poles. 180

aneuploidy A chromosome abnormality in which an individual's cells carry too many or too few copies of a particular chromosome. 228

animal Multicelled consumer that develops through a series of stages and moves about during part or all of its life cycle. 8

anticodon Set of three nucleotides in a tRNA; base-pairs with mRNA codon. 155

antioxidant Substance that prevents oxidation of other molecules. 86

archaea *See* archaeon.

archaeon Member of archaea, the most recently discovered and less well-known lineage of single-celled organisms without a nucleus. More closely related to eukaryotes than to bacteria. 8

asexual reproduction Reproductive mode by which offspring arise from a single parent only. 178

atom Fundamental particle that is a building block of all matter. Consists of varying numbers of protons, neutrons, and electrons. 4, 24

atomic number Number of protons in the atomic nucleus; determines the element. 24

ATP Adenosine triphosphate. Nucleotide that has an adenine base, a ribose sugar, and three phosphate groups. Subunit of RNA; also has an important role as an energy-carrying coenzyme. 49, 87

ATP/ADP cycle Process by which cells regenerate ATP. ADP forms when ATP loses a phosphate group, then ATP forms again as ADP gains a phosphate group. 87

autosome Any chromosome other than a sex chromosome. 135

autotroph Organism that makes its own food using carbon from inorganic molecules such as CO_2, and energy from the environment. 101

bacteria *See* bacterium.

bacteriophage Type of virus that infects bacteria. 137

bacterium Member of bacteria, the most diverse and well-known lineage of single-celled organisms with no nucleus. 8

basal body Organelle at the base of a cilium or flagellum; derived from a centriole. 69

base Substance that accepts hydrogen ions in water. 32

base-pair substitution Mutation in which a single base pair changes. 158

bell curve Bell-shaped curve; typically results from graphing frequency versus distribution for a trait that varies continuously. 214

biodiversity Scope of variation among living organisms; includes genetic variation within species, variety of species, and variety of ecosystems. 8

biofilm Community of microorganisms living in a shared mass of secretions. 59

biology The scientific study of life. 3

biosphere All regions of Earth's waters, crust, and air where organisms live. 5

bond *See* chemical bond, covalent bond, hydrogen bond, ionic bond.

buffer system Set of chemicals that can keep the pH of a solution stable by alternately donating and accepting ions that contribute to pH. 32

C3 plant Type of plant that uses only the Calvin–Benson cycle to fix carbon. 110

C4 plant Type of plant that minimizes photorespiration by fixing carbon twice, in two cell types. 111

calcium pump Active transport protein that pumps calcium ions across a cell membrane. 93

Calvin–Benson cycle Second stage of photosynthesis: light-independent reactions that form sugars by fixing carbon from CO_2. Runs in the stroma of chloroplasts on the chemical energy of ATP and the energy of electrons donated by NADPH. 109

CAM plant Type of C4 plant that conserves water by fixing carbon twice, at different times of day. 111

cancer Disease that occurs when the uncontrolled growth of body cells that can invade other tissues physically and metabolically disrupts normal body functioning. 163

carbohydrate Molecule that consists primarily of carbon, hydrogen, and oxygen atoms in a 1:2:1 ratio. 42

carbon fixation Process by which carbon from an inorganic source such as carbon dioxide gets incorporated into an organic molecule. 109

catalysis The acceleration of a reaction by a molecule (such as an enzyme) that is unchanged by participating in the reaction. 82

cDNA DNA synthesized from an RNA template by the enzyme reverse transcriptase. 237

cell Smallest unit of life; at minimum, consists of plasma membrane, cytoplasm, and DNA. 4

cell cortex Mesh of cytoskeletal elements that reinforces a plasma membrane. 68

cell cycle A series of events from the time a cell forms until its cytoplasm divides; comprises G1, S, G2, and mitosis. **178**

cell junction Structure that connects a cell to another cell or to extracellular matrix. **71**

cell plate After nuclear division in a plant cell, a disk-shaped structure that forms a cross-wall between the two new nuclei. **182**

cell theory Theory that all organisms consist of one or more cells, which are the basic unit of life; all cells come from division of preexisting cells; and all cells pass hereditary material to offspring. **55**

cellulose Polysaccharide that is a major structural material in plants. **42**

cell wall Semirigid but permeable structure that surrounds the plasma membrane of some cells. **58**

central vacuole Large, fluid-filled vesicle in many plant cells. **65**

centriole Barrel-shaped organelle from which microtubules grow. **69**

centromere Constricted region in a eukaryotic chromosome where sister chromatids are attached. **134**

charge Electrical property of matter. Particles with opposite charges attract, and those with like charges repel. **24**

chemical bond An attractive force that arises between two atoms when their electrons interact; can be ionic or covalent. **28**

chemoautotroph Organism that makes its own food using carbon from inorganic sources such as carbon dioxide, and energy from chemical reactions. **112**

chlorophyll a Most common photosynthetic pigment in plants, photosynthetic protists, and bacteria. **102**

chloroplast Organelle of photosynthesis in the cells of plants and many protists. Its two outer membranes enclose the thylakoid membrane and stroma. **66, 105**

chromatin Collective term for a cell's DNA and its associated proteins. **63**

chromosome A structure that consists of a molecule of double-stranded DNA and associated proteins; carries part or all of a cell's genetic information. **63, 134**

chromosome number The sum of all chromosomes in a cell of a given type. **134**

cilium Short, movable structure that projects from the plasma membrane of some eukaryotic cells. **68**

cleavage furrow Indentation where a contractile ring is pinching an animal cell in two during cytoplasmic division. **182**

clone Genetically identical copy of an organism. **133**

cloning vector A DNA molecule that can accept foreign DNA and get replicated inside a host cell. **237**

codominant Refers to two alleles that are both fully expressed in heterozygotes and neither is dominant over the other. **210**

codon In mRNA, a sequence of three nucleotides that codes for an amino acid or stop signal during translation. **154**

coenzyme An organic cofactor. **86**

cofactor A metal ion or an organic molecule that associates with an enzyme and is necessary for its function. **86**

cohesion Property of a substance that arises from the tendency of its molecules to resist separating from one another. **30**

community All species in a particular region. **5**

compound Molecule that consist of atoms of more than one element. **28**

concentration The number of molecules or ions per unit volume. **32, 90**

concentration gradient Difference in concentration between adjoining regions of fluid. **90**

condensation Enzymatic reaction in which two molecules become bonded together; water also forms. **40**

consumer Organism that gets energy and nutrients by feeding on tissues, wastes, or remains of other organisms; a heterotroph. **6**

continuous variation Range of small differences in a shared trait. **214**

control group In an experiment, a group of individuals who are not exposed to the independent variable being tested. **13**

covalent bond Chemical bond in which two atoms share a pair of electrons. **29**

critical thinking Judging the quality of information before allowing it to guide one's beliefs and actions. **12**

crossing over Process in which homologous chromosomes exchange corresponding segments during prophase I of meiosis. Gives rise to new combinations of parental alleles among offspring of sexual reproducers. **194**

cuticle Secreted covering at a body surface. **70**

cytokinesis Cytoplasmic division. **182**

cytoplasm Semifluid substance enclosed by a cell's plasma membrane. **54**

cytosine *See* nucleotide.

cytoskeleton Dynamic framework of protein filaments that support, organize, and move eukaryotic cells and their internal structures. **68**

data Experimental results. **13**

deductive reasoning Logical process of using a general premise to draw a conclusion about a specific case. **12**

deletion Mutation in which one or more base pairs are lost. **158**

denature To unravel the shape of a biological molecule. **48**

deoxyribonucleic acid *See* DNA.

dependent variable In an experiment, a variable that is presumably affected by the independent variable under investigation. **13**

development Multistep process by which the first cell of a new individual becomes a multicelled adult. **7**

differentiation Process by which cells of a multicelled organism become specialized. **164**

diffusion Spontaneous spreading of molecules or ions in a liquid or gas. **90**

dihybrid cross Cross between two individuals identically heterozygous for two genes; for example *AaBb* × *AaBb*. **208**

diploid Having two of each type of chromosome characteristic of the species (2*n*). **134**

disaccharide Polymer of two sugar subunits. **42**

DNA Deoxyribonucleic acid. Nucleic acid that consists of two chains of nucleotides (adenine, guanine, thymine, and cytosine) twisted into a double helix. Carries hereditary information. **7, 49**

DNA cloning Set of procedures that uses living cells to make many identical copies of a DNA fragment. **236**

DNA library Collection of cells that host different fragments of foreign DNA, often representing an organism's entire genome. **238**

DNA ligase Enzyme that seals gaps or breaks in double-stranded DNA. **141**

DNA polymerase DNA replication enzyme that uses a DNA template to assemble a complementary strand of DNA from free nucleotides. **140**

DNA profiling Identifying an individual by analyzing the unique parts of his or her DNA. **242**

DNA replication Process by which a cell duplicates its DNA before it divides. **140**

DNA sequence Order of nucleotide bases in a strand of DNA. **139**

DNA sequencing Method of determining the order of nucleotides in a fragment of DNA. **240**

dominant Refers to an allele that masks the effect of a recessive allele paired with it. 205

dosage compensation Theory that the inactivation of one of the two X chromosomes in the cells of females equalizes X chromosome gene expression between the sexes. 168

duplication Chromosomal structure change in which a section is repeated. 226

ecosystem A community interacting with its environment through a one-way flow of energy and cycling of materials. 5

egg Mature female gamete, or ovum. 196

electron Negatively charged subatomic particle that occupies orbitals around an atomic nucleus. 24

electron transfer chain Array of enzymes and other molecules that accept and give up electrons in sequence, thus releasing the energy of the electrons in usable increments. 85

electron transfer phosphorylation Process in which electron flow through electron transfer chains sets up a hydrogen ion gradient that drives ATP formation. Also called chemiosmosis. 106

electronegativity Measure of the ability of an atom to pull electrons away from other atoms. 28

electrophoresis Technique that separates DNA fragments by size. 240

element A pure substance that consists only of atoms with the same number of protons. 24

emergent property A characteristic of a system that does not appear in any of the system's component parts. 4

endergonic Describes a reaction that requires a net input of free energy to proceed. 80

endocytosis Process by which a cell takes in a small amount of extracellular fluid by the ballooning inward of its plasma membrane. 94

endomembrane system Series of interacting organelles (endoplasmic reticulum, Golgi bodies, vesicles) between nucleus and plasma membrane. 64

endoplasmic reticulum (ER) Organelle that comprises a continuous system of sacs and tubes extending from the nuclear envelope. Smooth ER makes lipids and breaks down carbohydrates and fatty acids; rough ER modifies polypeptides made by ribosomes on its surface. 64

energy The capacity to do work. 6, 78

enhancer Binding site in DNA for proteins that enhance the rate of transcription. 164

entropy Measure of how much the energy of a system has become dispersed. 78

enzyme Protein or RNA that speeds up a chemical reaction without being changed by it. 40

epigenetic Refers to heritable changes in gene expression that are not the result of changes in DNA sequence. 172

epistasis Effect in which a trait is influenced by multiple genes. Also called polygenic inheritance. 211

eugenics Idea of deliberately improving the genetic qualities of the human race. 248

eukaryote Organism that consists of one or more cells that characteristically have a nucleus. 8

eukaryotic flagella *See* flagellum.

evaporation Transition of a liquid to a gas. 30

exergonic Describes a reaction that ends with a net release of free energy. 80

exocytosis Process by which a cell expels a vesicle's contents to extracellular fluid. 94

exon Nucleotide sequence that remains in an RNA after post-transcriptional splicing. 153

experiment A test designed to support or falsify a prediction. 13

experimental group In an experiment, a group of individuals who are exposed to an independent variable. 13

extracellular matrix (ECM) Complex mixture of cell secretions that supports cells and tissues; also has roles in cell signaling. 70

fat Lipid that consists of a glycerol molecule with one, two, or three fatty acid tails. 44

fatty acid Organic compound that consists of a chain of carbon atoms with an acidic carboxyl group at one end. Carbon chain of saturated types has single bonds only; that of unsaturated types has one or more double bonds. 44

feedback inhibition Mechanism in which a change that results from some activity decreases or stops the activity. 84

fermentation Metabolic pathway that breaks down carbohydrates to produce ATP; does not require oxygen. Starts with glycolysis. 118

fertilization Fusion of two gametes to form a zygote. 191

first law of thermodynamics Energy cannot be created or destroyed. 78

flagellum Long, slender cellular structure used for motility. 59

fluid mosaic Model of a cell membrane as a two-dimensional fluid of mixed composition (of proteins and lipids). 88

frameshift Any type of mutation that causes the reading frame of mRNA codons to shift. 159

free radical Atom with an unpaired electron. Most are highly reactive and can damage the molecules of life. 27

functional group A group of atoms bonded to a carbon of an organic compound; imparts a specific chemical property to the molecule. 40

fungus Eukaryotic single-celled or multicelled heterotroph with cell walls of chitin; obtains nutrients by extracellular digestion and absorption. 8

gamete Mature, haploid reproductive cell; e.g., an egg or a sperm. 190

gametophyte A haploid, multicelled, gamete-producing body that forms in the life cycle of land plants and some algae. 196

gap junction Cell junction that forms a channel across the plasma membranes of adjoining animal cells. 71

gene DNA sequence that encodes an RNA or protein product. 150

gene expression Process by which the information in a gene becomes converted to an RNA or protein product. 150

gene therapy Treating a genetic defect or disorder by transferring a normal or modified gene into the affected individual. 248

genetic code Complete set of sixty-four mRNA codons. 154

genetic engineering Process by which deliberate changes are introduced into an individual's genome. 244

genetically modified organism (GMO) Organism whose genome has been modified by genetic engineering. 244

genome An organism's complete set of genetic material. 238

genomics The study of genomes. 242

genotype The particular set of alleles carried by an individual. 205

genus A group of species that share a unique set of traits; also the first part of a species name. 10

germ cell Immature reproductive cell that gives rise to haploid gametes when it divides. 191

glycogen Polysaccharide that serves as an energy reservoir in animal cells. 43

glycolysis Set of reactions in which glucose or another sugar is broken down to two pyruvate for a net yield of two ATP. First step of aerobic respiration, fermentation; occurs in cytoplasm. 118

Golgi body Organelle that modifies polypeptides and lipids; also sorts and packages the finished products into vesicles. 65

growth In multicelled species, an increase in the number, size, and volume of cells. In single-celled species, an increase in the number of cells. 7

growth factor Molecule that stimulates mitosis and differentiation. 184

guanine *See* nucleotide.

haploid Having one of each type of chromosome characteristic of the species. 191

heterotroph Organism that obtains energy and carbon from organic compounds assembled by other organisms. 101

heterozygous Having two different alleles of a gene. 205

histone Type of protein that associates with eukaryotic DNA and structurally organizes chromosomes. 134

homeostasis Set of processes by which an organism keeps its internal conditions within tolerable ranges. 7

homeotic gene Type of master gene; its expression controls the formation of specific body parts during development. 166

homologous chromosomes Chromosomes with the same length, shape, and set of genes. 179

homozygous Having identical alleles of a gene. 205

hybrid The offspring of a cross between two individuals that breed true for different forms of a trait; a heterozygous individual. 205

hydrocarbon Compound that consists only of carbon and hydrogen atoms. 40

hydrogen bond Attraction between a covalently bonded hydrogen atom and an electronegative atom taking part in a separate covalent bond. 30

hydrolysis Enzymatic reaction with water that causes a molecule to break into smaller subunits. 40

hydrophilic Describes a substance that dissolves easily in water. 31

hydrophobic Describes a substance that resists dissolving in water. 31

hypertonic Describes a fluid that has a high overall solute concentration relative to another fluid. 90

hypothesis Testable explanation of a natural phenomenon. 12

hypotonic Describes a fluid that has a low overall solute concentration relative to another fluid. 90

incomplete dominance Effect in which one allele is not fully dominant over another, so the heterozygous phenotype is between the two homozygous phenotypes. 210

independent variable Variable that is controlled by an experimenter in order to explore its relationship to a dependent variable. 13

induced-fit model The concept that substrate binding to an active site of an enzyme improves the fit between the two molecules. 82

inductive reasoning Arriving at a conclusion based on one's observations. 12

inheritance Transmission of DNA to offspring. 7

insertion Mutation in which one or more base pairs become inserted into DNA. 158

intermediate filament Stable cytoskeletal element that structurally supports cells and tissues. 68

interphase In a eukaryotic cell cycle, the interval between mitotic divisions when a cell enlarges, roughly doubles the number of its cytoplasmic components, and replicates its DNA. 178

intron Nucleotide sequence that is excised from an RNA during post-transcriptional splicing. Intervenes between exons. 153

inversion Structural rearrangement of a chromosome in which part of it becomes oriented in the reverse direction. 226

ion Atom that carries a charge because it has an unequal number of protons and electrons. 27

ionic bond Type of chemical bond that consists of a strong mutual attraction between oppositely charged ions. 28

isotonic Describes two fluids with identical solute concentrations. 90

isotopes Forms of an element that differ in the number of neutrons. 25

karyotype Image of an individual's complement of chromosomes arranged by size, length, shape, and centromere location. 135

kinetic energy The energy of motion. 78

knockout An experiment in which a gene is deliberately inactivated in a living organism. 167

Krebs cycle Cyclic pathway that, along with acetyl–CoA formation, breaks down pyruvate to carbon dioxide during aerobic respiration. In eukaryotes, occurs in mitochondrial matrix. 122

lactate fermentation Anaerobic carbohydrate breakdown pathway that produces ATP and lactate. 126

law of independent assortment During meiosis, members of a pair of genes on homologous chromosomes tend to be distributed into gametes independently of other gene pairs. 208

law of nature Generalization that describes a consistent natural phenomenon for which there is incomplete scientific explanation. 18

law of segregation The members of each pair of genes on homologous chromosomes end up in different gametes during meiosis. 207

light-dependent reactions First stage of photosynthesis: A series of reactions that collectively convert light energy to chemical energy of ATP and NADPH. A noncyclic pathway includes electron transfer phosphorylation and produces oxygen; a cyclic pathway does not. 105

light-independent reactions Second stage of photosynthesis: A series of reactions that collectively use ATP and NADPH to assemble sugars from water and CO_2. 105

lignin Organic compound that stiffens cell walls of vascular plants. 70

linkage group All genes on a chromosome. 209

lipid Fatty, oily, or waxy organic compound; e.g., a fat, steroid, or wax. 44

lipid bilayer Double layer of lipids arranged tail-to-tail; forms the structural foundation of all cell membranes. In eukaryotes, consists mainly of phospholipids. 45

locus Location of a gene on a chromosome. 205

lysosome Enzyme-filled vesicle that functions in intracellular digestion of waste, debris, and ingested particles. 65

mass number Total number of protons and neutrons in the nucleus of an element's atoms. 25

master gene Gene encoding a product that affects the expression of many other genes. 166

meiosis Nuclear division process that halves the chromosome number. Basis of sexual reproduction. 189

messenger RNA (mRNA) A type of RNA that carries a protein-building message. The mRNA of eukaryotes is modified after transcription. 150

metabolic pathway Series of enzyme-mediated reactions by which cells build, remodel, or break down an organic molecule. 84

metabolism All the enzyme-mediated chemical reactions by which cells acquire and use energy as they build and break down organic molecules. 40

metaphase Stage of mitosis at which the cell's chromosomes are aligned midway between poles of the spindle. 180

metastasis The process in which cancer cells spread from one part of the body to another. 185

microfilament Reinforcing cytoskeletal element; a fiber of actin subunits. 68

microtubule Cytoskeletal element involved in cellular movement; hollow filament of tubulin subunits. 68

mitochondrion Organelle that produces ATP by aerobic respiration in eukaryotes. 66

mitosis Nuclear division mechanism that maintains the chromosome number. Basis of body growth and tissue repair in multi-celled eukaryotes; also asexual reproduction in some plants, animals, fungi, and protists. Occurs in stages: prophase, metaphase, anaphase, telophase. 178

mixture An intermingling of two or more types of molecules. 31

model Analogous system used for testing hypotheses. 12

molecule A group of two or more atoms joined by chemical bonds. 4, 28

monohybrid cross Cross between two individuals identically heterozygous for one gene; for example $Aa \times Aa$. 206

monomers Molecules that are subunits of polymers. 40

monosaccharide Simple sugar that can be used as a monomer of polysaccharides. 42

motor protein Type of energy-using protein that interacts with cytoskeletal elements to move the cell's parts or the whole cell. 68

multiple allele system Gene for which three or more alleles persist in a population. 210

mutation Permanent change in the nucleotide sequence of DNA. 142

neoplasm An accumulation of cells that have lost control over their cell cycle. 184

neutron Uncharged subatomic particle in the atomic nucleus. 24

nondisjunction Failure of sister chromatids or homologous chromosomes to separate during nuclear division. 228

nuclear envelope Nuclear membrane. A double membrane that constitutes the outer boundary of the nucleus. Pores in the membrane control which substances can cross. 62

nucleic acid Single-stranded or double-stranded chain of nucleotides joined by sugar–phosphate bonds; e.g., DNA, RNA. 49

nucleic acid hybridization Base-pairing between DNA or RNA from different sources. 238

nucleoid Region of cytoplasm where the DNA is concentrated inside a bacterium or archaeon. 58

nucleolus In a cell nucleus, a dense, irregularly shaped region where ribosomal subunits are assembled. 63

nucleoplasm Viscous fluid enclosed by the nuclear envelope. 63

nucleosome A length of DNA wound twice around a spool of histone proteins. Basic unit of structural organization of eukaryotic chromosomes. 134

nucleotide Organic compound that consists of a five-carbon sugar (e.g., ribose, deoxyribose) bonded to one or more phosphate groups and a nitrogen-containing base (e.g., adenine, guanine, cytosine, thymine, uracil) after which it is named. Various types function as coenzymes, signaling molecules, and monomers of nucleic acids. 49

nucleus Of a eukaryotic cell, a double-membraned organelle that encloses the cell's DNA. 8, 54 Of an atom; core region occupied by protons and neutrons. 24

nutrient Substance that an organism needs for growth and survival, but cannot make for itself. 6

oncogene Gene that helps transform a normal cell into a tumor cell. 184

operator Part of an operon; a DNA binding site for a repressor. 170

operon Group of genes together with a promoter–operator DNA sequence that controls their transcription. 170

organ In multicelled organisms, a grouping of tissues engaged in a collective task. 5

organ system In multicelled organisms, set of organs engaged in a collective task. 5

organelle Structure that carries out a specialized metabolic function inside a cell. E.g., nucleus, mitochondrion. 54

organic Describes a compound that consists primarily of carbon and hydrogen atoms. 38

organism An individual that consists of one or more cells. 4

osmosis The diffusion of water across a selectively permeable membrane in response to a concentration gradient. 90

osmotic pressure Amount of turgor that prevents osmosis into cytoplasm or other hypertonic fluid. 91

oxidation–reduction reaction See redox reaction.

passive transport Mechanism by which a concentration gradient drives the movement of a solute across a cell membrane through a transport protein. Requires no energy input. 92

pattern formation Process by which a complex body forms from local processes during embryonic development. 166

PCR See polymerase chain reaction.

pedigree Chart of family connections that shows the appearance of a trait through generations. 220

peptide bond A bond between the amine group of one amino acid and the carboxyl group of another. Joins amino acids in proteins. 46

periodic table Tabular arrangement of the elements by atomic number. 24

peroxisome Enzyme-filled vesicle that breaks down amino acids, fatty acids, and toxic substances. 65

petal Unit of a flower's corolla; often adapted to attract animal pollinators. 486

pH A measure of the number of hydrogen ions in a fluid. 32

phagocytosis "Cell eating"; an endocytic pathway by which a cell engulfs particles such as microbes or cellular debris. 94

phenotype An individual's observable traits; product of genotype, epigenetics, and environmental influences. 205

phospholipid A lipid with a phosphate group in its hydrophilic head, and two nonpolar fatty acid tails; main constituent of eukaryotic cell membranes. 45

phosphorylation Transfer of a phosphate group from one molecule to another. 87

photoautotroph Photosynthetic autotroph; obtains carbon from carbon dioxide and energy from light. 112

photolysis Any process by which light energy breaks down a molecule. 106

photorespiration Reaction in which rubisco attaches oxygen instead of carbon dioxide to ribulose bisphosphate. 111

photosynthesis Metabolic pathway by which most autotrophs use light energy to make sugars from carbon dioxide and water. See light-dependent reactions, light-independent reactions. 6, 101

photosystem Cluster of pigments and proteins that work as a unit to convert light energy to chemical energy in photosynthesis. Two types (I and II) occur in thylakoid membranes. 106

pigment An organic molecule that can absorb light of certain wavelengths. 102

pilus Protein filament that projects from the surface of some bacteria and archaea. 59

pinocytosis Endocytosis of bulk materials. 95

plant A multicelled producer, typically photosynthetic and adapted to life on land. 8

plasma membrane A cell's outermost membrane. 54

plasmid Of many bacteria and archaea, a small ring of DNA replicated independently of the chromosome. 58

plasmodesmata Cell junctions that connect the cytoplasm of adjacent plant cells. 71

plastid Category of double-membraned organelle in plants and algal cells. Different types specialize in storage or photosynthesis; e.g., chloroplast, amyloplast. 66

pleiotropic Refers to a gene that influences multiple traits. 211

polarity Separation of charge into distinct positive and negative regions. 28

polymer Molecule that consists of multiple monomers. 40

polymerase chain reaction (**PCR**) Method that uses heat-tolerant *Taq* polymerase to rapidly generate many copies of a specific section of DNA. Used in DNA profiling, among other applications. 238

polypeptide Chain of amino acids linked by peptide bonds. 46

polyploid Having three or more of each type of chromosome characteristic of the species. 228

polysaccharide Polymer of monosaccharides; e.g., cellulose, starch, or glycogen. 42

population A group of organisms of the same species that live in a specific location and breed with one another more often than they breed with members of other populations. 5

potential energy Stored energy. 79

prediction Statement, based on a hypothesis, about a condition that should exist if the hypothesis is correct. 12

primary wall The first cell wall of young plant cells. 70

primer Short, single strand of DNA that base-pairs (hybridizes) with a complementary DNA sequence. 140, 239

prion Infectious protein. 48

probability Chance that a particular outcome of an event will occur; depends on the total number of possible outcomes. 17

probe Short fragment of DNA labeled with a tracer; designed to hybridize with a nucleotide sequence of interest. 238

producer Organism that makes its own food using energy and simple raw materials from the environment; an autotroph. 6

product A molecule that remains at the end of a reaction. 80

promoter A DNA sequence to which RNA polymerase binds. 152

prophase Stage of mitosis during which chromosomes condense and become attached to a newly forming spindle. 180

protein Organic compound that consists of one or more polypeptide chains. 46

protist General term for member of one of the eukaryotic lineages that is not a fungus, animal, or plant. 8

proton Positively charged subatomic particle that occurs in the nucleus of all atoms. 24

proto-oncogene Gene that, by mutation, can become an oncogene. 184

pseudopod "False foot." A temporary protrusion that helps some eukaryotic cells move and engulf prey. 69

Punnett square Diagram used to predict the genetic and phenotypic outcome of a cross or mating. 206

pyruvate Three-carbon end product of glycolysis. 118

radioactive decay Process by which an atom emits energy and/or subatomic particles when its nucleus spontaneously disintegrates. 25

radioisotope Isotope with an unstable nucleus. 25

reactant A molecule that enters a reaction. 80

reaction Process of molecular change. 40

receptor protein Plasma membrane protein that triggers a change in the cell's activities when it binds to a particular substance outside of the cell. 89

recessive Refers to an allele with an effect that is masked by a dominant allele on the homologous chromosome. 205

recognition protein Plasma membrane protein that identifies a cell as belonging to self (one's own tissue or body). 89

recombinant DNA A DNA molecule that contains genetic material from more than one organism. 236

redox reaction Oxidation–reduction reaction, in which one molecule accepts electrons (it becomes reduced) from another molecule (which becomes oxidized). 85

repressor Regulatory protein that blocks transcription. 164

reproduction Processes by which individuals produce offspring. 7

reproductive cloning Technology that produces genetically identical individuals. 144

restriction enzyme Type of enzyme that cuts specific nucleotide sequences in DNA. 236

reverse transcriptase An enzyme that synthesizes a strand of cDNA using mRNA as a template. 237

ribonucleic acid *See* RNA.

ribosomal RNA (**rRNA**) Type of RNA that, together with structural proteins, composes ribosomal subunits. As part of ribosomes, catalyzes formation of peptide bonds. 150

ribosome Organelle of protein synthesis. Has two subunits that consist of rRNA and proteins 58

RNA Ribonucleic acid. Nucleic acid, a polymer of nucleotides (adenine, guanine, cytosine, and uracil). Various types have roles in gene expression and controls over it. Most are single-stranded; some are enzymatic. 49

RNA polymerase Enzyme that carries out transcription. 152

rubisco Ribulose bisphosphate carboxylase. Carbon-fixing enzyme of the Calvin–Benson cycle. Operates inefficiently under high oxygen conditions (*see* photorespiration). 109

salt Ionic compound that releases ions other than H^+ and OH^- when it dissolves in water. 31

sampling error Difference between results derived from testing an entire group of events or individuals, and results derived from testing a subset of the group. 16

saturated fatty acid Fatty acid that contains no carbon–carbon double bonds in its hydrocarbon tail. 44

science Systematic study of the observable world. 12

scientific method Making, testing, and evaluating hypotheses. 13

scientific theory Hypothesis that has not been disproven after many years of rigorous testing. 18

second law of thermodynamics Energy tends to disperse spontaneously (the entropy of a system tends to increase). 78

secondary wall Lignin-reinforced wall that forms inside the primary wall of a plant cell. 70

semiconservative replication Describes the process of DNA replication, which produces two copies of a DNA molecule; one strand of each copy is new, and the other is a strand of the original DNA. **141**

sequence *See* DNA sequence.

sex chromosome Member of a pair of chromosomes that differs between males and females. **135**

sexual reproduction Reproductive mode by which offspring arise from two parents and inherit genes from both. **189**

shell model Model that helps us visualize how electrons populate atoms. **26**

short tandem repeats In chromosomal DNA, sequences of 4 or 5 bases repeated multiple times in a row. **243**

single-nucleotide polymorphism (SNP) One-base DNA sequence variation carried by a measurable percentage of a population. **235**

sister chromatid One of two attached DNA molecules of a duplicated eukaryotic chromosome. **134**

solute A dissolved substance. **31**

solution Homogeneous mixture. **31**

solvent Substance that can dissolve other substances. **31**

somatic Relating to the body. **190**

somatic cell nuclear transfer (SCNT) Reproductive cloning method in which genetic material is transferred from an adult somatic cell into an unfertilized, enucleated egg. **144**

somatosensory cortex *See* primary somatosensory cortex.

species Unique type of organism; designated by genus name and specific epithet. **10**

specific epithet Second part of a species name; differentiates the species from other members of the genus. **10**

sperm Mature male gamete. **196**

spindle Dynamically assembled and disassembled network of microtubules that moves chromosomes during nuclear division. **180**

sporophyte Diploid, spore-producing body in the life cycle of land plants and some algae. **196**

starch Polysaccharide that serves as an energy reservoir in plant cells. **42**

statistically significant Refers to a result that is statistically unlikely to have occurred by chance. **17**

steroid Type of lipid with four carbon rings and no fatty acid tails. **45**

stomata Closable gaps formed by guard cells in plant epidermis; when open, they allow water vapor and gases to diffuse across plant surfaces. Singular, stoma. **110**

stroma Semifluid matrix between the thylakoid membrane and the two outer membranes of a chloroplast. The light-independent reactions take place in it. **105**

substrate A molecule that is specifically acted upon by an enzyme. **82**

substrate-level phosphorylation A reaction that transfers a phosphate group from a substrate directly to ADP, thus forming ATP. **120**

surface-to-volume ratio A relationship in which the volume of an object increases with the cube of the diameter, and the surface area increases with the square. Constrains cell size and shape. **55**

taxon Ranked group of organisms that share a unique set of features; each comprises a set of the next lower taxon. E.g., domain, kingdom, phylum, etc. **10**

taxonomy Science of systematically naming and classifying species. **10**

telomere Noncoding, repetitive DNA sequence at the end of a chromosome. **183**

telophase Stage of mitosis during which the two sets of chromosomes arrive at opposite spindle poles and decondense as new nuclei form. **180**

temperature Measure of molecular motion, which increases with heat. **30**

testcross Method of determining genotype by tracking a trait in the offspring of a cross between an individual of unknown genotype and an individual known to be homozygous recessive. **206**

theory, scientific *See* scientific theory.

therapeutic cloning The use of SCNT to produce human embryos for research purposes. **145**

thylakoid membrane A chloroplast's highly folded inner membrane system; forms a continuous compartment in the stroma. The light-dependent reactions occur at the thylakoid membrane. **105**

thymine *See* nucleotide.

tight junctions Arrays of fibrous proteins that join epithelial cells and collectively prevent fluids from leaking between them. **71**

tissue In multicelled organisms, specialized cells organized in a pattern that allows them to perform a collective function. **4**

tracer Substance with a detectable component, such as a molecule labeled with a radioisotope. **25**

trait Physical, biochemical, or behavioral characteristic of an organism or species. **10**

transcription RNA synthesis; process by which an RNA is assembled from nucleotides using a DNA template. **150**

transcription factor Regulatory protein that influences transcription; e.g., an activator or repressor. **164**

transfer RNA (tRNA) A type of RNA that delivers amino acids to a ribosome during translation. **150**

transgenic Refers to a genetically modified organism that carries a gene from a different species. **244**

transition state Point during a reaction at which substrate bonds reach their breaking point and the reaction will run spontaneously to completion. **82**

translation Process by which a polypeptide chain is assembled from amino acids in the order specified by an mRNA. **150**

translocation, chromosomal Structural change of a chromosome in which a broken piece gets reattached in the wrong location. **226**

transport protein Protein that passively or actively assists specific ions or molecules across a cell membrane. **89**

transposable element Segment of chromosomal DNA that can spontaneously move to a new location. **159**

triglyceride A fat with three fatty acid tails. **44**

tumor A neoplasm that forms a lump in the body. **184**

turgor Pressure that a fluid exerts against a wall, membrane, or other structure that contains it. **91**

unsaturated fatty acid Fatty acid that has one or more carbon–carbon double bonds in its hydrocarbon tail. **44**

uracil *See* nucleotide.

vacuole A membrane-enclosed, saclike organelle filled with fluid; functions to isolate or dispose of waste, debris, or toxins. **65**

variable In an experiment, a characteristic or event that differs among individuals or over time. **13**

vesicle Small vacuole; different kinds store, transport, or degrade their contents. **64**

wavelength Distance between the crests of two successive waves. **102**

wax Water-repellent mixture of lipids with long fatty acid tails bonded to long-chain alcohols or carbon rings. **45**

X chromosome inactivation Shutdown of one of the two X chromosomes in the cells of female mammals. **168**

xenotransplantation Transplantation of an organ from one species into another. **247**

zygote Diploid cell formed by fusion of two gametes; the first cell of a new individual. **191**

Art Credits and Acknowledgments

TABLE OF CONTENTS **Page i** top, Bob Jensen Photography; middle, Jeff Vanuga/Corbis; bottom, John Easley. **Page iii** top, Bob Jensen Photography; bottom, John Easley. **Page v** top, NASA; bottom right, © Exactostock/ SuperStock. **Page vi** from left, © R. Calentine/ Visuals Unlimited; © Dylan T. Burnette and Paul Forscher; Hemoglobin models: PDB ID: 1GZX; Paoli, M., Liddington, R., Tame, J., Wilkinson, A., Dodson, G.; Crystal structure of T state hemoglobin with oxygen bound at all four haems. *J.Mol.Bio.*, v256, pp. 775–792, 1996. **Page vii** from left, Professors P. Motta and T Naguro / Photo Researchers, Inc.; Andrew Syred / Photo Researchers, Inc. **Page viii** from left, © Jose Luis Riechmann; Ed Reschke; Image courtesy of Carl Zeiss Micro-Imaging, Thornwood, NY; Moravian Museum, Brno. **Page ix** from left, Photo courtesy of The Progeria Research Foundation; Photo courtesy of MU Extension and Agricultural Information.

CHAPTER 1 **Page 2** © Sergei Krupnov, www.flickr.com/photos/7969319@N03; **Learning Roadmap**, from top, © Mauritius/ SuperStock; © Umberto Salvagnin, www.flickr.com/photos/kaibara; Dr. Marina Davila Ross, University of Portsmouth; From Meyer, A., Repeating Patterns of Mimicry. PLoS Biology Vol. 4, No. 10, e341 doi:10.1371/journal.pbio.0040341; © Antje Schulte; © Raymond Gehman/ Corbis; NASA. **1.1** Tim Laman / National Geographic Stock. **1.2** (3-4) © Umberto Salvagnin, www.flickr.com/photos/kaibara; (5) California Poppy, © 2009, Christine M. Welter; (6) Lady Bird Johnson Wildflower Center; (7) Michael Szoenyi / Photo Researchers, Inc.; (8) Photographers Choice RF/ SuperStock; (9) © Sergei Krupnov, www.flickr.com/photos/7969319@N03; (10) © Mark Koberg Photography; (11) NASA. **1.3** above, © Victoria Pinder, www.flickr.com/photos/vixstarplus. **1.4** Dr. Marina Davila Ross, University of Portsmouth. **1.5** (A) Clockwise from top left, Dr. Richard Frankel; Photo Researchers, Inc.; © Susan Barnes; www.zahnarzt-stuttgart.com; (B) left, ArchiMeDes; right, © Dr. Harald Huber, Dr. Michael Hohn, Prof. Dr. K.O.Stetter, University of Regensburg, Germany; (C) Protists, left, © Lewis Trusty/ Animals Animals; right, clockwise from top left, M I Walker / Photo Researchers, Inc.; Courtesy of Allen W. H. Bé and David A. Caron; © Emiliania Huxleyi photograph, Vita Pariente, scanning electron micrograph taken on a Jeol T330A instrument at Texas A&M University Electron Microscopy center; Oliver Meckes/ Photo Researchers, Inc.; © Carolina Biological Supply Company; Plants, left © John Lotter Gurling / Tom Stack & Associates; right © Edward S. Ross; Fungi, left, © Robert C. Simpson/ Nature Stock; right, © Edward S. Ross; Animals, from left, © Tom & Pat Leeson, Ardea London Ltd.; Thomas Eisner, Cornell University; © Martin Zimmerman, Science, 1961, 133:73-79, © AAAS.; © Pixtal/ SuperStock. **1.6** from left, Joaquim Gaspar; © kymkemp.com; Nigel Cattlin/Visu-als Unlimited, Inc.; Courtesy of Melissa S. Green, www.flickr.com/photos/henkimaa; © Gordana Sarkotic. **1.8** From Meyer, A., Repeating Patterns of Mimicry. PLoS Biology Vol. 4, No. 10, e341 doi:10.1371/journal.pbio.0040341. **1.9** (A) Volker Steger/ Photo Researchers, Inc.; (B) Cape Verde National Institute of Meteorology and Geophysics and the U.S. Geological Survey; (C) © Roger W. Winstead, NC State University; (D) Photo by Scott Bauer, USDA/ARS. **1.10** © Superstock. **1.11** (A) © Matt Rowlings, www.eurobutterflies.com; (B) © Adrian Vallin; (C) © Antje Schulte. **1.12** © Bruce Beehler/ Conservation International. **1.13** © Gary Head. **Table 1.3** © Raymond Gehman/ Corbis. **1.15** Courtesy East Carolina University. **Page 20**, Section 1.1, Tim Laman/ National Geographic Stock; Section 1.2, California Poppy, © 2009, Christine M. Welter; Section 1.3, © Victoria Pinder, www.flickr.com/photos/vixstarplus; Section 1.4, Courtesy of Allen W. H. Bé and David A. Caron; Section 1.5, From Meyer, A., Repeating Patterns of Mimicry. PLoS Biology Vol. 4, No. 10, e341 doi:10.1371/journal.pbio.0040341; Section 1.6, © Roger W. Winstead, NC State University; Section 1.7, © Adrian Vallin; Section 1.8, Gary Head; Section 1.9, © Raymond Gehman/ Corbis. **Page 21** Scientific Paper; Adrian Vallin, Sven Jakobsson, Johan Lind and Christer Wiklund, Proc. R. Soc. B (2005 272, 1203, 1207). Used with permission of The Royal Society and the author.

CHAPTER 2 **Page 22**, © Vicki Rosenberg, www.flickr.com/photos/roseofredrock; **Learning Roadmap**, from top, © Victoria Pinder, www.flickr.com/photos/vixstarplus; © Michael S. Yamashita/ Corbis; © Bill Beatty/ Visuals Unlimited; © Herbert Schnekenburger; W. K. Fletcher/ Photo Researchers, Inc. **2.1** left, Michael Grecco/ Picture Group; inset, Theodore Gray/ Visuals Unlimited, Inc; right, © Kim Westerskov, Photographer's Choice/ Getty Images. **2.4** Brookhaven National Laboratory. **Page 27** © Michael S. Yamashita/ Corbis. **2.8** (B) left, Gary Head; center, © Bill Beatty/ Visuals Unlimited. **Page 30** © Herbert Schnekenburger. **2.11** (C) www.flickr.com/photos/roseofredrock. **2.12** © JupiterImages Corporation. **Page 33** left, © Jared C. Benedict, www.flickr.com/photos/redjar; right, W. K. Fletcher/ Photo Researchers, Inc. **Page 34**, Section 2.1, © Kim Westerskov, Photographer's Choice/ Getty Images; Section 2.3, © Michael S. Yamashita/ Corbis; Section 2.5, © Herbert Schnekenburger; Section 2.6, W. K. Fletcher/ Photo Researchers, Inc.

CHAPTER 3 **Page 36**, © JupiterImages Corporation; **Learning Roadmap**, from top, © JupiterImages Corporation; © JupiterImages Corporation; From the collection of Jamos Werner and John T. Lis. **3.1** © ThinkStock/ SuperStock. **3.2** (B) © JupiterImages Corporation. **3.4** (A) © National Cancer Institute; (B) Hemoglobin models: PDF ID: 1GZX; Paoli, M., Liddington, R., Tame, J., Wilkinson, A., Dodson, G., Crystal structure of T state hemoglobin with oxygen bound at all four hems. J.Mol.Bio., v256, pp. 775–792, 1996. **3.5** (A) © Exactostock/ SuperStock. **3.7** left, **3.8** (C) © JupiterImages Corporation. **3.13** right, Tim Davis/ Photo Researchers, Inc. **Page 45** left, Kenneth Lorenzen. **Page 47** left, From: *Structure of the Rotor of the V-Type Na+-ATPase from Enterococcus hirae* by Murata, et al. Science 29 April 2005: 654-659. DOI:10.1126/science.1110064. Used with per-mission; right, This lipoprotein image was made by Amy Shih and John Stone using VMD and is owned by the Theoretical and Computational Biophysics Group, NIH Resource for Macromolecular Modeling and Bioinformatics, at the Beckman Institute, University of Illinois at Urbana-Champaign. Labels added to the original image by book author. **3.17** (A) © Lily Echeverria/Miami Herald; (B) Sherif Zaki, MD PhD, Wun-Ju Shieh, MD PhD; MPH/ CDC. **Page 49** © JupiterImages Corporation. **Page 50**, Section 3.1, © ThinkStock/ SuperStock; Section 3.3, © Exactostock/ SuperStock; Section 3.4, © JupiterImages Corporation; Section 3.5, Kenneth Lorenzen; Section 3.6, From: *Structure of the Rotor of the V-Type Na+-ATPase from Enterococcus hirae* by Murata, et al. Science 29 April 2005: 654-659. DOI:10.1126/science.1110064. Used with permission; Section 3.7, Sherif Zaki, MD PhD, Wun-Ju Shieh, MD PhD; MPH/ CDC.

CHAPTER 4 **Page 52**, © Dylan T. Burnette and Paul Forscher; **Learning Roadmap**, from top, Courtesy of © Johannes Kästner, Universität Stuttgart; R. Calentine/ Visuals Unlimited; Astrid & Hanns-Frieder Michler/ Photo Researchers, Inc.; Don W. Fawcett/ Visuals Unlimited; © ADVAN-CELL (Advanced In Vitro Cell Technologies; S.L.) www.advancell.com; Michael Clayton/ University of Wisconsin, Department of Botany. **4.1** left, © JupiterImages Corporation; right, Stephanie Schuller/ Photo Researchers, Inc. **4.3** Courtesy of © Johannes Kästner, Universität Stuttgart. **4.6** (A–B, D–E) © Jeremy Pickett-Heaps, School of Botany, University of Melbourne; (C) © Prof. Franco Baldi. **4.7** Whale, © Dorling Kindersley/ the Agency Collection/ Getty Images; Giraffe, © Ingram Publishing, SuperStock; Man, © JupiterImages Corporation; Dog, © Pavel Sazonov/ Shutterstock.com; Guinea Pig, © Sascha Burkard/ Shutterstock.com; Frog, Ant, photos.com; Louse, Edward S. Ross; Eukaryotic cell, Bruce Runnegar, NASA Astrobiology Institute; HPV particle, CDC. **4.9** (A) Rocky Mountain Laboratories, NIAID, NIH; (B) © R. Calentine/ Visuals Unlimited; (C) Cryo-EM image of Haloquadratum walsbyi, isolated from Australia. Courtesy of Zhuo Li (City of Hope, Duarte, California, USA), Mike L. Dyall-Smith (Charles Sturt University, Australia), and Grant J. Jensen (California Institute of Technology, Pasadena, California, USA); (D–E) © K.O. Stetter & R. Rachel, Univ. Regensburg. **4.10** © Dennis Kunkel Microscopy, Inc./ Phototake. **Page 60**, Astrid & Hanns-Frieder Michler/ Photo Researchers, Inc. **4.11** (A) Dr. Gopal Murti/ Photo Researchers, Inc.; (B) M.C. Ledbetter, Brookhaven National Laboratory. **4.13** right, © Kenneth Bart. **4.14** (A) Don W. Fawcett/ Visuals Unlimited; (B) © Martin W. Goldberg, Durham University, UK. **Page 63**, Andrew Syred/ Photo Researchers, Inc. **4.15** from left, © Kenneth Bart; (#2,3) Don W. Fawcett/ Visuals Unlimited; (#4) Micrograph, Gary Grimes. **4.16** Courtesy of Conner's Way Foundation, www.connersway.com. **Page 67**, © David T. Webb. **4.17** Micrograph, Keith R. Porter. **4.18** (A) © Dr. Ralf Wagner, www.dr-ralf-wagner.de; (B) Dr. Jeremy Burgess/ Photo Researchers, Inc. **4.19** (D) © Dylan T. Burnette and Paul Forscher. **4.21** (A) right, Don W. Fawcett/ Photo Researchers, Inc. **4.23** George S. Ellmore. **4.24** © ADVANCELL (Advanced In Vitro Cell Technologies; S.L.) www.advancell.com. **Page 72**, Above, © R. Llewellyn/ SuperStock, Inc.; Section 4.1, © JupiterImages Corpora-